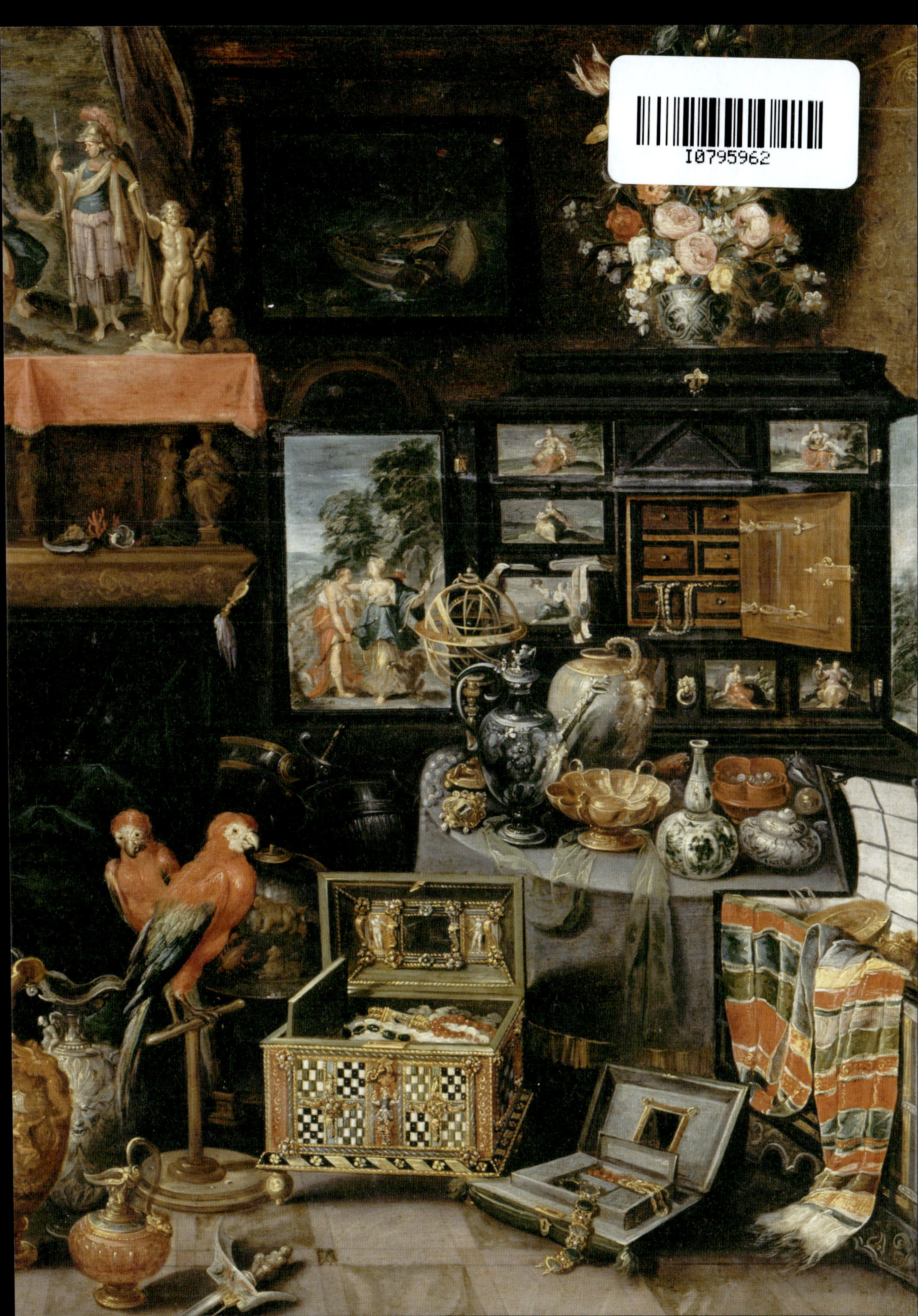

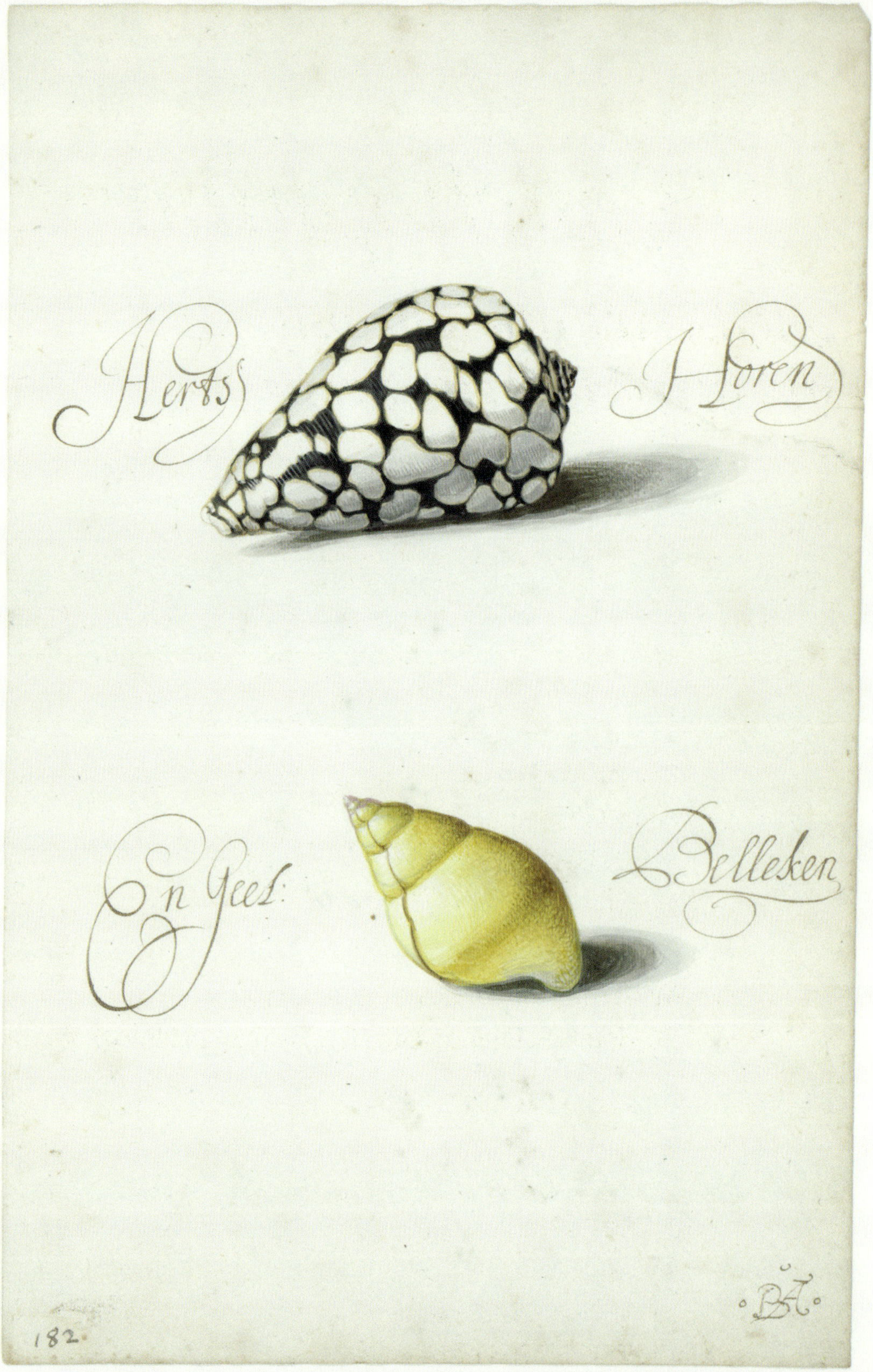
Hertss Horen
En Geel: Belleken
182

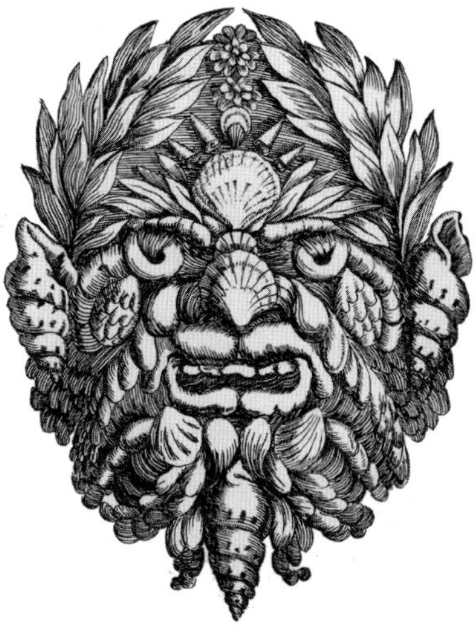

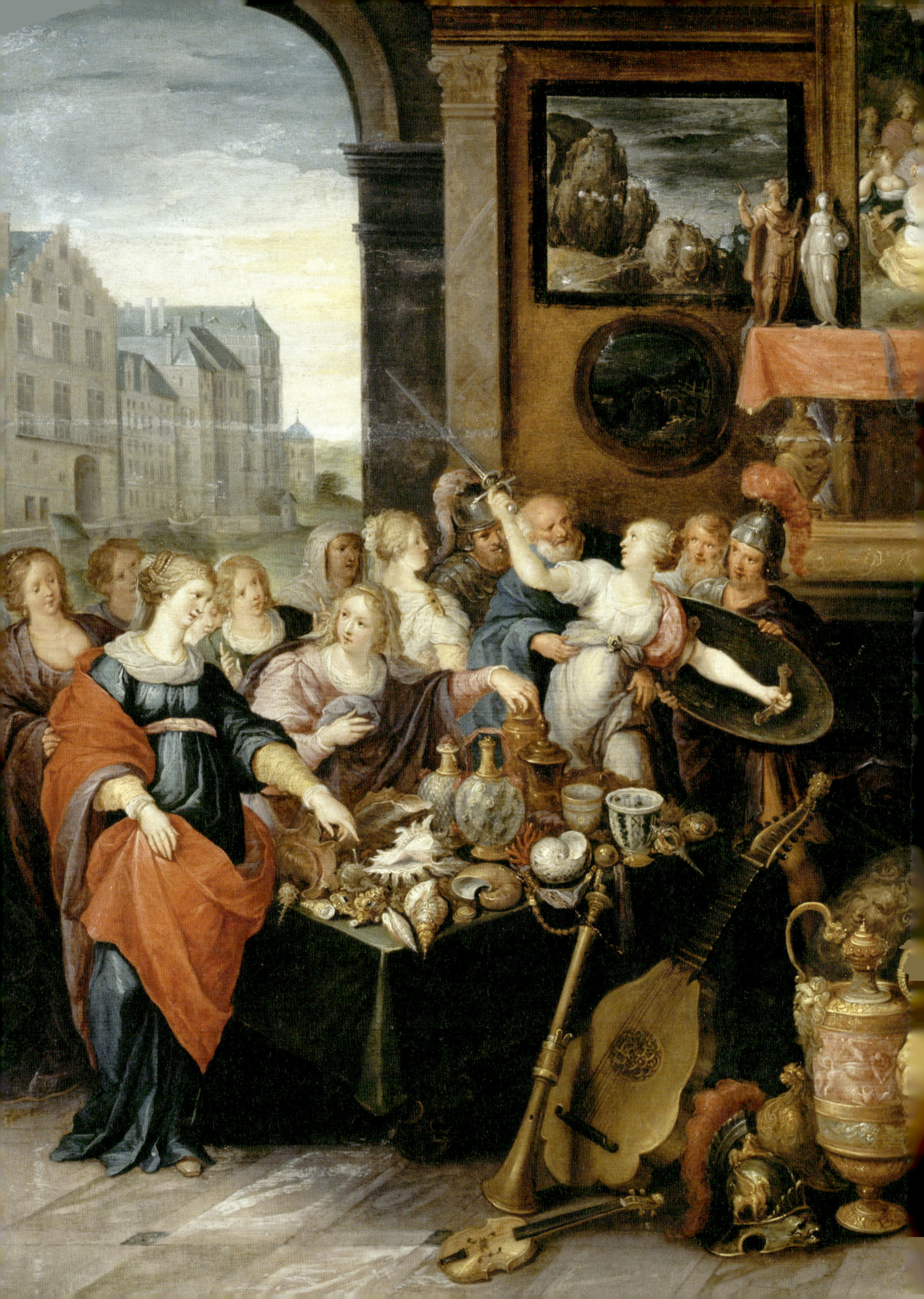

# *Conchophilia*

## *Shells, Art, and Curiosity in Early Modern Europe*

Marisa Anne Bass
Anne Goldgar
Hanneke Grootenboer
Claudia Swan

With contributions by
Stephanie S. Dickey, Anna Grasskamp, and Róisín Watson

PRINCETON UNIVERSITY PRESS
PRINCETON AND OXFORD

Published by Princeton University Press, 41 William Street, Princeton, New Jersey 08540
99 Banbury Road, Oxford OX2 6JX

press.princeton.edu

Cover illustrations (front) Jacques Linard, *Still Life with Shells and Coral* (detail), 1640. Oil on canvas, 53.5 × 62.2 cm. Musée des Beaux-Arts, Montreal [1999.149]; (back) Shell goblet with dragon foot, mid-seventeenth century. Shells and gilded silver, 26 × 19 × 14 cm. Landesmuseum Württemberg, Stuttgart [KK hellblau 13].

Endpapers: ID 105853579/ © 18042011e | Dreamstime.com

Illustrations in front matter: p. i, detail of fig. 52; pp. ii–iii, detail of fig. 26; p. iv, reproduction of fig. 80; pp. viii–ix, detail of fig. 47

Illustrations in back matter: p. 176, detail of fig 40; p. 178, detail of fig. 12; pp. 200–201, detail of fig. 13



The Library of Congress has cataloged the cloth edition as follows:

Names: Bass, Marisa, 1981–| Dickey, Stephanie. | Goldgar, Anne, writer of introduction. | Grasskamp, Anna. | Grootenboer, Hanneke. | Swan, Claudia. | Watson, Róisín.
Title: Conchophilia : shells, art, and curiosity in early modern Europe / Marisa Anne Bass, Anne Goldgar, Hanneke Grootenboer, and Claudia Swan ; with contributions by Stephanie S. Dickey, Anna Grasskamp, and Róisín Watson.
Description: Princeton : Princeton University Press, [2021] | Includes bibliographical references and index.
Identifiers: LCCN 2020038356 (print) | LCCN 2020038357 (ebook) | ISBN 9780691215761 (hardback) | ISBN 9780691220246 (ebook)
Subjects: LCSH: Shells. | Collectors and collecting—Europe—History.
Classification: LCC QL402 .C66 2021 (print) | LCC QL402 (ebook) | DDC 594—dc23
LC record available at https://lccn.loc.gov/2020038356
LC ebook record available at https://lccn.loc.gov/2020038357

British Library Cataloging-in-Publication Data is available

First Paperback Printing, 2023
Paper ISBN: 978-0-691-24859-2
Cloth ISBN: 978-0-691-21576-1

Published with the assistance of the Frederick W. Hilles Publication Fund of Yale University

Design by Jenny Chan / Jack Design
Typeset by Tina Henderson

This book has been composed in Jenson, Avenir Next, and Mazius Display.

Printed in Italy

# *Contents*


1 Introduction: For the Love of Shells *Anne Goldgar*

*Part I*
*Surface Matters*

21 Chapter 1: The Nature of Exotic Shells *Claudia Swan*

49 Chapter 2: Shells, Bodies, and the Collector's Cabinet *Anna Grasskamp*

*Part II*
*Microworlds of Thought*

75 Chapter 3: Shell Life, or the Unstill Life of Shells *Marisa Anne Bass*

103 Chapter 4: Thinking with Shells in Petronella Oortman's Dollhouse *Hanneke Grootenboer*

*Part III*
*The Multiple Experienced*

127 Chapter 5: Shells and Grottoes in Early Modern Germany *Róisín Watson*

155 Chapter 6: Shells, Prints, and the Discerning Eye *Stephanie S. Dickey*

177 Acknowledgments
179 Notes
190 Bibliography
202 Contributors
205 Index
214 Illustration Credits

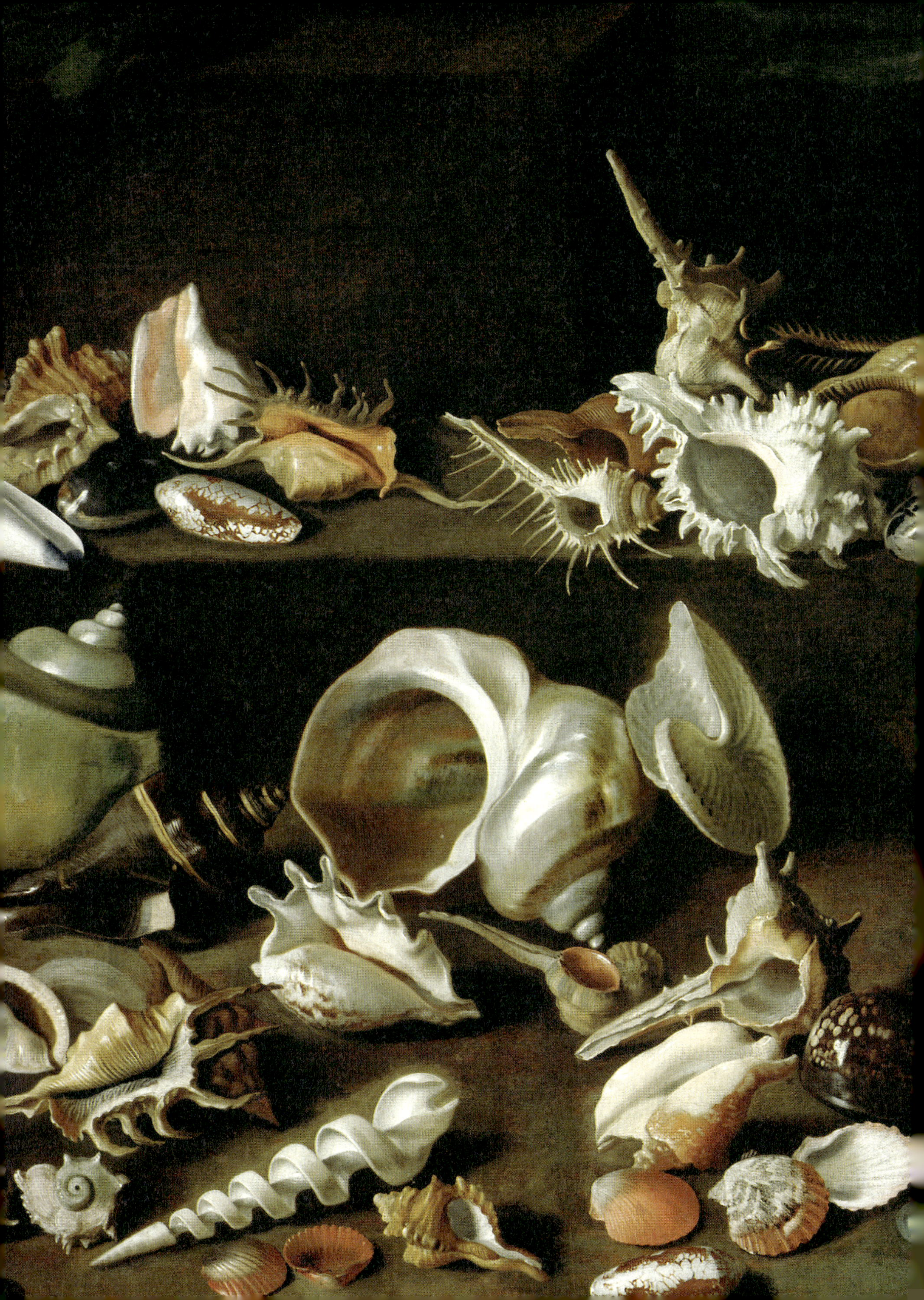

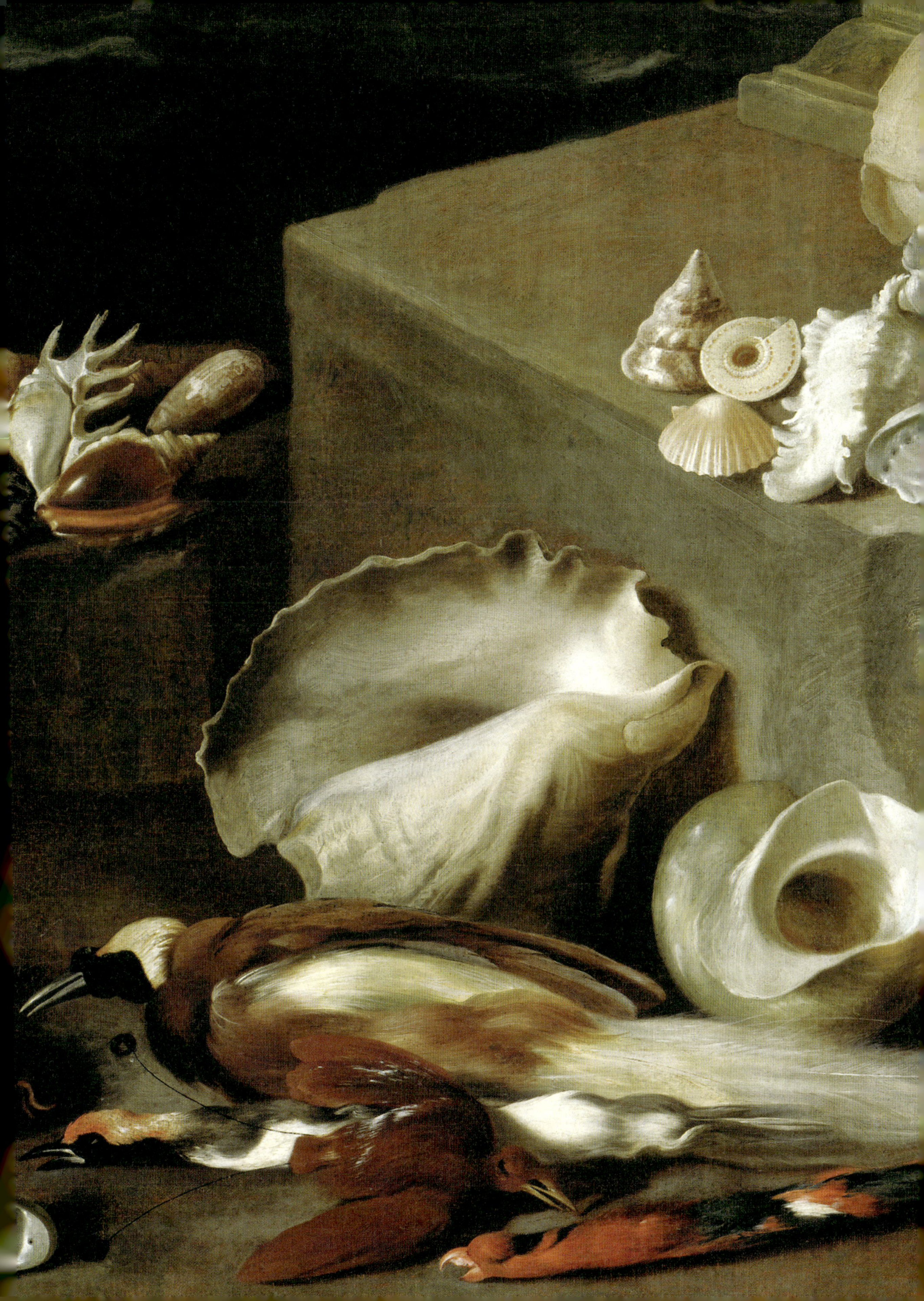

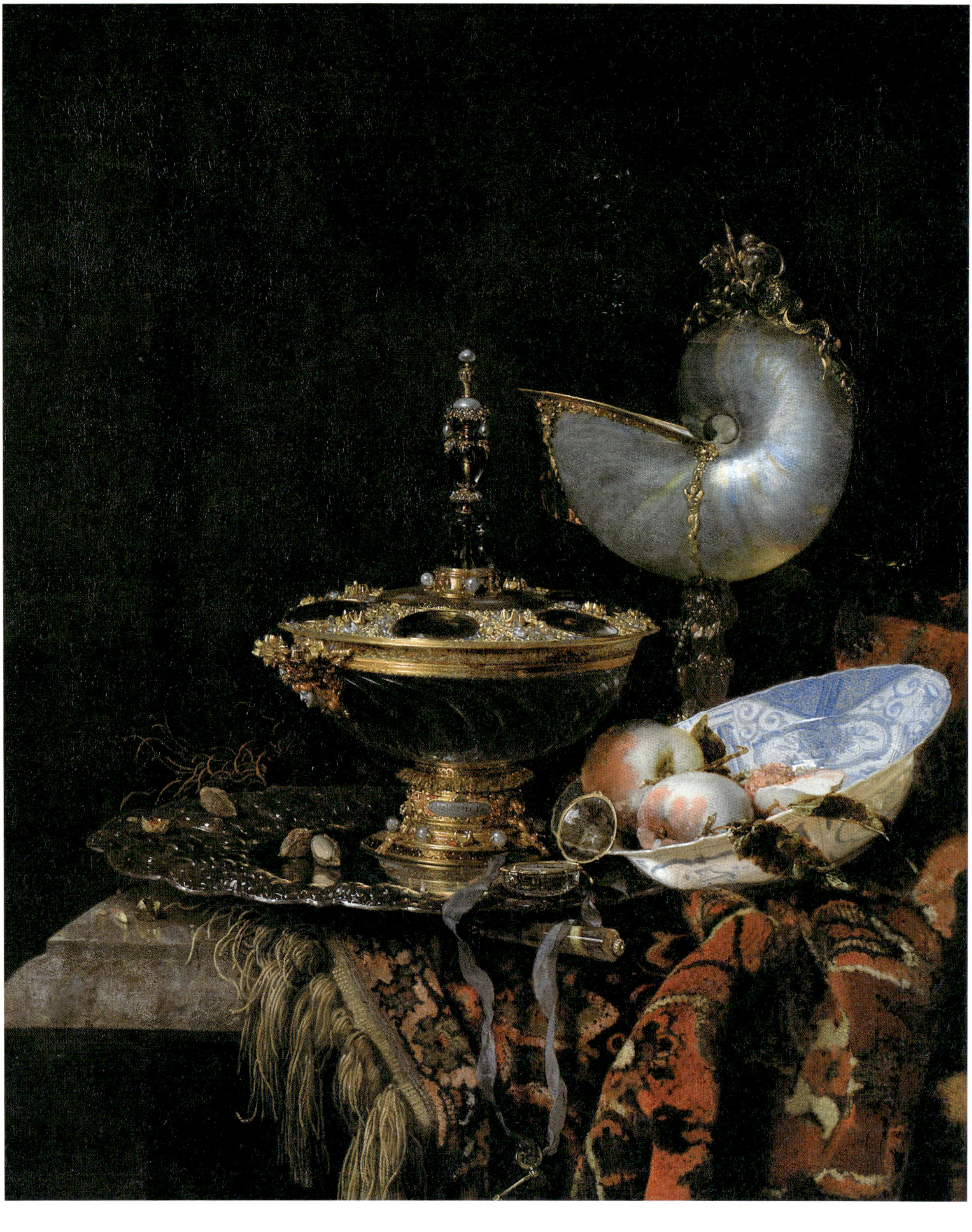

Anne Goldgar

# Introduction
# For the Love of Shells

FIG. 1. Willem Kalf. *Pronk Still Life with Nautilus Cup*, 1678. Oil on canvas, 68 × 56 cm. Statens Museum for Kunst, Copenhagen [KMS 1513].

*Conchophilia* ought to be a word, but it isn't. We may have conchophile—although even this the *Oxford English Dictionary* fails to recognize—but for the phenomenon of a passionate engagement with shells there seems to have been no word until my fellow authors and I conceived it as the title of this project in 2017. Yet conchophilia was a striking phenomenon in early modern Europe, occupying aristocrats and apothecaries, scholars and tradesmen, men and women, connoisseurs, art lovers, and enthusiasts for interior design. The exoticism of the most fashionable shells excited many, coming as they did from such distant shores as the Pacific and the West Indies. Their variegated colors and intriguing forms inspired pleasure, curiosity, introspection, and wonder at the endless variety of God's creation. A different sensibility might have informed the collector who attempted to systematize a cabinet of shells into a series of patterned drawers, and the craftsman creating shell-studded furniture, but whether the shells were on the inside or the outside of a cabinet, both phenomena testify to a shared fascination with these beautiful objects.[1]

When we talk about conchophilia in the early modern period, we inevitably talk not only about the shells, but also about the people who interacted with them. One such person was François Valentyn (1666–1727). On the first Wednesday afternoon of each month, he and other shell collectors in Dordrecht met to admire and discuss each other's collections, to see what each had newly acquired, and to clarify points about the shells. The club began in the winter of 1720 in the home of Bartholomeus van Segwaart, who had the largest and finest cabinet in Dordrecht, and in August 1723 its members decided to name themselves "the *Liefhebbers* of Neptune's Cabinet." It was an epithet, according to Valentyn, intended to encourage others along the same path. "We take great pleasure in this," he wrote, "besides which, it is a clear proof that we are serious about this business."[2]

*Liefhebber* is a Dutch word essentially meaning lover, amateur, devotee, or connoisseur, whether of art, poetry, flowers, or, as we see here, shells. It was an appropriate term for Valentyn, and for many other collectors and admirers of shells across Europe between the mid-sixteenth century and the end of the eighteenth century.[3] Valentyn's own interest in shells, and unusual advantage as a shell collector, came with his travels abroad. In 1685, as a newly qualified Reformed minister, he was sent to the island of Ambon in the East Indies, where he served a community largely consisting of indigenous people, learned to speak Malay, and gained firsthand knowledge of the archipelago and its many beautiful shells. There he also met Georgius Eberhard Rumphius (1627–1702), the blind author of a treatise devoted to sea creatures and shells; Valentyn later reworked and drew from parts of Rumphius's volume.[4] Having spent 1698 to 1705 back in the Netherlands, he returned there for good in 1713 after a disagreement with the church over a posting to Banda. Back in Dordrecht, he was able to meet a variety of other collectors across the Dutch Republic.[5] In his *Treatise on Seashells and Sea Creatures*, published posthumously in 1754, Valentyn thought it important to list fellow lovers of shells in Zeeland and in cities such as Amsterdam, Haarlem, Delft, The Hague, Zwolle, and Rotterdam, as well as the six members of his local shell society. In each case, he described the particularly wonderful specimens each collector possessed, providing a personal geography of his experience of conchophilia in the 1720s.[6] As for the naysayers among his contemporaries who did not consider the admiration of sea creatures a worthy pastime, Valentyn dismissed them. In the words of the Latin proverb, he wrote, "one can have no love for a thing of which one has no understanding (*ignoti nulla cupido*)."[7]

In this book we talk about shells in their relationship with people: people like Valentyn, who collected shells, but also people who collected them on the beaches of Indonesia or the Caribbean, who polished and changed them, who etched images on them, who created elaborate decorative metal mounts for them, who painted pictures of them, who wondered about them or used them to think about other things. As we see from the material turn in historical studies, for example, we could do the same for any material object, most obviously objects of craft, such as paintings, furniture, jewelry, pottery; but this is also true of natural specimens such as shells, minerals, flowers, and plants. In the case of both artificial and natural objects, we can consider their materiality, discuss their location in space, and examine the conditions of their making, their use, and their representation. But even when we look at materiality, we continue to think about the social. The social life of things, the biography of objects, always encompasses the role material objects played in the lives of people who interacted with them.[8] To understand the passion for shells, we must also consider the people who handled them: the people who traded in them, the people who consumed and took pleasure in them, the people who labored over them or even died for them.

The love of shells was not new to early modernity; indeed, it was age-old. Shells have long fascinated through their variety, their beauty, their shapes, colors, and luminescence. Shells have been used throughout history as decoration, ornament, ritual objects, and objects of play. They have served as currency across the globe. Shells offer historical evidence of taste, commerce, and patterns of habitation and civilization. Archaeologists have used shell middens, vast deposits of waste shells in coastal waters, to understand prehistoric food consumption and movements of population.[9] Shells were the first objects clearly used as personal decoration: shells with holes bored in them to create jewelry have been found on sites from as early as the Middle Stone Age.[10] Shells and shell craft feature in cultures the world over, from Goa to the Trobriand Islands to the indigenous peoples of the North American coasts.[11]

Ancient writers also showed enthusiasm for shells. Aristotle studied shells as the domiciles of aquatic animals, but not without observing the qualities of their forms. His description of the paper nautilus (*Argonauta argo*) as a wonderous sailing vessel captured the imagination of his readers well into the early modern era, when nautilus shells were avidly collected in Europe, sometimes artfully carved, and transformed into vessels for elaborate display (fig. 1).[12] Pliny the Elder not only attempted to classify a diversity of shells in his *Natural History*; he also noted the use value of specimens like the murex, the famous source of the purple dye employed to color the robes of the Roman elite. The ancient orator Cicero advocated shell collecting, whether actual or figurative, as a means to achieve serenity in the midst of political turmoil, while the historian Suetonius reports that Caligula, in a fit of madness, ordered his soldiers to gather seashells along the shore as spoils of war.[13] Gardens and grottoes adorned with shells attest to the ancient Roman love affair with these specimens as artful and evocative decoration.[14] The Renaissance, with its new refocusing on the ancient world, saw a revival of shells and shell motifs in grotto architecture and gardens from Italy to northern Europe.[15] The Ciceronian imagery of collecting shells as a form of repose was in turn resurrected by scholars from the great Netherlandish humanist Desiderius Erasmus (c. 1466–1536) to the Dutch poet and playwright Joost van den Vondel (1587–1697).[16] At the same time, the early modern fervor for shells as prized possessions from foreign lands was founded on a long history of trade and, ultimately, imperial and colonial exploitation.

Conchophilia, then, has a long and complicated history, and can be examined in the way so many histories of material objects have done. The history of natural history has, moreover, investigated interest in shells as part of the more general exploration of important topics such as the history of collecting, the rise of classification, the conceived relationship between art and nature, and the understanding of the relationships among colonial trade, scientific inquiry, and material culture. But the curiosity aroused by shells, this book argues, is something unique, regardless of the time or place in which it manifests itself, and one that deserves to be treated as such. Indeed, there is something particular about shells as liminal things, objects that hover between life and death, and that throw our thoughts into far places, not least those from whence they came. Shells are things that were once inhabited by creaturely life, that once belonged to an ecosystem distant from humankind before finding their way into the artist's studio, or the drawers of a collector's cabinet. Not only were glistening, polished nautilus shells transformed into drinking vessels in sixteenth- and seventeenth-century Europe; shells in the early modern period (and beyond) were also vessels in both a literal and a figurative sense. Shells invite and contain a range of thoughts and reflections by virtue of their diverse forms, colors, and surfaces. As the essays in this volume lay out, these thoughts are culturally specific. The ways people in the early modern period thought about shells elucidate other aspects of the culture in which the shells were gathered, altered, collected, used, and admired. A beautiful shell delighted the eyes, but the love of shells arose from something deeper: an interest in their formation, in the thoughts they could arouse, and in their potential for reenlivenment, whether among a circle of learned friends, in the depths of a grotto, or within the realms of a painting. Building on foundational work by such scholars as Karin Leonhard, Leopoldine van Hogendorp Prosperetti, Emma Spary, Beth Fowkes Tobin, and Anna Marie Roos, this book aims to explore the love of shells as a cultural phenomenon, examining the cultural resonances and uses of objects and images produced around shells across early modern Europe, as well as through consideration of the global networks that brought them to the continent.[17]

Because of the early emergence of enthusiasm for shells in the Low Countries, and the optimal circumstances there for acquiring them, Dutch interaction with shells plays a particularly prominent role in the pages that follow. By the seventeenth century, to be a cultured person in the Netherlands had come to mean owning shells—even if just a few. When the Portuguese Jewish merchant Francisco Gomes da Costa went bankrupt in Amsterdam in 1646, we can observe, in the inventory of his possessions made by notaries, attributes of his cultured lifestyle: paintings, including five floral still lifes, a breakfast scene, and portraits; a mirror with a gilded frame; purple silk curtains; Spanish chairs with velvet seats; a porcelain mustard pot; a Turkish carpet; and two shells.[18]

The contents of Da Costa's best room should immediately tell us something about his understanding of luxury. His acquisitions suggest his sense of what was fashionable in the 1640s, particularly in their emphasis on the exotic: the purple (Chinese? Japanese? Persian?) silk curtains, the (Chinese or Japanese) porcelain mustard pot, the Turkish carpet. If he had had more money, he might well have had a parrot, a monkey, a Japanese lacquerwork cabinet, or any number of other exotic goods, including still more shells.[19] As for the many who

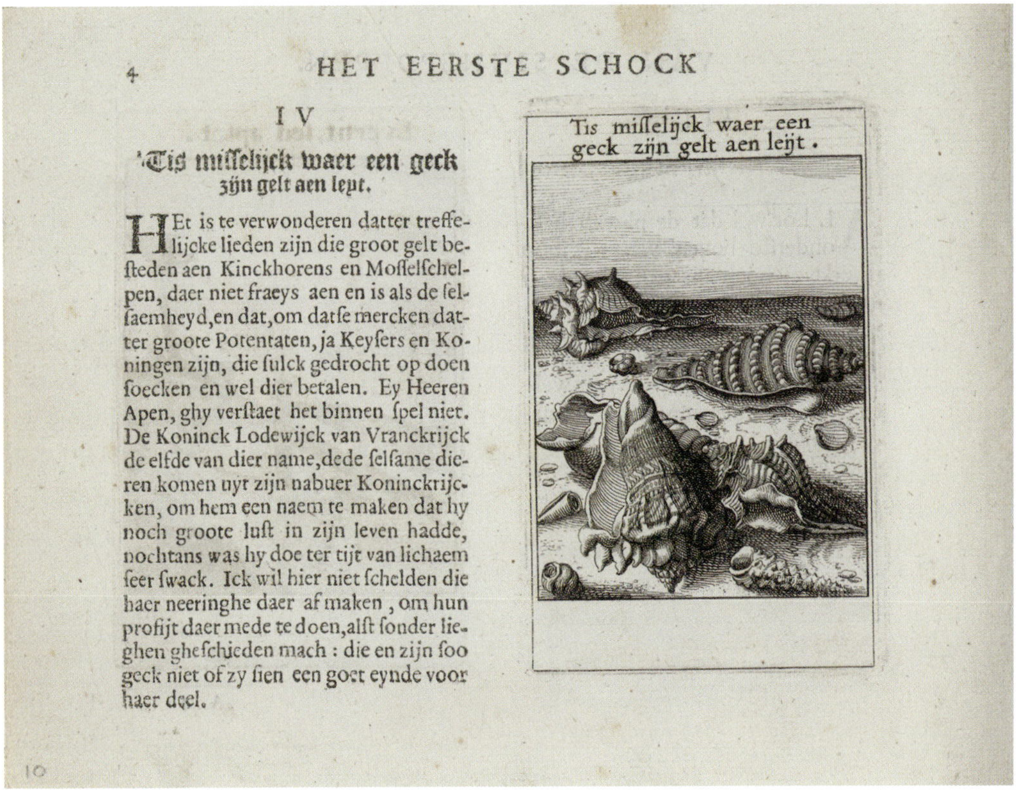
4 HET EERSTE SCHOCK

IV

'Tis miſſelijck waer een geck zijn gelt aen leyt.

HEt is te verwonderen datter treffelijcke lieden zijn die groot gelt beſteden aen Kinckhorens en Moſſelſchelpen, daer niet fraeys aen en is als de ſelſaemheyd,en dat,om datſe mercken datter groote Potentaten, ja Keyſers en Koningen zijn, die ſulck gedrocht op doen ſoecken en wel dier betalen. Ey Heeren Apen, ghy verſtaet het binnen ſpel niet. De Koninck Lodewijck van Vranckrijck de elfde van dier name,dede ſelſame dieren komen uyt zijn nabuer Koninckrijcken, om hem een naem te maken dat hy noch groote luſt in zijn leven hadde, nochtans was hy doe ter tijt van lichaem ſeer ſwack. Ick wil hier niet ſchelden die haer neeringhe daer af maken , om hun profijt daer mede te doen,alſt ſonder lieghen gheſchieden mach : die en zijn ſoo geck niet of zy ſien een goet eynde voor haer deel.

Tis miſſelijck waer een geck zijn gelt aen leijt .

10

**FIG. 2.** Roemer Visscher, "Tis misselijck waer een geck zijn gelt aen leijt," in *Zinnepoppen*, 1614 (Bk. 1, no. 4). Rijksmuseum, Amsterdam [BI-1893-3539-10].

focused more specifically on the collection of *artificialia* and *naturalia*, nearly all collectors in seventeenth-century Amsterdam had shells at the core of their collections.[20] With the regular voyages of the Dutch East India Company (VOC) and later West India Company (WIC) ships to the tropics, the seventeenth-century Netherlands in particular saw a great increase in the importation of large and colorful shells from the Indonesian archipelago, the Moluccas, the Cape of Good Hope, the Antilles, and other sites of Dutch colonial trade. To have such shells was a marker of fashion, and consequently a subject of ridicule for moralists. Roemer Visscher, for example, famously criticized both the fashion for buying tulips and that for buying shells in his 1614 emblem book *Sinnepoppen*, remarking about shells that they were merely children's toys, and that "it is crazy what some people spend their money on" (fig. 2).[21]

The increase in trade was intimately connected with shells, whether for good or for ill. The greater access to unusual and exotic items fueled the interest in shells. Already in the sixteenth century, before the founding of the VOC in 1602, there was a great demand for exotic plants, animals, and other goods, as humanistic curiosity joined together with fashion.

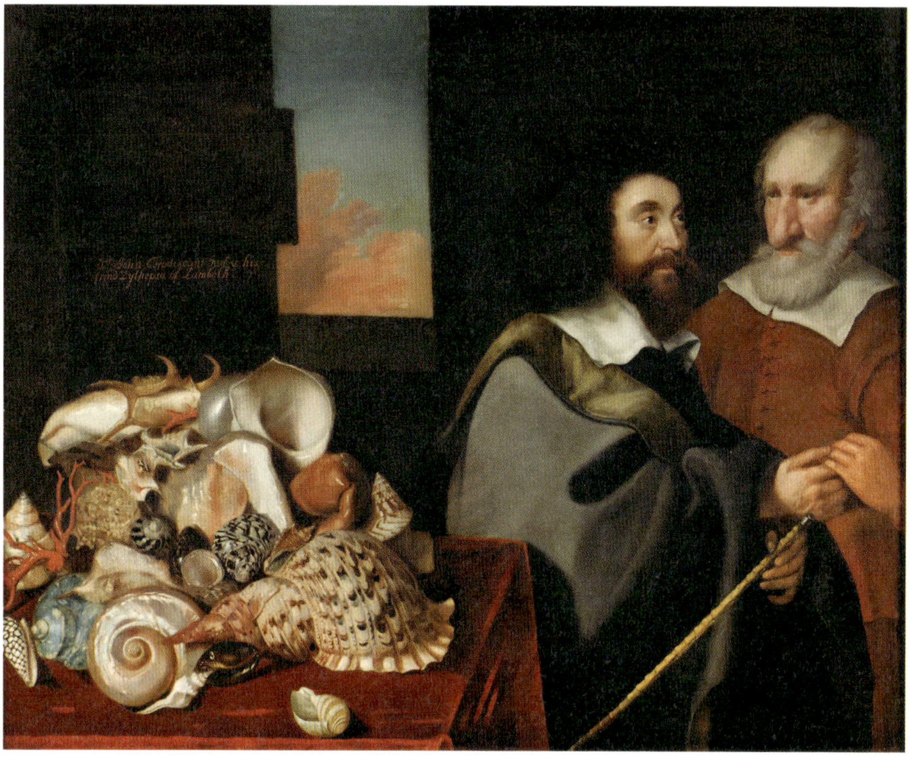

**FIG. 3.** Thomas de Critz (attributed to), *John Tradescant the Younger with Roger Friend and a Collection of Exotic Shells*. Oil on canvas, 107 × 132 cm. Ashmolean Museum, Oxford [WA1898.11].

Some collectors met East India ships; in Amsterdam and other Dutch towns, merchants and collectors sold curiosities and natural history objects from the East. As we see in part I of this volume, however, this commercial expansion was founded on colonial power relationships that manifested themselves in this case both in the exploitative extraction of shells and in the labor involved in their preparation for sale. Shells were intimately connected with slavery, both in the employment of slaves—the *dienaers* ("servants") Rumphius described, who gathered shells and cleaned and polished them, were actually enslaved—but also in their purchase.[22] Toby Green has recently highlighted in the title of his history of West Africa, *A Fistful of Shells*, that shells played a large part in funding the slave trade. Shell money, cowrie shells from the Maldives, was used as currency in India and Africa in this period, as well as becoming important in ritual practice in some West African societies.[23] The Portuguese, and later the Dutch and the English, took advantage of this monetary use of shells, transporting vast quantities of cowrie shells across the world to finance their purchase and transportation of enslaved African people.[24] Especially in the period from the 1630s, growing Dutch ascendancy in the slave trade also brought wealth to the Netherlands that made possible the purchase of more luxury and exotic goods, not least shells.[25] The same Dutch commercial networks that circulated enslaved peoples and the goods that helped to purchase them also brought these luxury goods into the European commercial theater. We are conscious of this today, and, in a very different way, so were those in the sixteenth and seventeenth centuries interested in shells and other exotica, who often elaborated in writing and in craft on the aesthetic desirability of both bodies and goods from far-flung lands.[26] In art, particularly in painting, artists often sought to emphasize the contrast between shining black skin and the surface of shells and mother-of-pearl, suppressing the reality of slavery by seeking aesthetic fulfillment.[27]

Dutch primacy in world trade contributed to a particularly Dutch enthusiasm for shells, which in turn fueled the fashion for shells in eighteenth-century Paris after the art dealer Edmé-François Gersaint began importing whole collections of shells to feed a growing market.[28] However, the love of shells was certainly not exclusive to the Low Countries. The earliest scholarly works on shells were German and French books from the 1550s by such scholars as Adam Lonitzer, Conrad Gesner, Pierre Belon, and Guillaume Rondelet.[29] The first printed work devoted entirely to shells was Italian, by the Jesuit Filippo Buonanni; and the Duchess of Portland, whose fabulous collection was sold at auction to great acclaim in London in 1786, is perhaps the most famous of all early modern shell collectors.[30] Shell grottoes were built across early modern Europe, and still-life paintings in which shells feature prominently were also popular in France and Italy, not just in the Low Countries. Shells mounted to form elaborate drinking vessels were prized possessions of elite German patrons as well as wealthy Dutchmen. The collecting of shells may have been especially fervent in the Netherlands, but the impulse to collect and admire them was commonplace among European social and intellectual elites (fig. 3).

This interest in shells was driven by their sensual qualities, although such aesthetic interest merged in sometimes complex ways with the priorities and analytic categories of natural philosophy.[31] Georgius Eberhard Rumphius, living for the greatest part of his life in Ambon, was fascinated not just by the beauty of shells; he was also interested in their animal occupants, as well as their part in the culture of the indigenous peoples of the region.[32] As Rumphius's modern translator and editor, E. M. Beekman, has pointed out, this manner of appreciating marine animals inside their shells rarely reflected the interest of collectors back in Europe to whom Rumphius's work was eventually marketed, and who had little interest in anything but the shells: "Where one expects a conchology, one is surprised by a malacology. These pages were conceived with a different intention than the one which caused it to be marketed."[33] With the exception of Martin Lister and his daughters, whose *Historiae conchyliorum* of 1685 deviated from the norm both by examining the ecology of mollusks and in its focus on local shells, most admirers of shells dwelled instead on their aesthetics and their meanings.[34] Where beauty was concerned, we can note a particular enthusiasm at this time for items that were or appeared to be marbled: shells, agate, marble, marbled paper, striped or speckled tulips.[35] In 1611, Philip van Borsselen published an elegaic poem about shells and

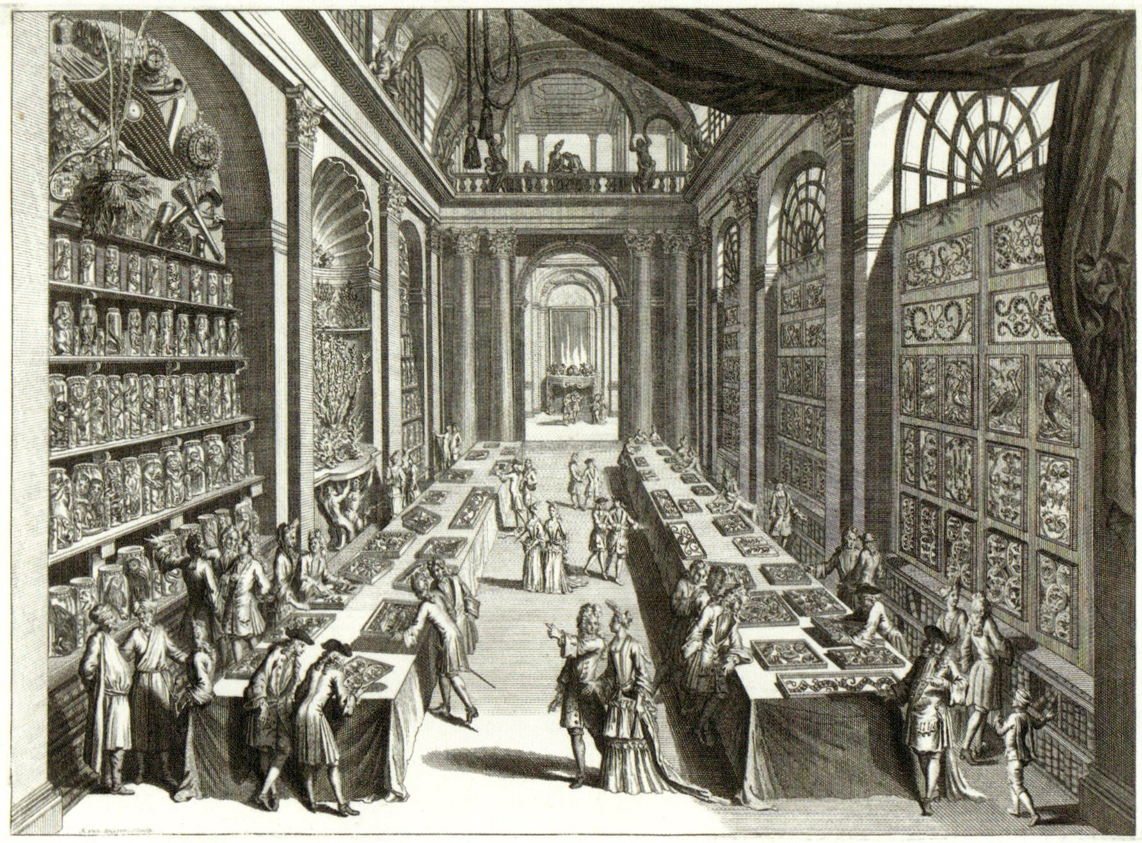

**FIG. 4.** Andries van Buysen (after Romeyn de Hooghe), *Visitors in the Natural History Cabinet of Levinus Vincent in Haarlem*, c. 1706, engraving, 22 × 31.5 cm. Rijksmuseum, Amsterdam [RP-P-BI-5324].

shell collecting in honor of the collection of his brother-in-law, Cornelis van Blijenburgh, comparing shells to many other natural objects that shone, glinted, or were shot through with color.[36] That shape, color, and luster were key to the popularity of shells is evident both in the writings of collectors and in contemporary auction catalogues. Valentyn, for example, identified shells by name when writing of the treasures possessed by his fellow collectors, but he also made sure to remark on their appearance in detail. He described an item in the collection of Johan de la Faille as follows: "a highly rare Shell, named by [de la Faille] *Cedo Nulli*, almost as big as an Orange *Admiraal*, but somewhat shorter and pressed together, and somewhat thicker, having small yellow, white, and brownish bands in a chain fashion over the whole Shell, which struck me as extraordinarily rare, and beautiful, being an unusual piece, which no one else possesses."[37] Even the choice of the name *Cedo Nulli* bespeaks the learned play entailed in the collection and appreciation of shells. The phrase, famously adopted by Desiderius Erasmus as a sly and boastful personal motto, declares that de la Faille's rare shell "yields to none."[38] Moreover, the emotive poetic language used by those writing about shells in the period underscores the way shells evoked affect and creativity. In writing about the nautilus, for example, Rumphius describes the shiny second layer, revealed when one polishes the shell, as "the beautiful color of Mother-of-Pearl, though showing more green and red, giving off a reflection like a rainbow. The partitions are also of a Substance that is like Mother-of-Pearl, but much more beautiful, smoother, and whiter, shining like silver."[39]

The aesthetic reflected here is one valuing rarity and variety: speckles, stripes, unusual shapes and colors, luminosity and sheen. The stress in the poem by van Borsselen is on creation and workmanship, even though the workman, in this case, is God. Indeed that is partly the point. Van Borsselen's title contains the phrase "In praise of the Creator of all things" (*Tot lof van den Schepper aller dinghen*). The aesthetic qualities of shells were appreciable in part because they reflected the contemporary taste in painting—variety and appeal to the eye—but also because, unlike painting except in the most indirect sense, they owed their seemingly infinite shapes and colors not to humankind, but to the divine creator. Van Borsselen's expression of natural theology—an appreciation of and interest in nature because of its reflection of the creation of God—was typical of the period, and a major impetus for the collection and representation of shells and natural curiosities more generally.[40]

Levinus Vincent (1658–1727), a merchant and one of the greatest Dutch collectors of natural history objects of the late seventeenth and early eighteenth centuries, expressly appealed to natural theology in his published descriptions of his collection, which was open to the public for a fee at different points in Haarlem and Amsterdam (fig. 4). For Vincent, and for his wife, Johanna van Breda, a talented embroiderer, a collection was a work of art similar to embroidery, artfully constructed by the collectors and pleasing to the eye; but it was intended as well to show off the works of God to the public. Indeed, for them the main point of having a collection was to bring the wonders of God to the attention of the public, and to confirm in their more pious viewers their understanding of God through his infinite variety and order. Vincent wrote that the collection, originally begun by his brother-in-law, Anthony van Breda, was very necessary because of its ability to convey "the magnificence and almightiness of the great Creator of the Universe."[41] Similarly, one of the other great collectors in Amsterdam in the same period, the apothecary Albertus Seba (1665–1736), founded his collection on principles of natural theology. He wrote in the preface to the first volume of the description of his collection that "the glory of God is my great goal. Because without such a goal, in what way would the knowledge of the works of Creation serve us? Should it hold any other use than to make us celebrate the perfections of the supreme Being?"[42]

Yet even as scholars, poets, and collectors dwelled on the wonder of divine creation, the makers of shell objects and manipulators of shells were also keen to assert the role of human craftsmanship.[43] In the early modern period, a particular thrill for observers of art and nature came from the way natural and artificial could come together in unexpected ways. One of these was to "improve" the natural object by framing it or impressing it with ingenious works of human artistry.[44] As Claudia Swan and Róisín Watson discuss below in their respective essays, shells in their natural state were generally less favored in Europe, with shell polishers in both the Indies and Europe stripping off outer layers to reveal the beauties underneath. To understand shells in the early modern era, we need to understand the socioeconomic connections with trade and imperialism, including the networks that brought shells to European shores and resulted in their being reworked and placed in new contexts, in order to think more expansively about the social and aesthetic lives of these particular things.

How these objects were changed tells us much about the taste of the period. Some large shells were etched with pastoral or mythological scenes, as we find mentioned in the illustrations of Seba's collection as well as in contemporary auction catalogues.[45] In similar fashion, shells were often displayed in cabinets in elaborate designs, portraying faces, garlands, flowers, and other patterns. We know that Seba's collection included grotesque faces made of shells, and these *arcimboldesques* were yet another way to make art out of nature. Some of Vincent's displays were constructed using the embroidery of his wife, Johanna van Breda.[46] This type of image making, as Róisín Watson argues below, was similar to the creation of grottoes, with shells serving as building blocks for another form of creation and purpose. Shells were often arranged in symmetrical and aesthetically pleasing designs both in the drawers of shell cabinets and on the page, a practice that has sparked debate among historians of science over whether it is in fact correct to suggest that "artistic" renditions of a page of shells were fundamentally different from a supposedly more scientific portrayal of shells.[47] Recent scholarship has in fact suggested that we cannot draw sharp distinctions among these different modes of representation of shells in the eighteenth century, because the preferences of consumers—their desire for polite displays of taste—and the training of artists in this period shaped the representational choices made.[48] Indeed, eighteenth-century Parisian collections, including collections of natural history objects such as shells, were displayed in ways consonant with the taste and style of their domestic surroundings: they were often simply forms of domestic décor.[49] Collecting in this period was at least in part an activity shaped by notions of polite self-representation; even those primarily interested in the philosophical aspects of shells were themselves caught up in a world of self-fashioning that depended on certain notions of aesthetics. In all these cases, the choice of objects and the manner of display were related to a set of learned practices and attitudes toward aesthetic authority.

We can see, then, how particular objects and the way they were displayed were shaped by considerations of the social group. This also related to understandings of the individual. Collecting and display could be aspirational and self-defining, allowing status to depend on one's ability to acquire, but also to arrange and to discuss, the objects one had. There could be a competitive quality to these forms of shell collecting and display. For all of François

Valentyn's praise of the communal nature of his Dordrecht shell club, it is easy to notice in his account of the cabinets and possessions of his fellow collectors a sense of what they had and he did not, or vice versa. Valentyn even printed a table in which he listed various shells by type and indicated the number of them in the possession of other collectors, followed by the total for *Ik* ("I, myself"). Although he was not always the person with the most shells of a particular type, there seems to be no reason other than sheer competitiveness for him to construct such a table.

In some ways, one might see a similar performativity in the display of humanistic knowledge and appreciation indicated by the more contemplative uses of shells. Although not as obvious as the social behaviors of collectors, who visited each other's collections and relied upon their abilities to discuss objects to boost their status in the learned world, thinking through the internal, emblematic, and symbolic meanings of objects also entailed a certain performance. There was of course sincerity in the piety of adherents to a natural theology that directed people toward the study of nature as a means of honoring God's creation. But it was also the case that there was a form of self-display in understanding the symbolism of shells. To know that a spiraling shell might be a metaphor and an emblem for spiritual repose was to possess a high level of humanistic education.

There is display here, but as the essays in this volume emphasize, there is thought as well. We see in the early modern period people interacting with shells, through sociability and performance, but, as Hanneke Grootenboer argues in her essay, shells were at the same time objects to think with. Valentyn's club of six *liefhebbers* in Dordrecht discussed the qualities of their shells, compared and contrasted them, and, as Stephanie Dickey argues below, behaved like connoisseurs of art objects, especially prints.[50] They paid particular attention to small variations in what were otherwise conceived as multiples, and this in itself, as we can see from many of the early modern authors on shells, led them to a contemplation of religion and the wonders of God. That this religious wonder, an appreciation of the aesthetic qualities of shells, and the social and intellectual competitiveness of collecting occurred at the same time and in the same space is of a piece with the nature of early modern collecting. Shells were something to discuss with others—but also something to think about in private, part of a life of serene learning, of contemplative solitude broken periodically by quiet learned sociability.[51] Through their Ciceronian connotations, but also through their very variety of color and shape, shells gave rise in turn to a range of thoughts, whether about the best life for a humanist, the nature of humankind, the playfulness of nature, the nature of God, or the regularity of God's design. The complex but regular inner chambers of turban shells could be an inspiration for a contemplation of natural theology; but they could also inspire other forms of contemplation. The Fibonacci sequence or golden section—rediscovered by the mathematician and astronomer Johannes Kepler in 1611—was fascinating to naturalists and mathematicians, and the fact that the nautilus shell followed this spiral was an inspiration

FIG. 5. Cornelis de Man, *The Curiosity Seller*, 1660s. Oil on canvas, 60 × 51 cm. Private collection.

for contemplation of mathematics as well as of God's design. Spirals of shells could even be taken as a microcosm of a rotating universe.[52] In a more analogous fashion, the idea of "turning into oneself" like a shell, contemplating the nature of life and death, was a long-standing trope, as we see in different ways in the essays in this volume by Hanneke Grootenboer and Róisín Watson. Shells helped inspire deeper and more complex thoughts, like the chambers of a shell, even when they were part of a decorative design in a grotto or in a dollhouse. They could tell us something about the character of the person who owned them, for example as a pictorial strategy in one of the portraits made at the time of collectors with their shells; Grootenboer has discussed this elsewhere in reference to the Goltzius portrait of Jan Govertsz van der Aar.[53] And, as Marisa Anne Bass argues here, paintings of shells in this period could be elaborate contemplations of the relationships among life, death, and potential reanimation. The liminal character of shells returns to the fore in these representations of shells: Were they alive or dead? What was their relationship with the hand of the artist, or the hand of God? Shells were not just another subject for still-life painting or other forms of artistry: they had resonances of their own, as commodities, as things with multiple lives, as objects that obeyed their own rules.

Shells appeal to the eye, but also to the other senses, in particular to the sense of touch. Despite our usual understanding of the genteel enjoyment of shells as primarily visual, Anna Grasskamp suggests here that conchophilia also included the sensual, and indeed sexual, excitement aroused by the luster, shape, smoothness, or prickliness of shell specimens. The sensuality of shells invites us not only to look, but to touch and feel. It even invites us to listen to the sound of the sea in the shell, as Cornelis de Man illustrated in his *The Curiosity Seller* of c. 1675 (fig. 5). Early modern shell grottoes were similarly designed to highlight the material delights of the shell, to confound viewers with glinting light, smoothness, and repetition, which would in turn give rise to other thoughts and feelings. Some of these could be philosophical, others sensual, as shell grottoes could be sites of festivity and debauchery, as well as of contemplation and mourning. Indeed, the materiality of shells, with their soft white smoothness and suggestive shapes, called handlers' attention to shells as substitutes for bodies. These could be sexual bodies, and the rarity and exoticism of shells created analogies for the sexualized understandings of foreign and colonized peoples in the same period. Shells appealed to touch, but they also called to mind thoughts of conquest, whether sexual or political. Indeed, these two were intertwined, at the very least in a metaphorical sense.

*Conchophilia* is divided in three parts. The first, "Surface Matters," begins with an essential question: How did shells from distant locales come to the shores of Europe, and what reflections could be made in the early modern period—and what reflections can we make now—on the resonances these processes call up? As Claudia Swan argues, in order to understand the passion for shells in this period, we need to understand how shells were gathered, manipulated, and transformed even before they arrived on European shores, as

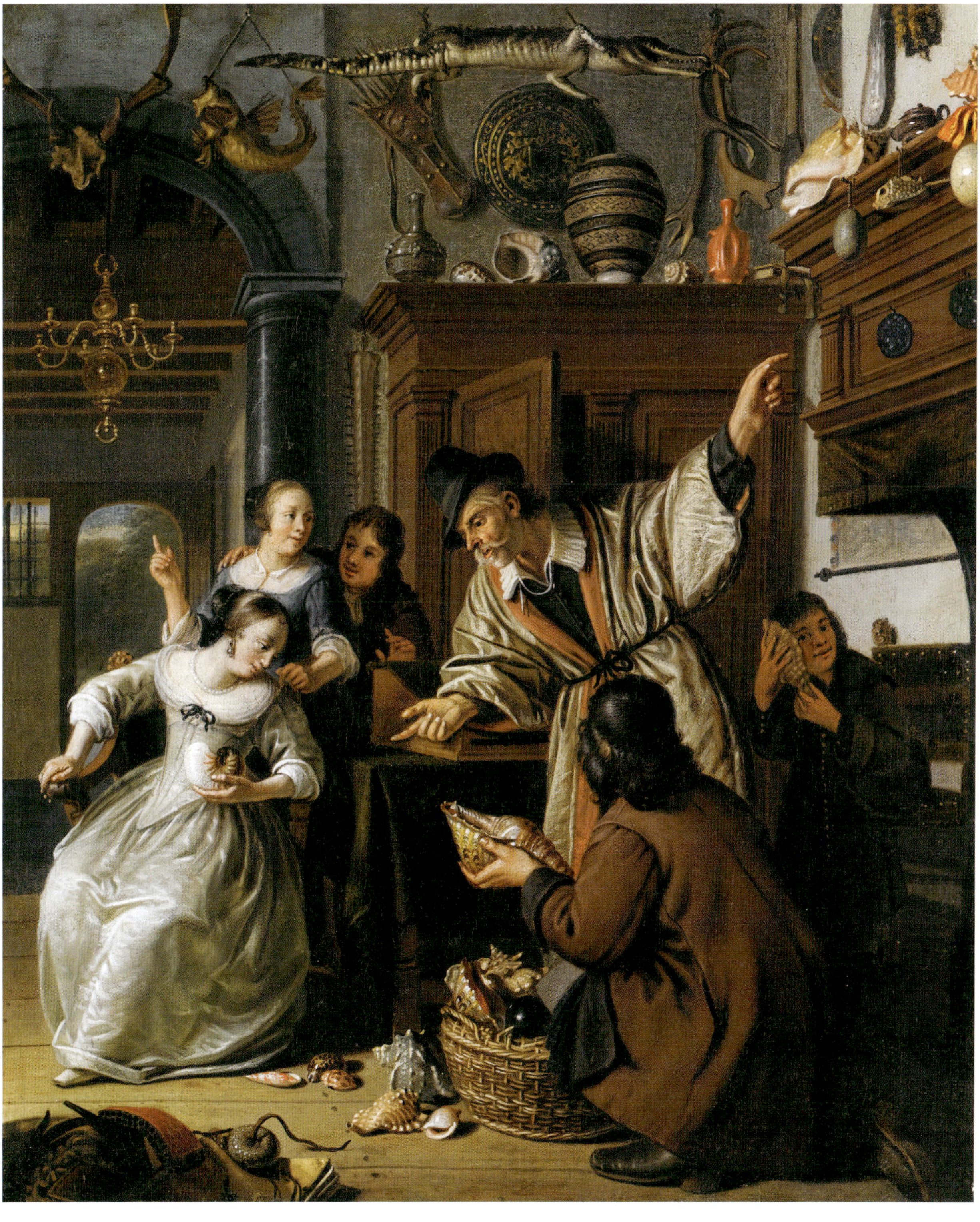

well as how they were handled once they arrived. Shells were aesthetic objects, as we have seen, and Swan discusses the role shells played in European society and the sensuality of the experiences collectors had with them. But shells were also commercial items, part of a world of trade and money, in their most extreme form as shell money or wampum in colonial exchanges and in the slave trade. Shells and their traders and collectors, we can see now, were implicated in a colonial project whose resonances are far different today from what they were in the early modern period. By exploring the "great effort and grief" required to render shells and their surfaces appealing to European buyers (who scarcely appreciated the labor involved), Swan asks us to look again at shells in the early modern context, to think beyond the notion of their status as natural wonders. This means thinking, first, about their sensual appeal and the ways this was enhanced by the work of many hands to delight a European aesthetic. It also means thinking about the biography of these objects: what lives they lived in the sea, who retrieved them, who altered their appearance, and with what methods and labor. Ultimately it also makes us think about other ways shells and bodies interacted: in the slave trade, in the labor of enslaved bodies, as well as through the sensuality of European bodies coming into contact with newly smooth and luminescent shells.

Anna Grasskamp in turn takes up the sensual appeal of shells deriving both from their exotic origins and from their provocative associations with the sexual characteristics of the female body. Through their shapes, colors, and translucent qualities, shells described in inventories as "Indian" played a prominent role in the frequent interaction between the exotic and the erotic. The desirability of shells was undeniably shaped by a sense of their foreignness, as indeed was the case with other prized commodities like porcelain, with which Grasskamp demonstrates that shells were felt to have a particular affinity. Yet their use in elaborate drinking vessels to conjure fantasies surrounding the bodies of mythological beings and the non-European "other" reveals a singular aspect of the associative allure of shells as natural objects. The mixture of vision, touch, allusion, and fantasy in both nautilus cups and paintings of shells and marine subjects brings to the fore an important way of thinking with shells about the nature of colonial exploitation, and a darker side to the desire to possess shells on the part of early modern collectors.

The second section of this book, "Microworlds of Thought," takes us from the bodily to the intellectual, while continuing the theme of shells as vessels of connotations and cultural resonances. The essays in this section consider the ways artists and collectors conceived of shells as objects to "think with." Shells were not just mute subjects of painting or static specimens in a collector's cabinet. They were also integral to early modern reflections on art, identity, and the relation between the human and nonhuman realms. The essays here explore the significance ascribed to shells by European audiences, particularly beyond traditionally discussed themes such as luxury or vanity. Early modern paintings of shells, as Marisa Anne Bass shows, often present their subjects less as prompts for intellectual repose than as uncanny things, which unsettle not only the compositions they inhabit but also the viewer's understandings of them. Thinking into shells—their histories, their forms, their associations—reveals, according to Bass, the fundamental inadequacy of "still life" as a term to describe what shells do within the realm of a two-dimensional representation. Shells implied movement, not least because they retained a connection with the living animals previously within them. A shell's potential for reanimation within the pictorial realm through techniques of compositional juxtaposition and association suggest that "shell life" is a category of its own.

Thinking with shells—using their shape and former function as shelters or homes—was, according to Hanneke Grootenboer, a way in which shells could inspire introspection. These naturally formed containers would not only trigger religious reflections on how the body was bound to the soul, but also provide a model of intimacy and interiority. Shells emerge here as philosophical objects, objects that echo the idea of "turning into oneself," encouraging thoughts about placement, displacement, and grief. Using the powerful example of the shell collection within Petronella Oortman's dollhouse, now in the Rijksmuseum, Grootenboer explores the way Oortman created a miniature world into which the spectator, and in particular Oortman herself, could spiral deeper and deeper into philosophical and personal associations, along the lines of the wentletrap, a famously rare shell in her collection.

The final part of the book, "The Multiple Experienced," focuses on shells not as individual objects, but in profusion and repetition, and how this could also resonate with themes we have already noted, of sensuality, contemplation, and sociability. Early modern audiences were particularly intrigued by the multiplicity of shells and yet their seemingly infinite variety in form. From classical times, these themes were exploited in architectural details, in garden design, and in the decoration of houses. While these uses of shells were about fashioning the exterior, they also were in their way about interiority and thought. The use of the natural to decorate the artificial, whether singly or in multiples, gave rise to the contemplation and delight of this evocation of the boundaries of man and nature. A single shell could be a source of wonder; how much more so the walls of a grotto encrusted with natural shells, with the luster and glint of their surfaces in candlelight? Róisín Watson's essay explores how shell-encrusted grottoes attached to palaces, used as spaces of entertainment as well as monuments of mourning, were also objects of knowledge and philosophy. Their very making required architectural and engineering knowledge and deep thinking about the properties of shells. The material properties of shells further aroused pleasure, but also evoked thoughts concerning the divine perfection of nature and the ephemerality of life. And given the value of shells, such multiplicity was also a kind of performance, a performance of political power and wealth, as well as an overwhelming play on the senses. Learning, contemplation, discussion, and social interaction could also work through an experience of the multiple; as a site of both social and individual experience, the shell-encrusted grotto was,

Watson argues, akin to the *Kunstkammern*—cabinets of curiosity popular in the period—in the way it encouraged discussion and introspection, while intertwining nature with technical proficiency. A grotto was also a shell in macrocosm, arousing curiosity and wonder as the spectator explored ever deeper underground, with complex architectural effects promoting thoughts ranging from wonder to grief.

This sense of the multiple as a conduit of thought and meaning is explored in a different way in the final essay in this volume, by Stephanie Dickey. Taking up the theme of shells as both commercial objects and objects of knowledge, Dickey examines the similarities of the eighteenth-century markets for shells and for prints, markets that converged in the figure of the Parisian dealer Edmé-François Gersaint, who dealt in both. Dickey returns us to the world of collectors like François Valentyn, who met on Wednesdays with his shell-collecting friends to make comparisons between their shells. Dickey considers shells not as individual objects to be explored philosophically, but as multiples whose variations were to be discussed endlessly, and about which discernment and judgment of tiny details carried not only social significance in a world of connoisseurs, but also market value. This analysis brings out further material for debates about the relationship between aesthetics and natural knowledge, since Gersaint was marketing shells specifically to an audience that already knew about art; further, through the comparison with prints, it also helps us understand more specifically how expensive shells acquired their particular value in Europe, a value they probably had not had in their places of origin. This essay delineates in a new way how early modern knowledge about shells interacted with knowledge about art, and how sociability, the market, and connoisseurship informed this knowledge.

When we talk about early modern shells, then, we are really talking about early modern people. Although Martin Lister and a few others were interested in shells for their ecological and malacological properties, for most Europeans interested in shells at this time, shells were metaphorical (and sometimes literal) vessels to be filled with meanings about their own society. Shells show us the ways people interacted with each other, using objects to discuss intellectual matters even as they competed for status. They show us people considering the nature of their world and its relationship to God, taking pleasure in the complex design that so clearly seemed to indicate the designing hand of the Creator. They show us people contemplating their own mortality, exploring their own sensuality, thinking about the nature of relationships with foreign lands and peoples, ruminating on the nature of life and the possibility of reanimation, judging objects singly and in multiplicity, and undertaking exciting explorations of the nature of value. Even in paintings of shells that are stunningly true to life, we still think about the hand of the artist. Someone, perhaps just out of sight, placed the shells in the pattern portrayed; someone bought the shells; someone brought the shells from the Indies to Europe; someone labored, perhaps under duress, to polish the shells and bring them to an unnatural luster; someone, at the earliest stage, retrieved the shells from the sea or shore in a far-off clime. Shells, then, speak particularly to a multiplicity of relationships between people and objects, and people and people in the early modern period, whether social, commercial, intellectual, theological, sensual, or sexual.

All these are themes that have been discussed in various disciplines that usually develop separately: for example, work on material and visual culture, the production of knowledge, the history of ideas, the realm of the senses, and the life of the mind. Yet in this book we see that it is possible for us to investigate the way these themes intertwine and connect by seeing them all through the intimate spiral of the shell. Shells were objects to think with in the early modern period, and, as we show in this book, they remain tools for scholarly analysis.

Thinking with shells indeed continues to be a theme in art and thought in the twentieth and twenty-first centuries. From Paul Valéry's 1936 work on the aesthetic qualities of the shell, to Gaston Bachelard's comparison of domestic space to shells, we see the way that the architecture and design of the shell continue to inspire and compel. Photographers like Dora Maar in her *Hand-Shell* of 1934 or Edward Weston's *Shells* series in the late 1920s amplify the surreal mystery or, in Weston's case, the fascination with the form of the shell. More recently, Kara Walker's *Fons Americanus* in the Tate Turbine Hall, with its themes of maritime conquest and colonialism, resonates with modern interpretations of the period discussed in this book, and with some of the analyses we have made here. Damian Hirst's 2017 installation *Treasures from the Wreck of the Unbelievable*, an exhibition of imagined goods retrieved from a fictional ancient shipwreck, including the sea creatures clinging to them, meditated, as one review put it, on "art for a post-truth world."[54] These reflections on and uses of shells in the modern world demonstrate that the age-old mystery of "shell life" continues to fascinate, and that, as we have already observed concerning the early modern period, shells are vessels for the meanings of the societies in which they are found. To explore the contents of these beautiful vessels in early modern Europe is the task of this book.

# Part I

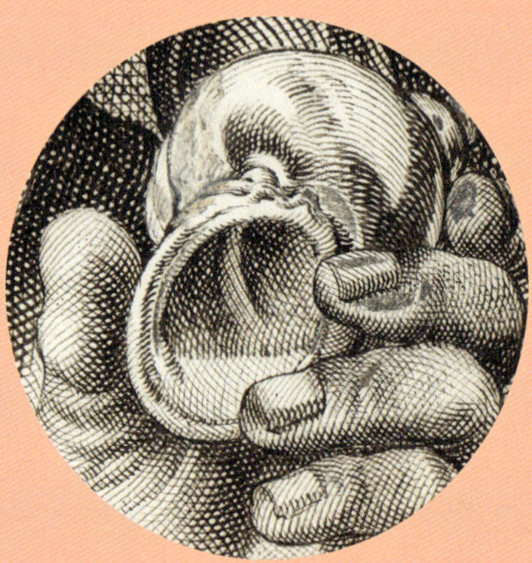

# SURFACE MATTERS

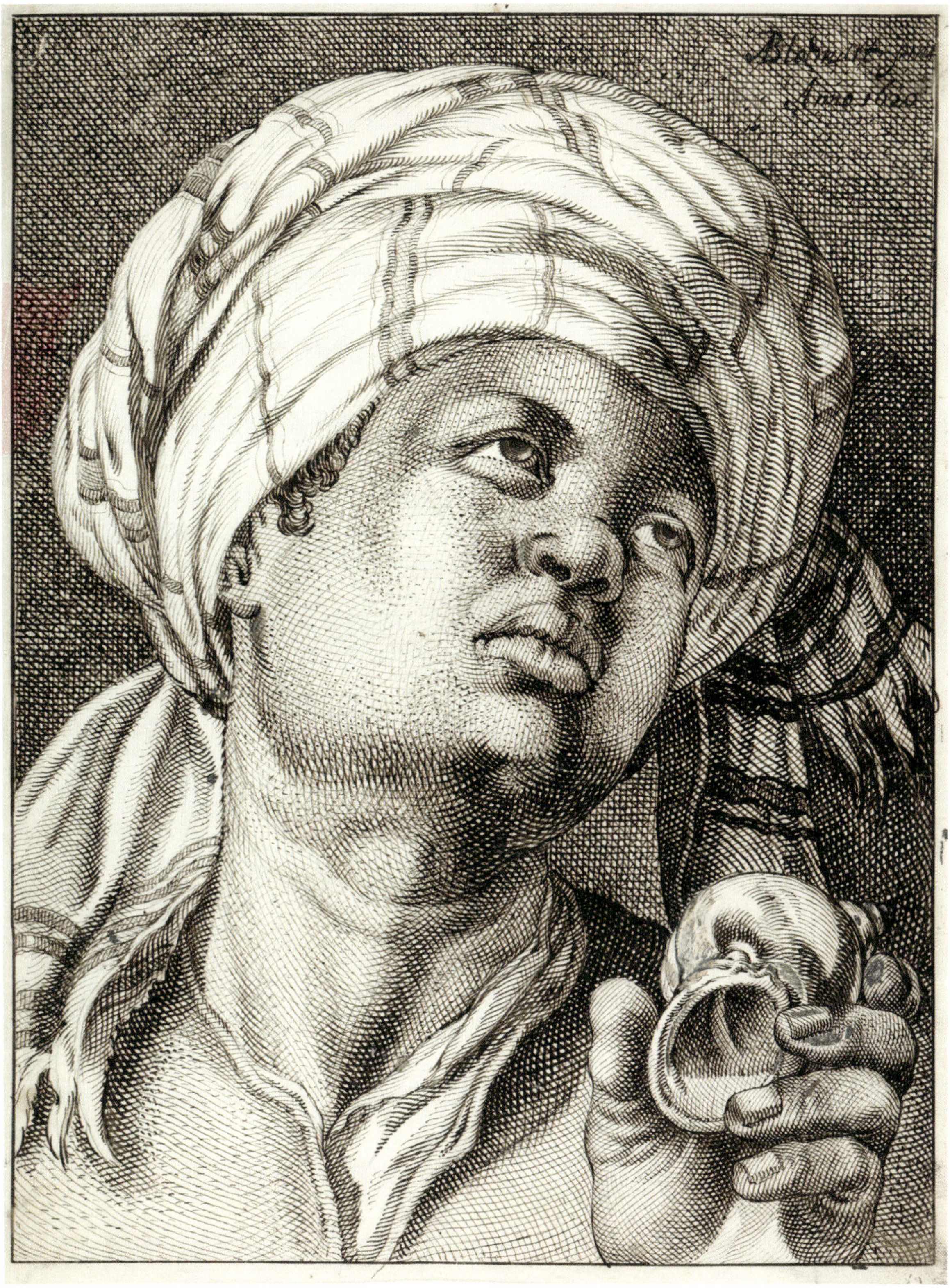

Why this lattice ornament? Why these fluted scales, these lumpy nodes?
Ignorance was, in the end, and in so many ways, a privilege:
to find a shell, to feel it, to understand only on some unspeakable level
why it bothered to be so lovely.
What joy he found in that, what utter mystery.

—Arthur Doerr, "The Shell Collector"

Claudia Swan

# Chapter 1
# The Nature of Exotic Shells

## HANDLING SHELLS

**FIG. 6.** Abraham Bloemaert, *Tronie of a Black Man with a Conch*, 1620. Pen and brown ink, 35.2 × 26.7 cm. Herzog Anton Ulrich-Museum Braunschweig, Kunstmuseum des Landes Niedersachsen [inv. no. Z.1816].

Shell collectors of all ages and levels of expertise share the joy that Arthur Doerr's blind malacologist experiences searching out and handling shells: taking them in hand, we encounter the mysteries of their forms. This chapter is about handling shells, through the lens of early modern shell collecting. Its focus, however, is on onerous forms of handling shells—catching, preparing, maneuvering these fluted and lumpy wonders.

Often, where the handling of shells is articulated in pictures, it is to express possession of and identification with the objects at hand. Early modern portraits emphasize collectors' and naturalists' attachment to their spiny, variegated, nacreous specimens; and shells and marine creatures serve as attributes to identify nymphs, Nereids, and the god of the seas, Neptune. In c. 1620–25 the Dutch artist Abraham Bloemaert drew a *Tronie of a Black Man with a Conch*, a figure he repeated in other media (fig. 6).[1] The sitter is not identified, and the presentation of the black-skinned man, headgear, and *Turbo marmoratus* (marbled turban shell) suggest that the image is an allegory of the other as seen through European eyes. The foreign shell, the source of mother-of-pearl, seems an emblem of the figure's exoticism. He wears a cotton turban typical of Melanesian Muslims and may thus present a shell from the region whence he came, but this image appears to be a study of foreign forms as such without regard to specific points of origin. The artist's handling of the man and his mollusk objectifies them both.[2]

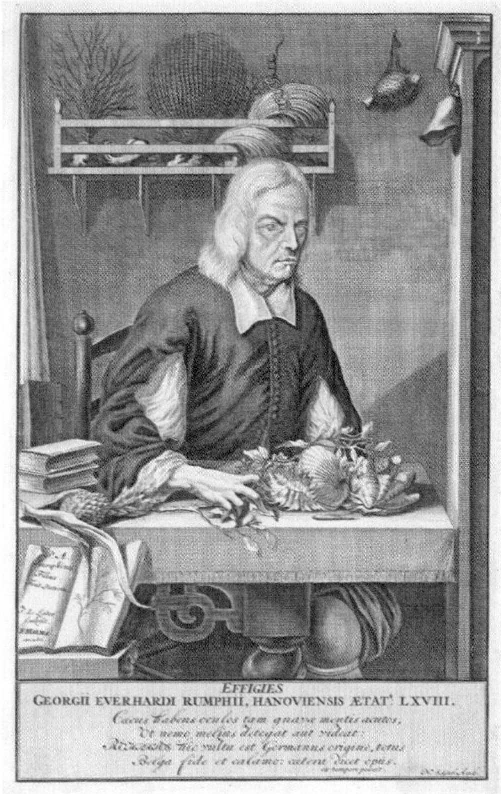


**FIG. 7.** J. de Later after Paulus Augustus Rumphius, *Portrait of Georgius Eberhard Rumphius at the Age of 68*. In Georgius Eberhard Rumphius, *D'Amboinsche rariteitkamer*, 1705. Etching and engraving, 34.7 × 22 cm. Rijksmuseum, Amsterdam [RP-P-1937-1064].

In ways this chapter will address, Black bodies and exotic shells share a history. Bloemaert's Black man handles a shell typical of those that were highly prized, avidly collected, and widely studied in the early modern era. Generally, when we speak of exotic shells depicted in painted still lifes, drawings, and scientific publications, we focus on such forms of labor as the artist's touch and the natural historian's classificatory zeal—labor that eclipses the work of procuring the shells in the first place. It is the latter form of labor, performed in the early modern era by Black bodies, that I aim to bring into view.

This chapter opens and concludes with observations on a signal early modern publication on shells by the German-born naturalist Georgius Eberhard Rumphius (1627–1702), *D'Amboinsche rariteitkamer* (The Ambonese curiosity cabinet). Rumphius, a longtime functionary in the Dutch East India Company (VOC), lived out his life on the island of Ambon in present-day Indonesia. His aim as an author was to write a natural history of the Moluccas. Rumphius, who suffered serial tragedies in Ambon, including the loss of his eyesight, was known even in his time as the "Pliny of the Indies" (fig. 7). His publications—*D'Amboinsche rariteitkamer* and an herbal, *Het Amboinsche kruidboek* (The Ambonese herbal)—convey information about morphology, ecology, biology, and local knowledge of Moluccan specimens that continues to be cited.[3]

Both of Rumphius's publications, which appeared posthumously, were intended for an audience of European collectors.[4] The very title of *The Ambonese Curiosity Cabinet* places the book in dialogue with collections of the time, whose curious contents redounded to the social, epistemological, and political credit of their owners. Rumphius gave the volume its title, he explains in the introduction, "because it describes those things, both from living and lifeless creatures, that, on account of their unusual shape, or because they are rare, are collected by *liefhebbers* as rarities."[5] Appreciation for rarities was expressed in a variety of ways in the early modern period, when sensual and philosophical wonder readily overlapped.

The elaborate frontispiece of *The Ambonese Curiosity Cabinet* depicts the sort of collection that informed Rumphius's efforts and where his book was to be consulted (fig. 8). Crabs, crayfish, sea urchins, shells, and whelks are strewn in the foreground, where two seminaked figures, whose skin is colored black in the hand-colored impression illustrated here, gather them in a large vessel. While these figures do not enter the chambers within, the things they are handling do. Transported by a white-skinned intermediary carrying a basket and a box,

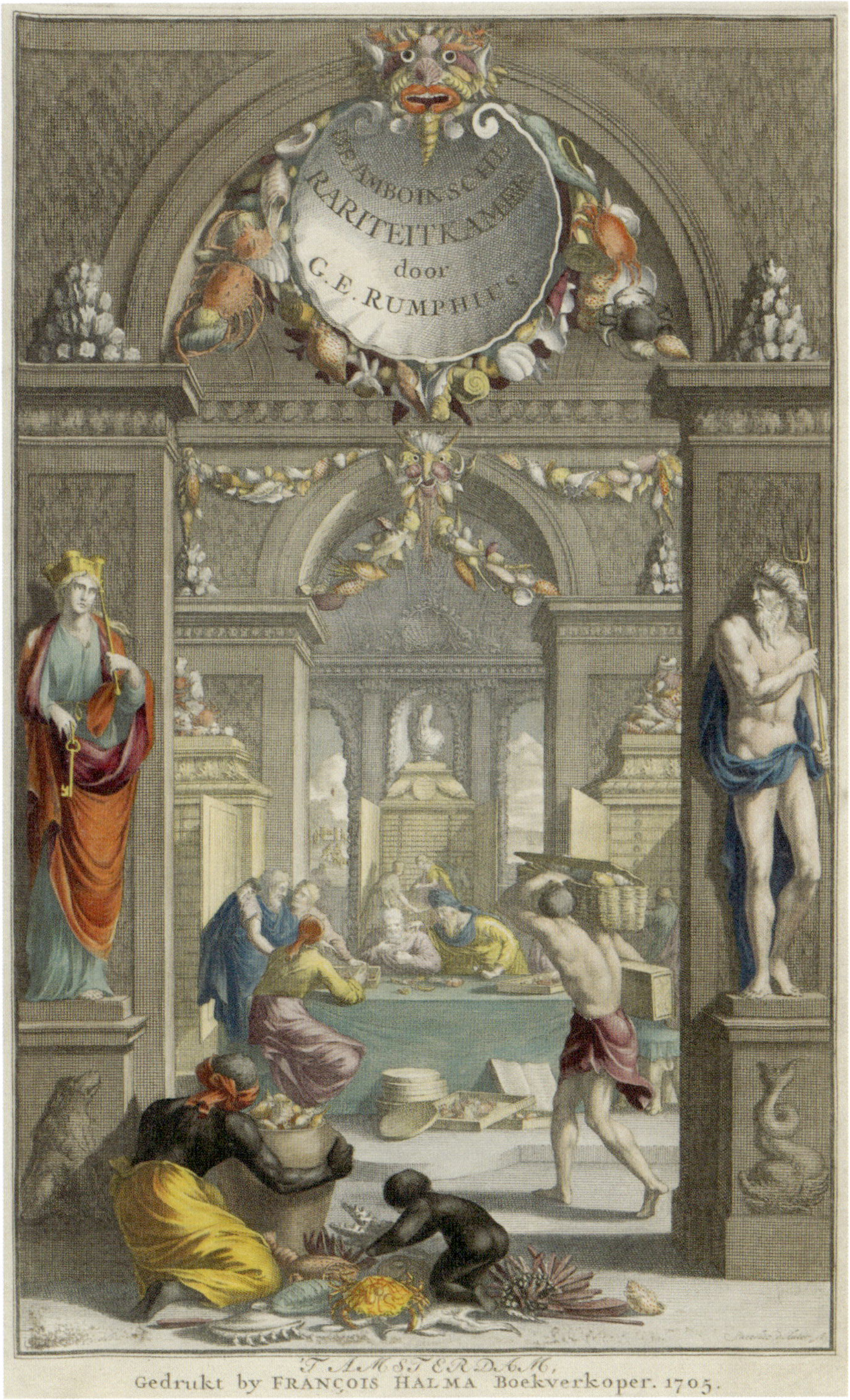


**FIG. 8.** J. de Later after Jan Goeree, Frontispiece. In Georgius Eberhard Rumphius, *D'Amboinsche rariteitkamer*, 1705. Hand-colored engraving, 39.5 × 25.7 cm. Universiteit van Amsterdam.

**FIG. 9.** *Nautilus major sive crassus* and its snail, Plate XVII. In Georgius Eberhard Rumphius, *D'Amboinsche rariteitkamer*, 1705. Hand-colored engraving. Universiteit van Amsterdam.

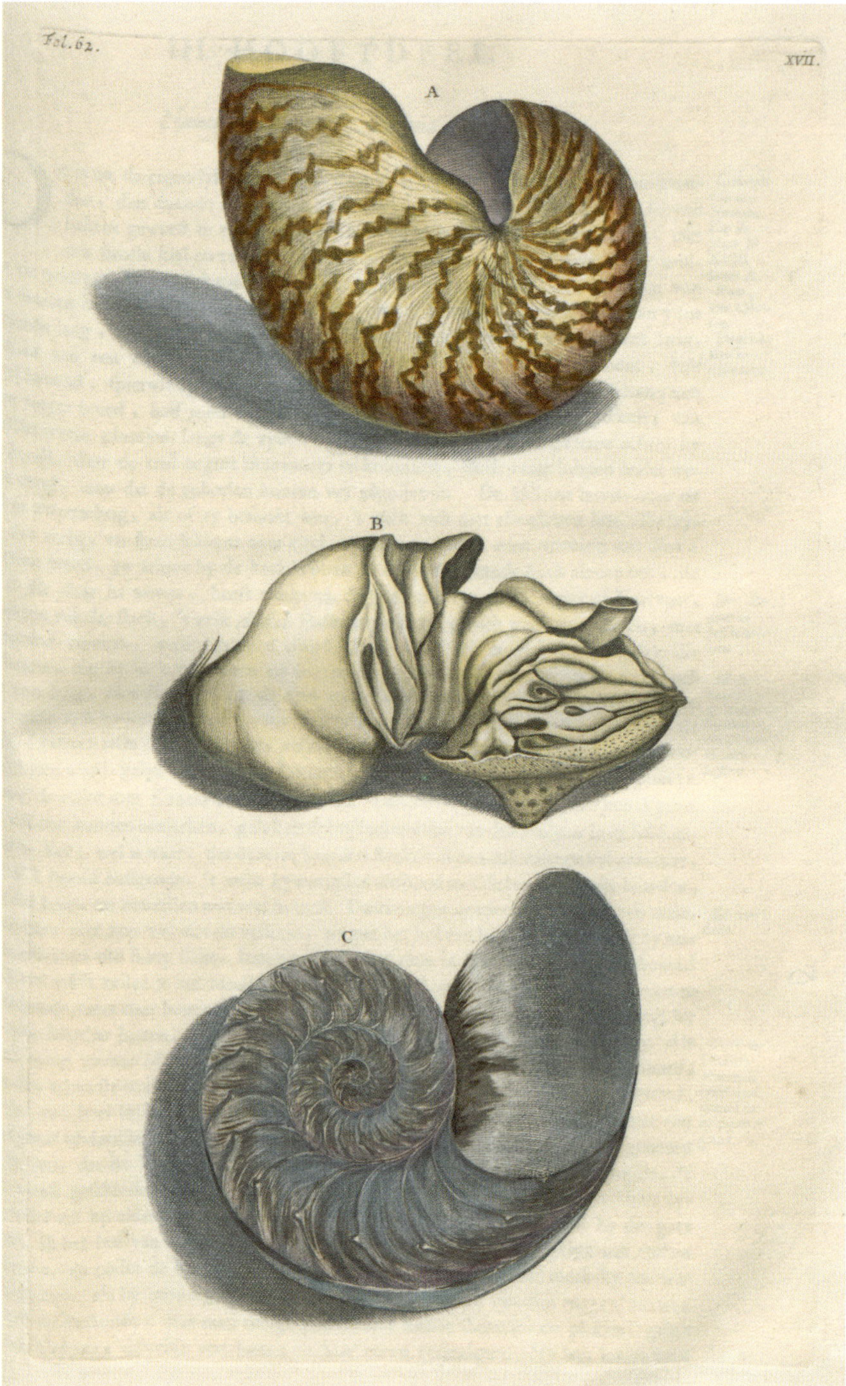

they reach the European men gathered around a table, surrounded by cabinets, examining shells. While the frontispiece alludes to the labor of gathering and shipping shells, it omits any specific reference to their provenience or the labor of procuring them. The European collectors, shown sorting and discussing their exotic objects, perform rituals of natural historical and, perhaps, aesthetic discernment without evident regard for where or how they were found. The respective modes of handling shells enacted by the figures trace the passage from exotic beaches and waters to European spaces of collecting, appreciation, and scientific observation, and labor is subsumed under the aegis of devotion to rarities. Sociable interactions among those inside the cabinet, built on the labor of others, undergird natural history and the culture of curiosity.[6]

Rumphius's *Ambonese Curiosity Cabinet* is divided into three books of which the second, on hard shellfish, is "the true cabinet of *liefhebberij*."[7] The publisher's preface—intended for an audience of Dutch collectors—lavishes attention on book 2, enjoining the reader to enter "the innermost chambers of this eastern cabinet of nature, in which so many and various sorts of whelks and shells, painted with the most variegated colors, are presented to you, that you will have reason to be gratified and amazed by them."[8] Rumphius's text regularly declares shells "beautiful," "finely painted," and "curious," and the publisher's preface to the reader describes the contents of the book as "the wonders the sea brings forth from her womb." But these natural wonders did not present themselves directly to their admirers, nor were they at their most beautiful when discovered, as they were often inhabited by the mollusks that produced them and likely to be slimy, mossy, covered in a sort of fleece (periostracum), or scratched by sea grit. Procuring shells required labor, without which the descriptive and taxonomic efforts of authors and collectors would have been stranded, and on which their pleasure depended.

The first necessary step in the process of collecting shells is to find and gather them. The most highly prized shells in early modern Europe were gathered at a very distant remove from the cabinets in which they were arranged by their European owners. The formidable *Turbo marmoratus* and the chambered nautilus shell or *Nautilus pompilius* (pearly nautilus), for instance, are native to foreign seas—the Indian and Pacific Oceans (fig. 9). Such shells had changed hands several times by the time they were appreciated by artists and collectors in Europe, and it is doubtful that the conditions of their discovery would have been known to their European owners or admirers. Someone gathered the shells, scrubbed them, and cleaned out their original inhabitants, and someone may have polished and even carved them, before they were shipped halfway around the globe and bought and sold in Europe, where they were studied, classified, and attentively portrayed.

In a plaintive chapter of *The Ambonese Curiosity Cabinet*, Rumphius writes of the "expenditure of time and effort (*tyd en moeite*) needed, if one wants to put a set of rarities together," declaring that "I know whereof I speak, for I have devoted a great deal of my free time to such things."[9] It cost Rumphius over twenty-five years to assemble a collection of 360 sorts of whelks and shells from the Moluccan region, with multiple examples of several sorts, that he sold to the Grand Duke of Tuscany, Cosimo III de' Medici (1642–1723) in 1682.[10] Cosimo passed those shells on to Ferdinando Cospi (1606–86) for the collection of *naturalia, artificialia,* and exotica he kept in Bologna, to which Cosimo made serial gifts of exotica.[11] Noting that he omitted to send Cosimo "common and vile" items, Rumphius stated in a letter that accompanied the shipment of his collection that he doubted whether "anything so rare, novel or strange has been seen by so many illustrious, learned, and excellent men in Italy."[12] "Time and effort" were converted, by the sale of Rumphius's shells, into economic value. And when Cospi arranged the shells in a decorative manner, they lost all relation to Rumphius's scientific considerations, as well as to the shells' origins in the Moluccas.

Rumphius's "time and effort" were surely not his alone. Rumphius lost his eyesight in 1670, more than a decade prior to the date of the sale of his collection to the Tuscan noble, so it is reasonable to assume that he had significant assistance in his labors.[13] An otherwise playful exchange with the retired governor-general of the East Indies who supported Rumphius in his endeavors, Johannes Camphuijs (1634–95), attests to who actually hunted specimens. Camphuijs, who lived in Batavia and had a small house on the tiny island of Edam, given to him by the VOC, wrote in the year he died to Rumphius of his penchant for searching for shells, whelks, and other sea creatures, which he said he had discovered during his stay on Edam the prior year. Whether Camphuijs himself combed the beaches, though, is highly doubtful. He was old and ill at the time. His letter states that his slaves, both boys and girls (*lyfeygenen, zoo jongens als meyden*) had found, on Edam and on the small island named after another Dutch city, Alkmaar, shells more beautiful and more curious than any to be found on Ambon. Rumphius responded that he doubted Batavian shells could outshine those local to Ambon and the Moluccas, and sent one hundred kinds to vie with Camphuijs's shells and defend the honor of the "Ambonese monarchy" over marine curiosities. In his letter, Rumphius does not credit Camphuijs, but his servants (*dienaers*), with discovering such things.[14]

## WONDERS OF NATURE, PRODUCTS OF HANDLING

Shells have long been celebrated as wonders of nature; these discarded exoskeletons are cherished as the mysterious products of natural forces. Aristotle held that shells derive their color from the mud whence they emerged: "Speaking generally, all the Testacea arise by spontaneous generation in mud, though they exhibit differences according as the mud differs: in slimy mud oysters grow, in sandy mud cockles and the others I mentioned; on the eroded hollows of rocks the tethya, barnacles, and the commoner kinds, such as limpets and nerites."[15] Early modern collectors and artists expressed admiration for a wider variety of coloring than what Aristotle described, and for the natural artistry of shells' morphology.

The sixteenth-century French ceramicist and author Bernard Palissy admired the colors of shells while proclaiming ignorance of their cause, and "exalt[ed] mollusks as natural craftsmen who channel water and earth deposits in order to paint rainbows on their calcified shells."[16]

In his encyclopedic verse poem dedicated to a shell collector, *Strande* (The beach, 1611), Philibert van Borsselen (1580–1627) compares shells with painted works of art, ivory turnings, and porcelain: nature turns the forms of shells, making her own art. In passages about gathering shells, van Borsselen specifies that they are selected for their strangeness of form, beautiful veining, and the variety of colors and shapes. Morphological intersections and aesthetic overlaps are especially vivid in the case of what he calls the *Concha murrha* (porcelain shell) and what we know as the *Nautilus pompilius*. These shells were transformed by craft into drinking vessels; "on a golden foot [they] are set at table" where, the poet writes, they shine like suns, wondrously containing the rainbow (fig. 10; see fig. 21, fig. 22). They surpass pure gold beakers set with gems, transparent crystal glasses, and noble porcelain with its pretty decorations: the sheen and iridescence, especially when heightened by artistic confection, renders shells noble bearers of heavenly light.[17]

Early modern authors and artists construed shells that are etched and carved or set in precious metal filigree as instances of nature's artistry compounded by human art (fig. 11), and many art historians have followed in suit. This chapter brings to the fore other features of the exotic shells prized at the time, especially the extent to which they themselves were the products of manual processes of culling, selection, and preparation. No shells ever sprang into the hands of their admirers. Acquisition, appreciation, and collection of shells require extraction—of their original inhabitants and from their places of origin. The association of shells with nature's artistry, which runs deep, occludes the extent to which shells are also products of handling, of considerable human labor and many hands.

## VALUES

Who, in early modern Holland, loved to handle shells and how were exotic shells valued? The Leiden-born merchant, patron of the arts, and collector Jan Govertsz van der Aar (1544/45–before 1615) was passionate about curious things, including unusual flowers and shells. In a 1597 letter to the naturalist and director of the Leiden University garden Carolus Clusius (1526–1609), from whom he hoped to obtain "something unusual," van der Aar admitted, "I desire things that are rare and appealing and I covet colorful flowers; the rarer they are, the more covetous I am."[18] In addition, he mentions his passion for things of the sea, embodied in the life-size seated portrait of him painted by Hendrick Goltzius (1558–1617) in 1603 (fig. 12). Within a year of its completion, the portrait was identified as "Jan Govertsz, residing in Haarlem, an admirer of shells, holding in his hand a Mother-of-Pearl, with other shells near him: this is in its entirety exceptionally pleasing, in manner and in

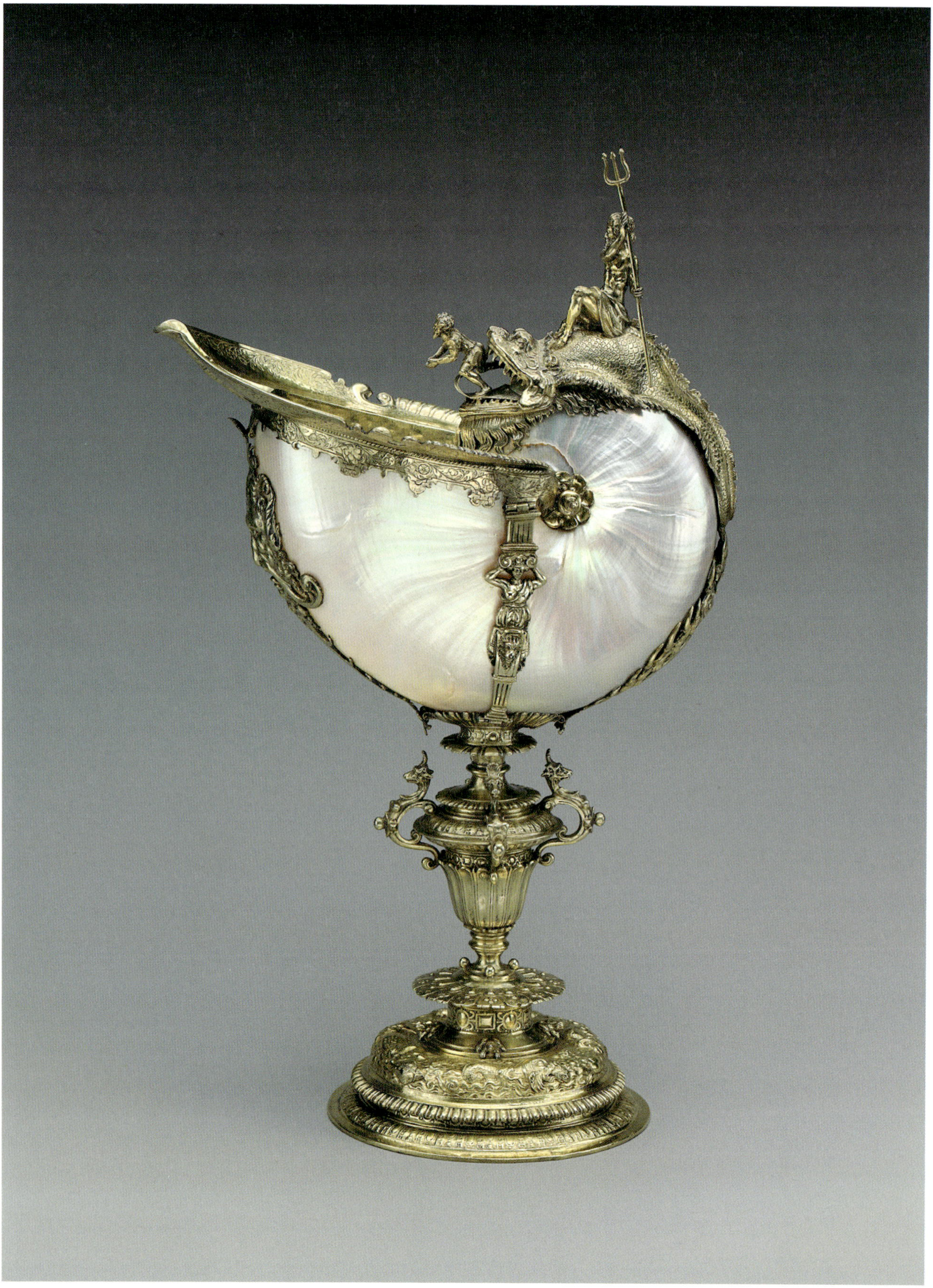

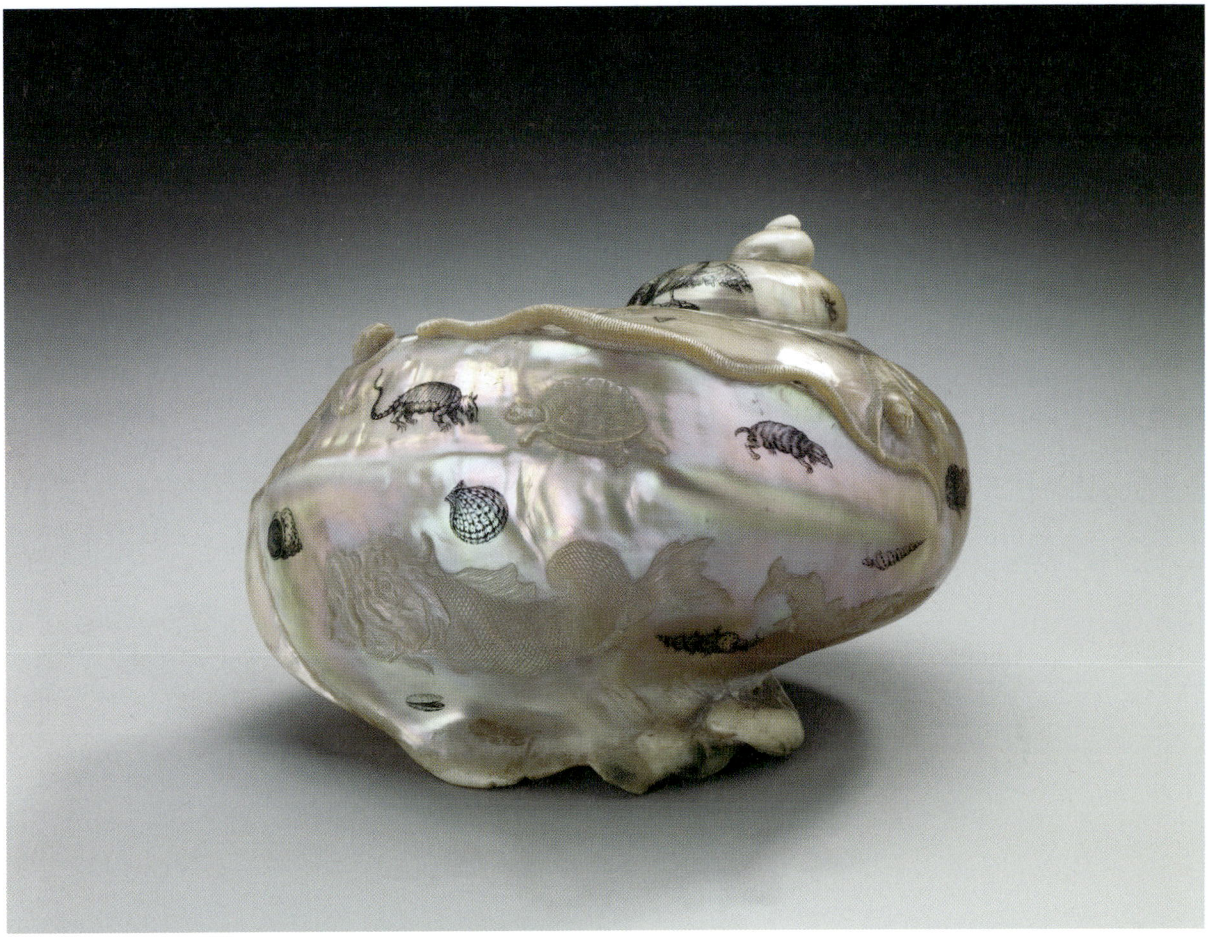

**FIG. 10.** Utrecht manufacture, Nautilus shell with gilded silver mounts, 1602. 27.9 × 16.8 × 10.8 cm. Metropolitan Museum of Art, New York, Gift of J. Pierpont Morgan, 1917 [17.190.604].

**FIG. 11.** Polished, incised, and inked *Turbo marmoratus*, with depictions of flying, crawling, swimming, and seated animals and shells, late seventeenth century. 8.7 × 15.5 × 12.8 cm (d). Amsterdam Museum [KA 18812].

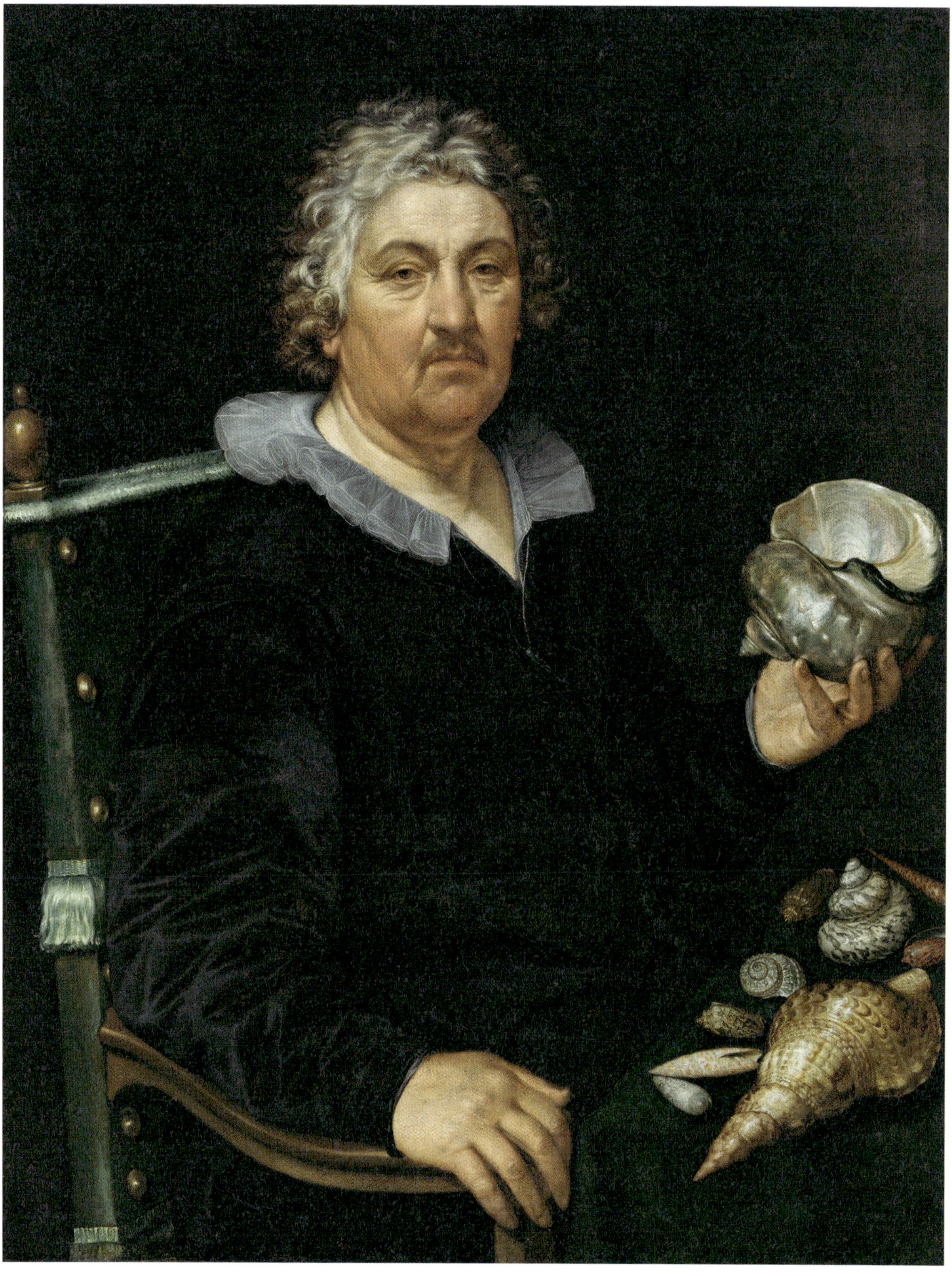

likeness."[19] Van der Aar is depicted seated in a stately chair covered in lush green velvet, his right arm and hand on the armrest. The painter has described a variety of shells, variously patterned and colored, arranged before the collector: a red-and-white speckled episcopal miter (*Mitra mitra*), a dark-red subulate auger shell (*Terebra subulata*), and the tremendous oaky brown Triton's trumpet (*Charonia tritonis*) are prominently placed at lower right. The small but rich collection includes cone shells (*Conus* species), a turban shell (*Turbo* species), a bat volute shell (*Cymbiola vespertilio*), and a minor harp shell (*Harpa amouretta*). While these are native to the Indo-Pacific region, the bulbous black-and-white volute is a West Indian top shell (*Livona pica*).[20] In his left hand van der Aar holds an exotic *Turbo marmoratus* nearly as large as his head; the coloring and forms of the nacreous mouth of the shell and its volutes, polished to a silvery opalescent sheen, rhyme with his locks of graying hair. Van der Aar is casually dressed: his jacket and shirt are open at the neck. The intimacy of the moment is simultaneously secured and ruptured by the sitter's direct gaze.[21] We interrupt the collector in his contemplative and sensual interaction with the nacreous curiosities before him, to join him.

**FIG. 12.** Hendrick Goltzius, *Portrait of the Merchant and Shell-collector Jan Govertsz van der Aar*, c. 1603. Oil on canvas, 107.5 × 82.7 cm. Museum Boijmans-van Beuningen, Rotterdam, [3450 (OK)]; Loan: P. & N. de Boer Foundation 1960.

Goltzius's painted portrait of Jan Govertsz van der Aar is one of a dozen extant portraits of the merchant that, taken together, attest to his commitment to the visual arts. In this formidable work and another he is depicted as a collector of shells from far and wide, compass points navigated by the Dutch.[22] Although van der Aar, who made a substantial investment in the Dutch East India Company at its inception in 1602, was a man of means and an important patron of the arts, no record of his collection is known to survive.[23] Clearly, he invested a great deal in his collection of unusual shells, and in commissioning a record of it from a renowned painter of the time. If he saw in his shells the artistry of nature, they also signal his wealth and access to "things that are rare and appealing" procured through overseas trade. He demonstrates their value as objects he owns and covets by handling them. The *Turbo marmoratus* in his hand stands in for the others laid out before him, which will surely next be or have just been handled.

Dutch conchophilia corresponds with a taste for other rarities—and the pictorial evidence abounds. The Middelburg painter Christoffel van den Berghe (c. 1590–1645) depicted a vase filled to overflowing with exotic and domestic flowers poised between delicate porcelain cups on one side and exotic shells on the other (fig. 13). These painted cups, nestling against one another in a manner that echoes the pairing of shells, underline associations between their consistency and formal qualities; they match cups recovered from the wreck of the Dutch East India company ship the *Witte Leeuw* (fig. 14), which sank near St. Helena, where it had stopped for provisioning en route back to the Netherlands from the East Indies in 1613. The picture is a fictional construct, most conspicuously where it brings together flowers that do not blossom contemporaneously, but it contains references to actual practices of transporting, cultivating, and collecting exotic goods. Of the sorts of shells recovered

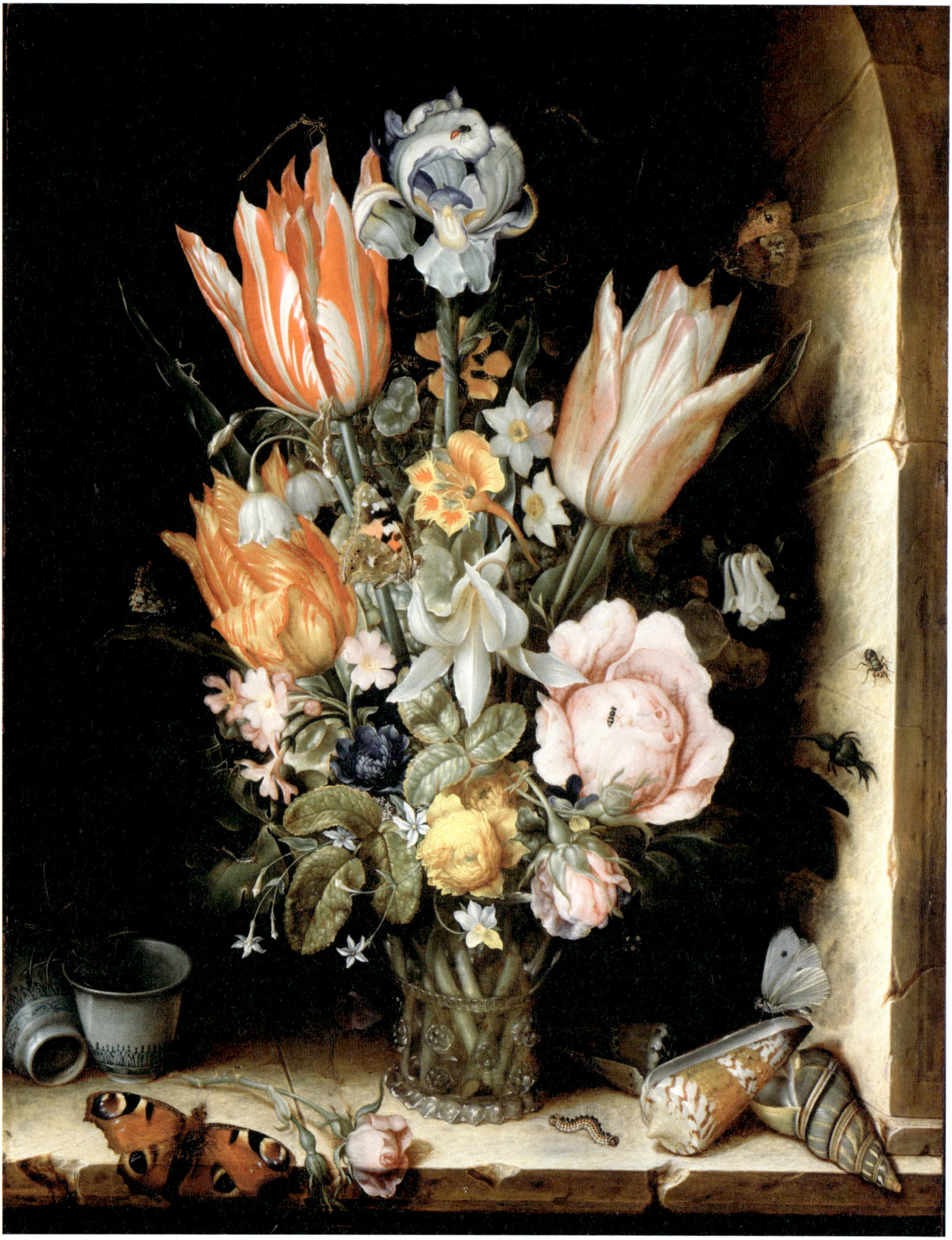

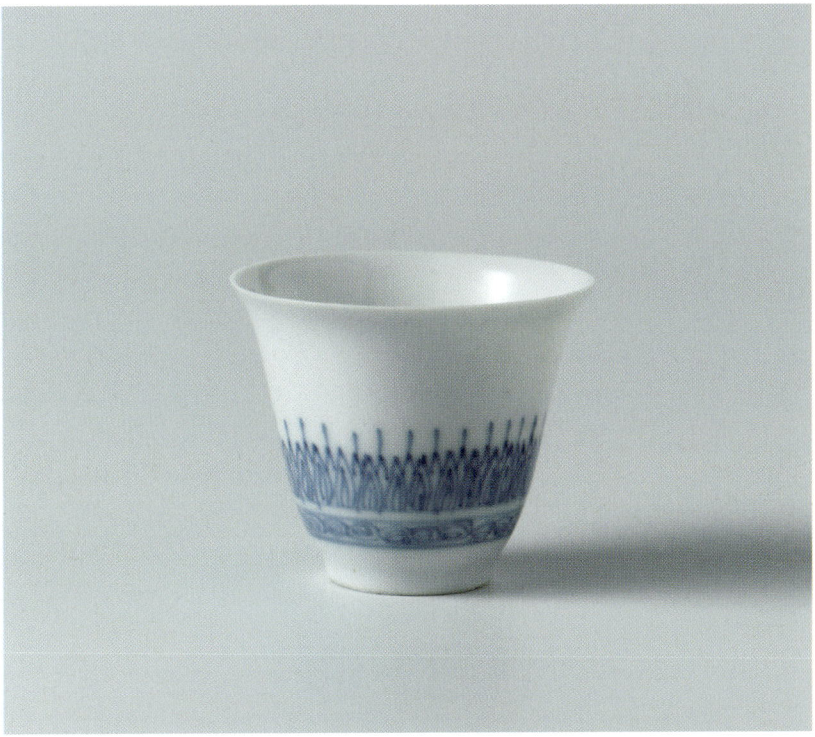

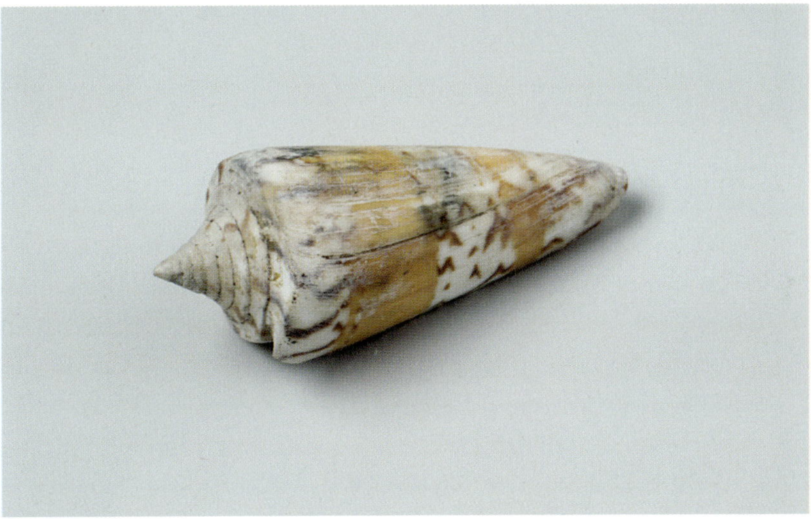

**FIG. 13.** Christoffel van den Berghe, *Still Life with Flowers in a Vase*, 1617. Oil on copper, 37.6 × 29.5 cm. Philadelphia Museum of Art. John G. Johnson Collection [648].

**FIG. 14.** Rice wine cup, remains of the *Witte Leeuw* (sunk 1613), Jingdezhen, Wanli-period, porcelain, 4.9 × 4 cm. Rijksmuseum, Amsterdam [NG-1977-134-W].

**FIG. 15.** *Conus generalis* shell, remains of the *Witte Leeuw* (sunk 1613), 4.8 × 2 cm. Rijksmuseum, Amsterdam [NG-1977-224-W].

from the wreck of the *Witte Leeuw* at least one (*Conus generalis*) is attentively depicted here (fig. 15). Remarkably, neither shells nor porcelain is listed on the official bill of lading for the *Witte Leeuw*. Van den Berghe's still life, painted four years after the *Witte Leeuw* went down, contains the sorts of specimens we know to have been transported as contraband. Procuring these highly valued exotic commodities from afar was complex and laborious and, with a cohort of collectors like van der Aar eagerly awaiting new arrivals, profitable.[24]

The Dutch humanist, diplomat, and collector Ernst Brinck (1582–1649) kept voluminous records of his travels, experiences, and observations. Brinck's surviving notebooks contain several pages of annotations under the rubric "Of sea shells" (*De conchis marinis*). Initial notations describe remarkable items in the most widely renowned Dutch collection of *naturalia* of its time, amassed by the city doctor Bernardus Paludanus (1550–1633) in the northern port town of Enkhuizen. Paludanus's collection was truly vast, containing in 1603 four thousand shells, in addition to as many as five thousand other maritime items. Brinck notes a shell bearing "naturally inscribed" Chaldean or Punic lettering (likely a *Conus litteratus*) that was of special interest to the philologist and historian Joseph Scaliger (1540–1609). Brinck also records small shells "variously colored, with pretty circles" for which Landgrave Maurice of Hesse-Kassel (1572–1632) offered Paludanus generous payment if the collector could procure a full set of twenty-four, for use as buttons on a doublet. Brinck mentions a variety of shells and various individuals who owned, acquired, or traded in shells—including a diplomat in Leiden; a *liefhebber* in Amsterdam; a poet, a queen, a jeweler, and a courtier.[25] Invariably, Brinck notes the value of the shells in question. The "inscribed" shell that so fascinated Scaliger fetched five guilders, and the diplomat in Leiden owned a collection of shells worth six thousand guilders. According to Brinck's notes, the Duke of Buckingham paid five thousand guilders for a single very large shell. The Amsterdam jeweler and merchant Jan van Wely (c. 1569–1616) sold shells to the Holy Roman Emperor Rudolf II (1552–1612), which enterprise earned him more than one hundred thousand rijcksdaelders, reports Brinck.[26] In his well-known emblems, Roemer Visscher captured this very market, in which nobility committed enormous sums to secure specimens for their collections and clever merchants could earn their living as agents of the desire for shells (see introduction, fig. 2).[27] Conchophilia was socially contagious—and lucrative.

The collections Brinck describes trading hands are now lost to the sands of time, but individual specimens of shells that were polished and whose forms were supplemented with metalwork and jewels and enamel do survive, as do works made of shells and pieces of shell. Brinck himself describes shells that had been variously crafted in his and other collections—a bracelet of shells from the Straits of Magellan; a marble table inlaid with shells and whelks; and "numerous beautiful tableaus of inlaid mother-of-pearl, as well as an entire chest of nacre filled with all sorts of beautiful shells of mother-of-pearl, cut with all sorts of beautiful figures" and "some beautiful mother-of-pearl shells mounted in silver."[28]

In his 1611 poem *The Beach*, Philibert van Borsselen invokes objects made in the East Indies with mother-of-pearl inlay from the *Nautilus pompilius*, mentioning tables and chests and gameboards. Like shells themselves, these were imported in increasing numbers over the course of the century and appear to have inspired local production in Amsterdam. The goldsmith Dirck van Rijswijck (1596–1680) specialized later in his career in inlaid tabletops and compositions featuring mother-of-pearl. His nacreous tableaus inspired by Namban lacquerwork include a beguiling example consisting of polished, incised, and inked shells and mother-of-pearl that depicts a variegated bouquet of flowers, two moths, and a snail, merging the forms of European still life with the lustrous effect of materials newly available in quantity (fig. 16).[29] Similarly, it was after midcentury that members of the Bellekin family gained renown for their carved shells—works whose artistry lay as much in the forms they rendered on the shells, with the use of a burin and ink or charcoal, as in their carvings, by which they elicited extraordinary patterning from the natural forms (see fig. 48).[30] A delicate medallion portrait of the stadholder of the United Provinces, Prince Maurits of Nassau, carved in cowrie shell illustrates how the layers of shells can be deployed to artistic effect (fig. 17). Comparison with a cowrie shell excavated from the marine wreck of the *Witte Leeuw* demonstrates that the artist drew out the forms of the man in profile from the contrast between layers of shell and their dusty rose and pale ivory tones; corrosion to the *Witte Leeuw* specimen has revealed the shell's natural palette (fig. 18).

Nature's beauteous marine bounty had both political and economic currency, as exemplified by accounts concerning Marie de' Medici, queen of France. Brinck refers twice to her having spent significant sums on shells: in Amsterdam he saw a box of "lovely" (*fraije*) shells whose owner had been offered 700 guilders for one shell for the queen; and Brinck notes that she purchased a rather shabby specimen for 450 guilders, also in Amsterdam. Marie de' Medici was in Amsterdam in 1638, so these episodes may date to that visit, during which she went shopping (incognito) for exotic fineries. Three decades earlier, she had received a significant gift of East Indian shells from the States General of the United Provinces of the Netherlands. The French statesman Pierre Jeannin (1542–1623) spent much of 1608 arranging French support for a truce between the United Provinces and Spain. During that time he also mediated the purchase by the Dutch government of a sizable collection of "Indian shells" from the humanist and collector Abraham Gorlaeus (c. 1549–1608) for the extravagant sum of 9,000 guilders, for presentation to the French court (see fig. 44).[31]

Exotic shells, transported at risk from afar, were highly valued for their aesthetic qualities, became precious market commodities, and could be politically expedient. That they were available in great quantities in the Netherlands was the consequence of burgeoning trade in the East and West Indies. There is evidence that specific sorts of shells served *as* currency in these circuits as well, as they had done in Asia previously. Of the small cowrie shell (*Mauritia arabica*) from the Maldives, Brinck notes that the VOC transported "an incredible

**FIG. 16.** Dirck van Rijswijck, *Tableau with Flower Still Life*, 1662. Black marble inlaid with engraved mother-of-pearl and breccia marble, 15.9 × 9.8 cm. Metropolitan Museum of Art, New York, Purchase: The Howard Bayne and Rogers Funds, 1986 [1986.21].

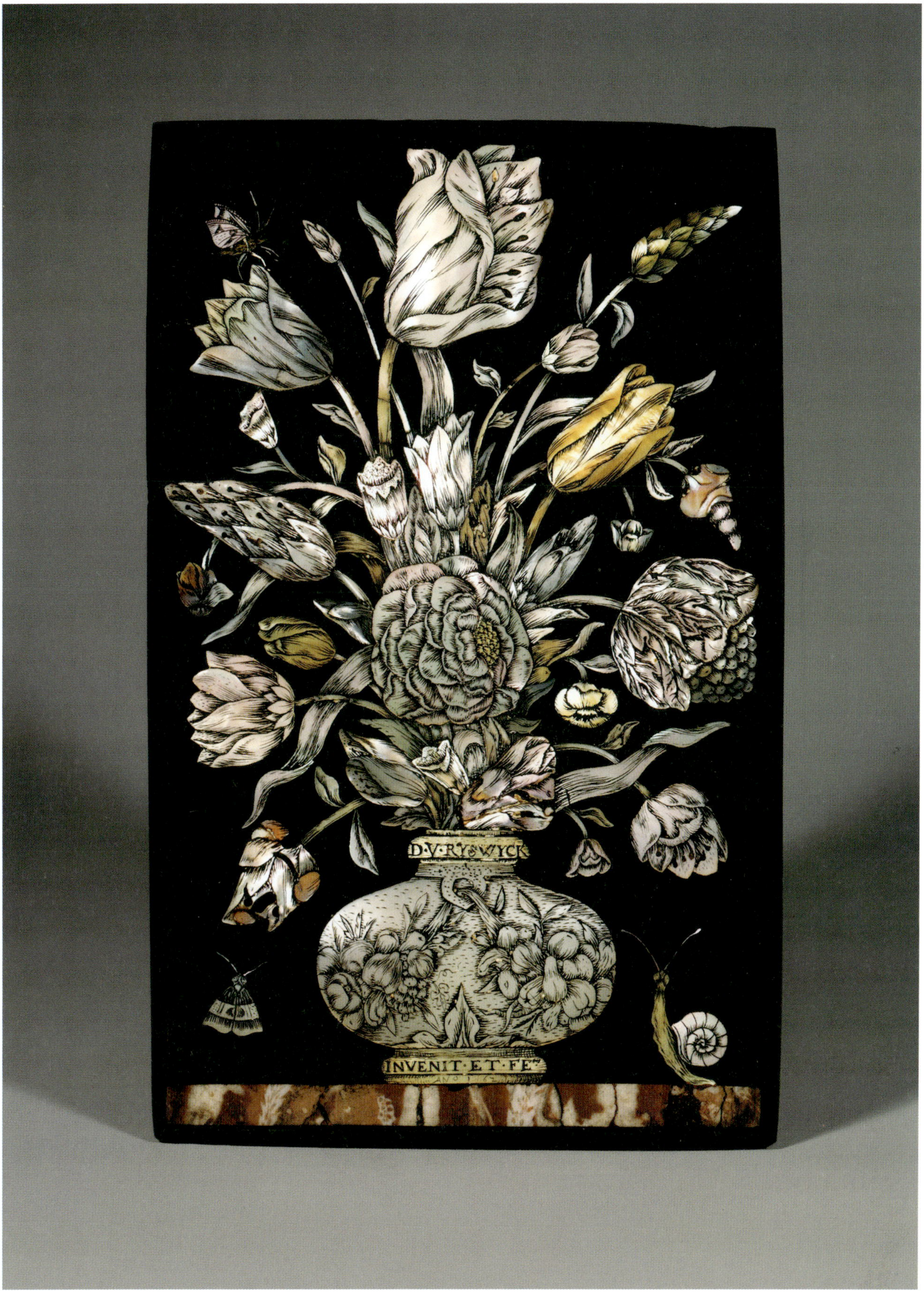
D·V·RYSWYCK
INVENIT·ET·FE

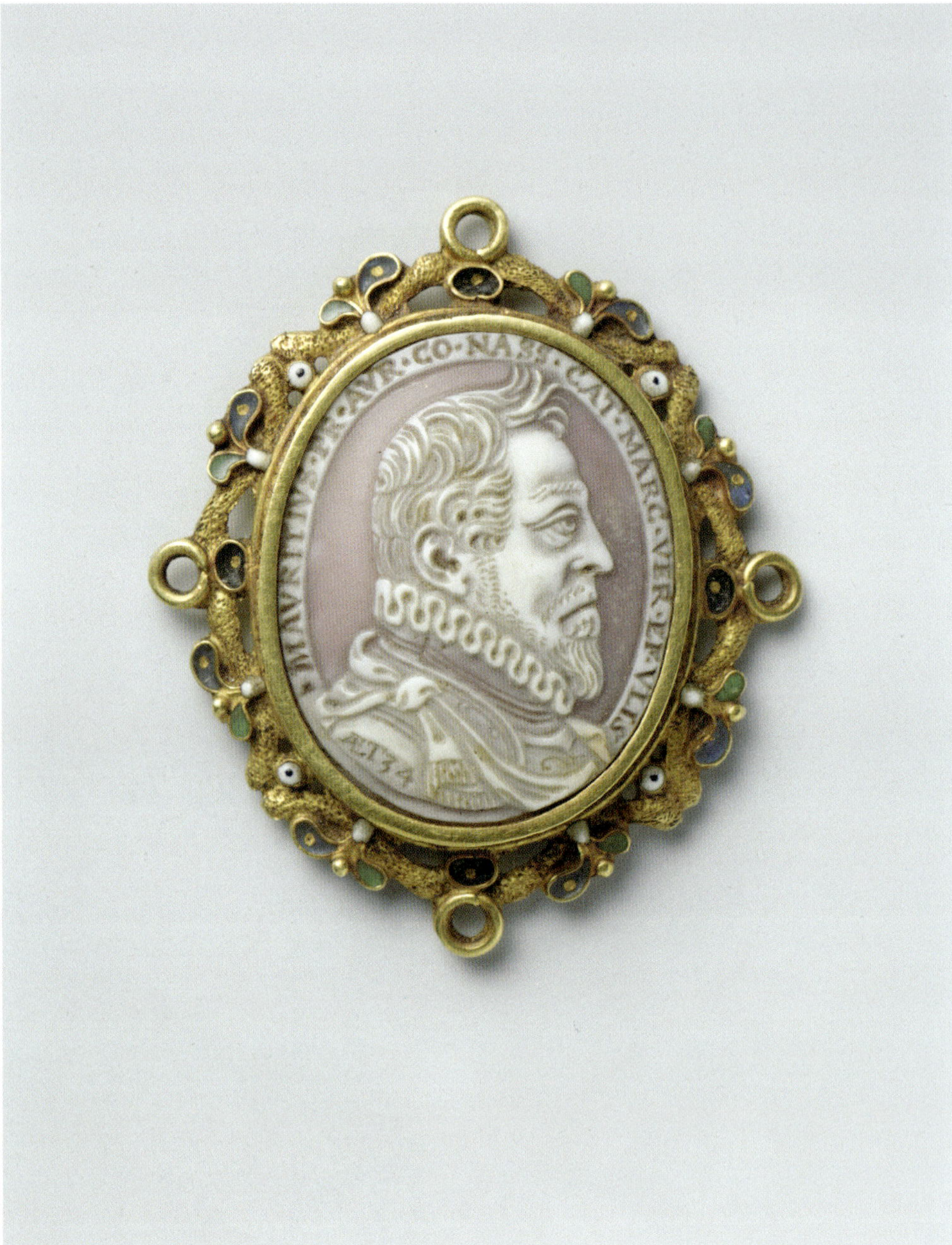

**FIG. 17.** Anonymous, *Cameo medallion with portrait of Prince Maurits, Stadholder*, c. 1600, Incised *Cypraea tigris* shell, in a filigree gold setting with enamel flowers, 3.4 × 3 × 0.5 cm. Rijksmuseum, Amsterdam [BK-NM-195].

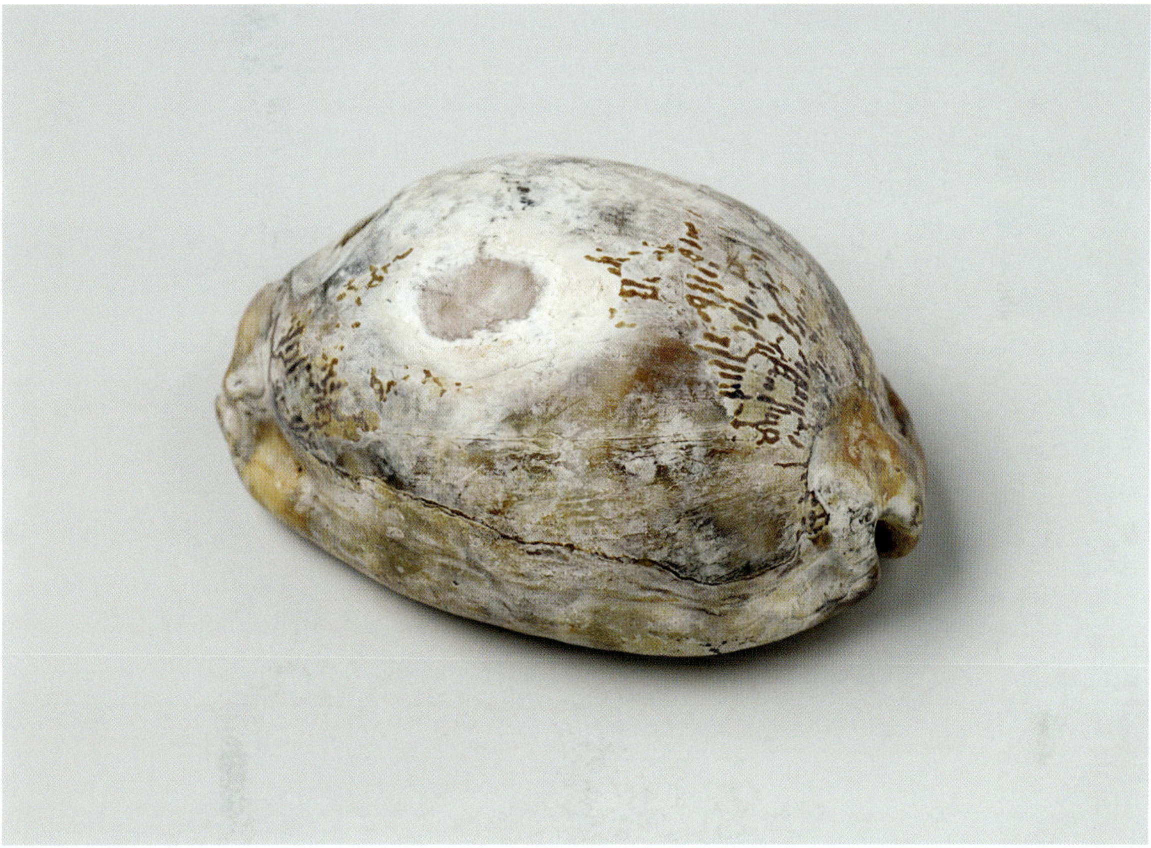

**FIG. 18.** *Cypraea arabica* (cowrie) shell, remains of the *Witte Leeuw* (sunk 1613), 5.3 × 3.6 × 2.9 cm. Rijksmuseum, Amsterdam [NG-1977-223-W].

FIG. 19. Cornelis Cornelisz van Haarlem, *Neptune and Amphitrite*. Oil on panel, 71 × 93 cm. Stichting P. en N. de Boer, Amsterdam.

quantity" in 1646; in 1648, eighty thousand pounds were imported to the Netherlands by the company; and in Brinck's final notation on shells, he writes that in 1649 (the last year of his life) forty thousand pounds arrived. Cowrie shells were beginning to be used, Brinck writes, to decorate horses' bridles. He also notes that so many were imported that they were then transported to Angola, "where they are distributed as currency and of great value. Thus, these shells are now also valid in commerce."[32] By the middle of the eighteenth century, Amsterdam was recognized as "the chief *European* Market for these Shells . . . where are spacious Warehouses of them, the *French* and *English* Merchants buying them up to send to *Afric*."[33] Brinck does not specify the nature of the commerce, but it is clear from other sources that European trade in the East sourced the cowries to conduct the slave trade.

## SENSUALITIES

Van der Aar touches, holds, offers to the viewer's gaze the ridged surface of a turban shell. His fingertips are poised to maximize receptivity, and the weight of the shell asks to be compared with that of others on the tabletop. Among collectors in early modern Holland appreciation of the sort embodied by van der Aar—a collector's interest—was a sensual matter. This is borne out by the 1628 sale in Leiden of the collection of Leiden pharmacist Christiaen Porret (1554–1627), which included myriad shells, under the rubric of *Sinnelickheden*, or sensualities. The title of the auction catalogue reads, in translation, "Exceptional Items or Rarities and Exceptional Sensualities / From Indian and other foreign locales conchs / shells / terrestrial and maritime creatures / minerals / and also strange animals; and some artfully made handicrafts and paintings."[34] While the Porret auction catalogue is a tantalizing invocation of an aesthetic of the exotic, it leaves us wondering how the thousands of items listed were sorted, viewed, cared for, handled. A 1603 inventory of Paludanus's collection lists 11 boxes of fish, shells, "strange things from the sea," corals, and other marine items divided into no fewer than 1,440 *small boxes*.[35] The experience of singularities of nature and the morphological variety at hand in his collection in Enkhuizen encouraged classificatory efforts and at the same time inspired Paludanus to describe his shells as bearing out the presence of the divine, of "whose wondrous almighty power the poet [Ovid] has rightly said: 'Divine power sports in human matters.'"[36]

Fascination and sensual engagement with the products of the sea is the subject of a number of early seventeenth-century Dutch paintings, which resonate with the prerogatives of collectors of their time. In one of them, painted by Cornelis Cornelisz van Haarlem (1562–1638) in 1617, Neptune is depicted in the company of a woman who may be the Nereid Amphitrite, his wife. Wearing a lopsided crown of four brightly patterned whelks, Neptune implores us to behold his shells—a hefty, luminescent *Turbo marmoratus* he holds in his left hand and an array of specimens laid out on a ledge on which he leans with his right arm (fig. 19).[37] The god of the seas and waters cradles in his right hand a flecked, ribbed, pointy

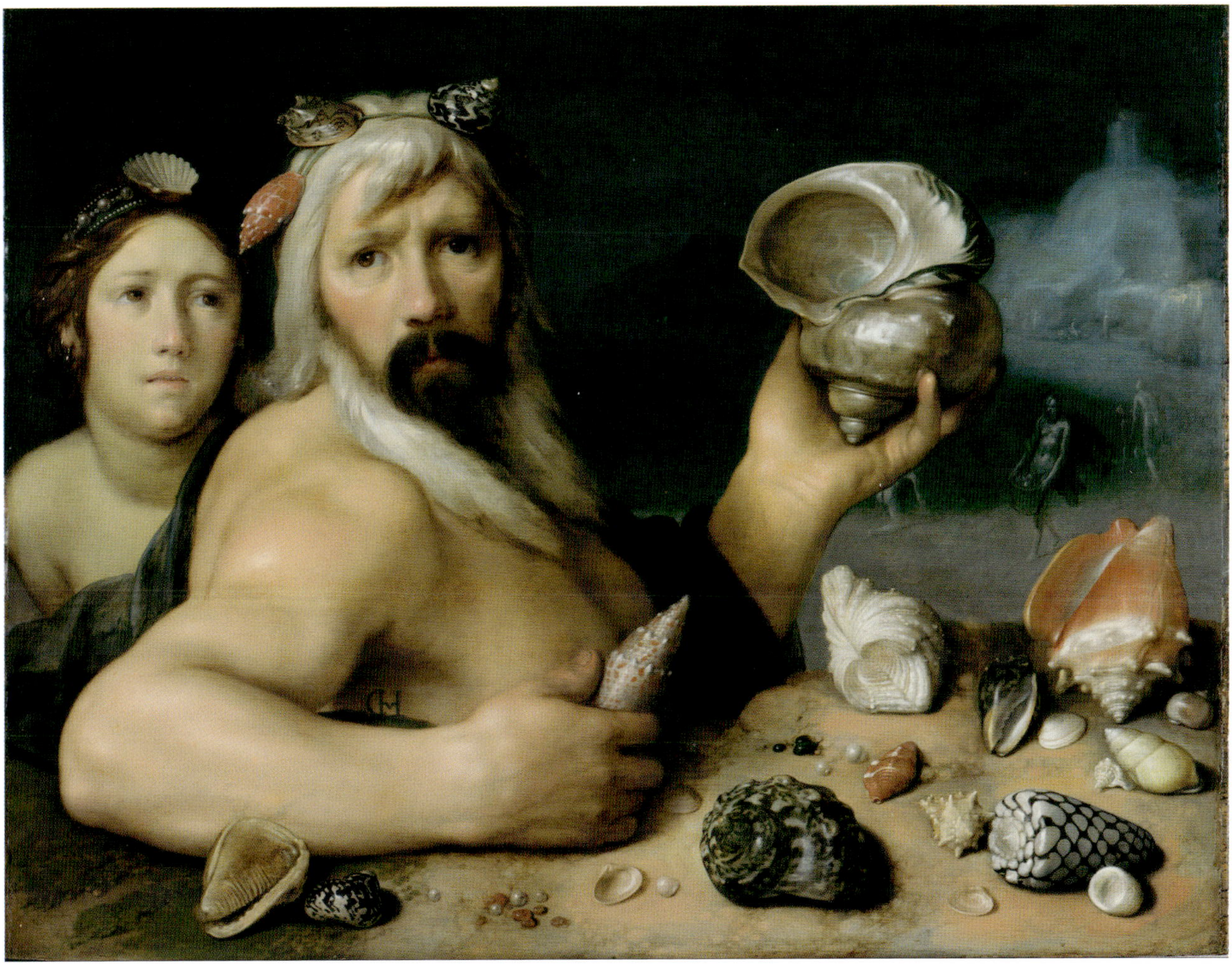

*Mitra papalis* (Papal miter), which seems to emerge from his midriff, erect. His physique and dark beard make him out to be youthful, although his long hair is white, flowing thickly past his shoulders. His female companion sports a headband embellished with pearls and a cockle, in addition to a gleaming pearl drop earring. She glances quizzically at a dangling red-and-white *Mitra stictica* (also known as the Pontifical miter), on Neptune's head, the inverse of the one he holds close to himself. The shells depicted are all highly prized varieties, most from the Indo-Pacific region, and scattered among them are lustrous pearls. In the hazy bluish background of the painting, naked figures wander about, one of them carrying a basket filled with round objects—shells gathered from the shoreline. What unifies the objects depicted with the figures and the viewer are the omnipresent references to beholding; we are everywhere incited to look, to observe the sheen and the luster, the whorls and keels. And the picture depicts the handling of shells—at the beach and in the summary space of a collection.

The colors, patterns, shapes, and textures of these shells range from smooth and pale yellow to gradually ribbed and brown, crass, ridged to knobby and striated, and from scalloped or round to conical or oval. Like the collector van der Aar, Neptune holds aloft a *Turbo marmoratus* that is in itself a study in shape, texture, and luminosity (see fig. 12). The lip of the shell is almost doughy, and seems to have been pressed open at its outer marbled green and white edge; the rim of the nacreous cavity gleams brightly; the blurry white highlights on the outer shell are the result of the shell having been polished (and the painter's craft). Perhaps more than any of the others, but certainly in combination with the Papal miter that Neptune points at it, the slippery, bulbous, firm *Turbo marmoratus* is sensual to the point of being sexy. Inside lies a world of sheen—the stuff of nacre (mother-of-pearl) for which *Turbo marmoratus* shells continue to be prized. Like most of the shells painted by Cornelisz, this one is unmarred, intact, and cleaned to the point of gleaming, bearing no traces either of former inhabitants or of their specific place of origin or provenience. Generally, this is how *Turbo marmoratus* shells were depicted. Notably, though, on the ledge below this specimen, directly in the middle, is an unscraped, unpolished *Turbo marmoratus*, closer in appearance to what the figures on the beach might find than what Neptune is holding.[38]

## *MOEITE EN VERDRIET*

The play of art and nature may have structured the aesthetic that supported the production of copious shell vessels, but where the pairing takes shells to be natural products, it rings hollow in light of the elaborate, extensive processes that collecting shells entailed.[39] In an Aristotelian sense and biologically speaking, shells are indeed shaped by nature. Merged with the forces of art by silversmiths who gave them bases, girded and chased them, and transformed the former homes of sea creatures into courtly drinking vessels, mounted shells are often described as plying the boundaries of art and nature. Shell vessels, like vessels crafted of emu or ostrich eggs, coconuts, or Seychelles nuts, are upheld as representatives of a fruitful commingling of art and nature. Early modern shell specimens were highly mediated, though—by processes of selection and manual preparation, and by trade and transport. As a result of those handling processes, shells reached collectors hollowed out and polished: they were reduced to form without apparent function. In this regard, they were like figured stones, deemed inscrutable from the perspective of the boundary between nature and art.[40] Here, I want to emphasize how worked, how de-natured, exotic shells were when they arrived in European hands.

Among the wonders of Rumphius's *Ambonese Curiosity Cabinet* is his sustained attention to the living creatures whose shells he catalogued.[41] In the course of his description of shells, Rumphius frequently acknowledges the local uses of the creatures within, up to and including how they taste and their medicinal uses. Writing of the snail that generates the *Turbo marmoratus*, which he calls the *Cochlea Lunaris major* (and, in Dutch, *Reuse-ooren*, or Giant's ears) Rumphius notes that "the natives make much of it, that is as food"; the best part is the tail, and the kings of Buton (an island located off the southeast peninsula of Sulawesi) reserve the snail for their own food. "We," he declares, "make more of the shell."[42]

Recall that both the collector Jan Govertsz van der Aar, portrayed by Goltzius, and Cornelisz's Neptune single out a *Turbo marmoratus* for show. Rumphius describes the *Turbo marmoratus* shell as "usually the size of two fists or more . . . with a large round mouth, which sticks out a bit at its lowest corner, and therefore looks like a little earlobe" (fig. 20). Neptune's gloriously iridescent specimen, its pearly surface shaped by a greenish lip, is paired in the painting by a knobbier, darker specimen of the same shell, on the table (see fig. 19). Rumphius explains that *Turbo marmoratus* shells have two thick skins, and specifies that the outer layer is "greyish, like *doodverf*, and rough and full of scratches, though here and there speckled like snakes, with black, brown, and sometimes also Spanish green. . . . The inner layer is a beautiful mother-of-pearl, not white, but showing all the colors of the rainbow—to wit, green, red, and blue."[43] Though the outer layer, which he describes by reference to a ground layer pigment used by Netherlandish artists (*doodverf*), is not entirely unappealing—it may be speckled and colorful—the value of the shell lies beneath that surface. The specimen Neptune holds in his hand is splendid by virtue of a great deal of handling; scraping and polishing have made it worthy of his favor.

The wholly nacreous version of the *Turbo marmoratus* Neptune holds aloft is the product of a laborious process. First, the mouth of the shell (which often has a natural cap, the operculum, which Rumphius calls the *Umbilicus Veneris*, or Venus's navel) should be stopped up with clay or wax, he writes, so that the interior is not affected; then the shell should be immersed in vinegar or spoiled rice for several weeks. "The vinegar must be changed 2 or 3 times, and also scour [the shell], until the rough inner layer falls away, and the entire shell becomes nacre; then, rub it all over with aqua fortis, and rinse it again with soapy water." Rumphius adds that the Japanese use broken bits of this shell as inlay in their

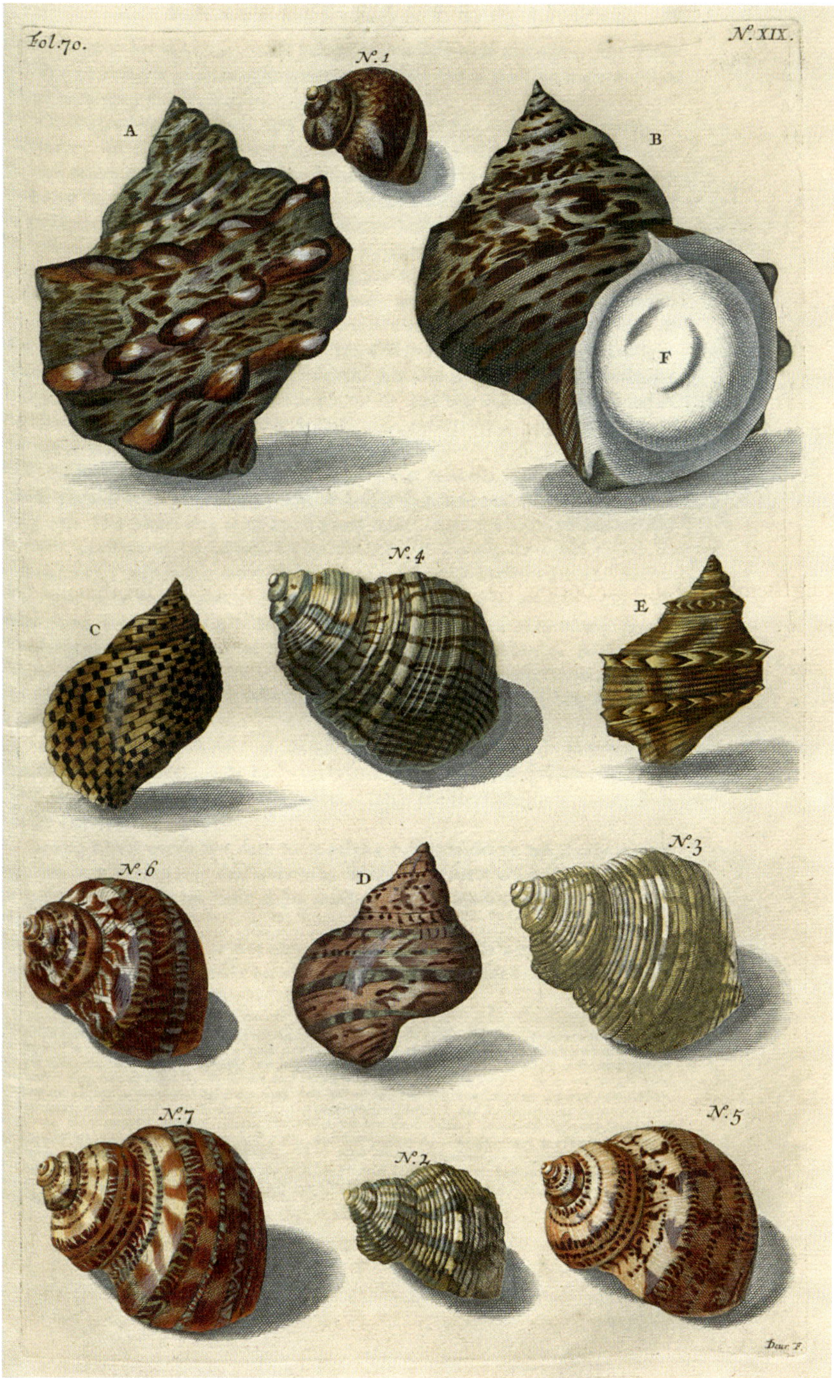


**FIG. 20.** *Cochlea lunaris major* (A and B; *Umbilicus marinus* labeled "F") and several kinds of *Cochlea lunaris minor* shells, Plate XIX. In Georgius Eberhard Rumphius, *D'Amboinsche rariteitkamer*, 1705. Hand-colored engraving. Universiteit van Amsterdam.

lacquerware cabinets and chests. Inlaying the mother-of-pearl in black lacquer "renders it gorgeous, displaying more beautiful colors than the mother-of-pearl itself."[44] Throughout the text of his *Ambonese Curiosity Cabinet* Rumphius refers to the preparation of specimens: many shells are described as having a slimy skin or other covering, or fleece, or cracks, and Rumphius regularly indicates how these should be removed and the shells preserved.

The final chapter of book 2 of *The Ambonese Curiosity Cabinet* is dedicated to explaining how Rumphius and his assistants handled Indian Ocean shells they procured and prepared for eager *liefhebbers* in the Netherlands. Whereas Aristotle held that shells take their colors from the mud whence they are spontaneously generated, Rumphius's text declares how artificial the qualities of collectable shells were. They were mediated, these wondrous objects, by processes encapsulated by Rumphius in the phrase, *moeite en verdriet* (trouble and tedium). "Our compatriots and friends in the Fatherland tend to be of the opinion that shells and rarities can be found on the beach or hauled from the sea as beautiful and clean as when we send them to them," Rumphius writes, "and that consequently the only trouble it requires is picking them up"—an assumption Cornelisz's painting animates. Clarifying that it is hardly worthwhile trying to refute the misconception, he wants to absolve those who live in the East Indies of accusations of rudeness "when we sometimes cannot fulfill our friends' requests or desires." Rumphius cites the time and effort, specifying "all the trouble and tedium (*moeite en verdriet*) one has to endure to clean them and make them presentable."[45]

The body of Rumphius's chapter "How to Gather and Clean Shells" consists of a litany of more than thirty specifications of the prevailing travails. The great majority of shells found on the beach are unworthy, as they have lost their sheen and color or are broken. Not all shells are found together, but in highly specific and disparate locales, and some only at sea. Searches should be conducted at night, under a full or new moon at low tide, and the best months are the first two rainy months of the year (on Ambon, May and June). Man-eating crocodiles make flat, sandy beaches perilous; spiny urchins lurk in marshy areas; and on rocky beaches one risks hurting one's feet by stepping on coral, burning sea slime, and slivers of stone. Having gathered shells, one must wash and prepare them. This is tricky, because some shells still contain their creatures. If this is the case, rain or sun or fresh water may harm the shell, while placing shells atop one another may cause the blood of one animal to mar the shell of another. Some may be cooked in seawater to remove the animal, but this can be a troublesome (and pungent) process.

Once the shells are cleaned, they are polished: rubbing with linen or sand is prescribed; grinding and scouring are recommended in some but not all cases; greasy or sweaty hands should not touch shells as they undergo preparation; and Rumphius suggests a variety of brushes, "rough and fine ones, thick and thin ones," made from the hairs of the palm tree. Throughout the process, beware the risk of predators: little crabs, he warns, "steal those shells they consider fitting for their bodies, leaving you with their old and broken houses."

Have you selected and prepared your shells? Do not pack them in cotton, which will pale their colors. More suitable are fine wood shavings or Chinese paper. Finally, foreseeing placement in the European collections they were destined for, he mentions that some shells sweat by nature and recommends wiping them carefully, to spare "the other things around them."[46]

Exotic shells were gathered under challenging conditions, exhaustively prepared, and then shipped halfway across the globe before they joined other rarities in European collections. In most respects their "natural" qualities were labored and highly artificial.

In an essay on early modern collecting titled "Wrought by No Artist's Hand," Martin Kemp writes of the "deliberate conflation of human and natural design" in shell vessels, where nature's (unsigned) designs are emulated by artists' hands in the figurative elements.[47] Emphasizing the chiasmic relationship in shell vessels between artistic and natural elements, Kemp calls such objects "cultural migrators," by which he means to highlight their fluidity, ambiguity, and diversity of meaning.[48] He also, mistakenly, I believe, posits that their inherent fluidity absolves them of implication in power relations, ideology, colonialism—which he calls "meta realities."[49]

When we assume that shells embody nature, adjoined to or adapted to the purposes of art, we risk missing the dynamics that are obscured by the notion of cultural migration. Who migrates what? From where to where and how? Refraining from iconographical impulses keen on tracing stability or instability of meaning in exotic shells and shell vessels, I have instead called attention to the *moeite en verdriet* involved in the preparation of these products of nature and artifice and, where possible, identified whose labor was involved. When coveted East Indian shells arrived on European shores, they had been cleansed of their original inhabitants and the undesirable residue of their accretion or provenience. Much modern historiography has continued to write the labor out—the bloodied feet and acid solutions Rumphius chronicles are long forgotten.[50]

The frontispiece of *The Ambonese Curiosity Cabinet* describes a one-way trajectory from the domain of nature, where shells are sourced, into the enclosed world of scholarly study, and the movement of exoskeletons from beach to cabinet via laboring human bodies (see fig. 8). The tripartite composition of Rumphius's frontispiece clearly distinguishes the labor of gathering shells from their study and appreciation. In the hand-colored copy of the volume in the University of Amsterdam library, black-skinned figures are juxtaposed with a group of sociable White men whose clothing and demeanor distinguish them from the work of procuring the shells they study. At either side of the architectural frame, statues of the figures Cybele and Neptune engage in a bemused face-off.[51] Far from including the laboring Black bodies, they signal a shift in tone from the forms of labor required to provide the specimens to the collectors to the forms of attention associated with the production of artworks and of natural history in and for collections. That shift registers power relations too, enacted in labor and marked by color: the foreground figures are not necessarily enslaved, but they might as well be. The glorification of the study of the natural world that Rumphius's text and its frontispiece advertise for shells and marine creatures is a steady feature of early modern collections of rarities of the era—spaces where abject labor is translated into high-minded handling.

Calling attention to the forms of handling and labor involved in the production of shells and shell vessels asks us to consider the processes and the practices and the persons that early modern European shell lovers overlooked and oppressed in the interest of sustaining an aesthetic regard for hard-won luxury commodities. Van der Aar the collector (see fig. 12), Cornelisz's Neptune (see fig. 19), and de Gheyn's shaggy god of the sea (see fig. 25) embody the keen curiosity and sensual engagement with shells typical of the era and familiar from a range of sources touched on in this chapter. This luxury pursuit did not capture nature so much as it invested laboriously prepared objects, transported from afar, with local aesthetic and economic values. Doing so meant leaving the labor—the *moeite en verdriet*—out of the picture.

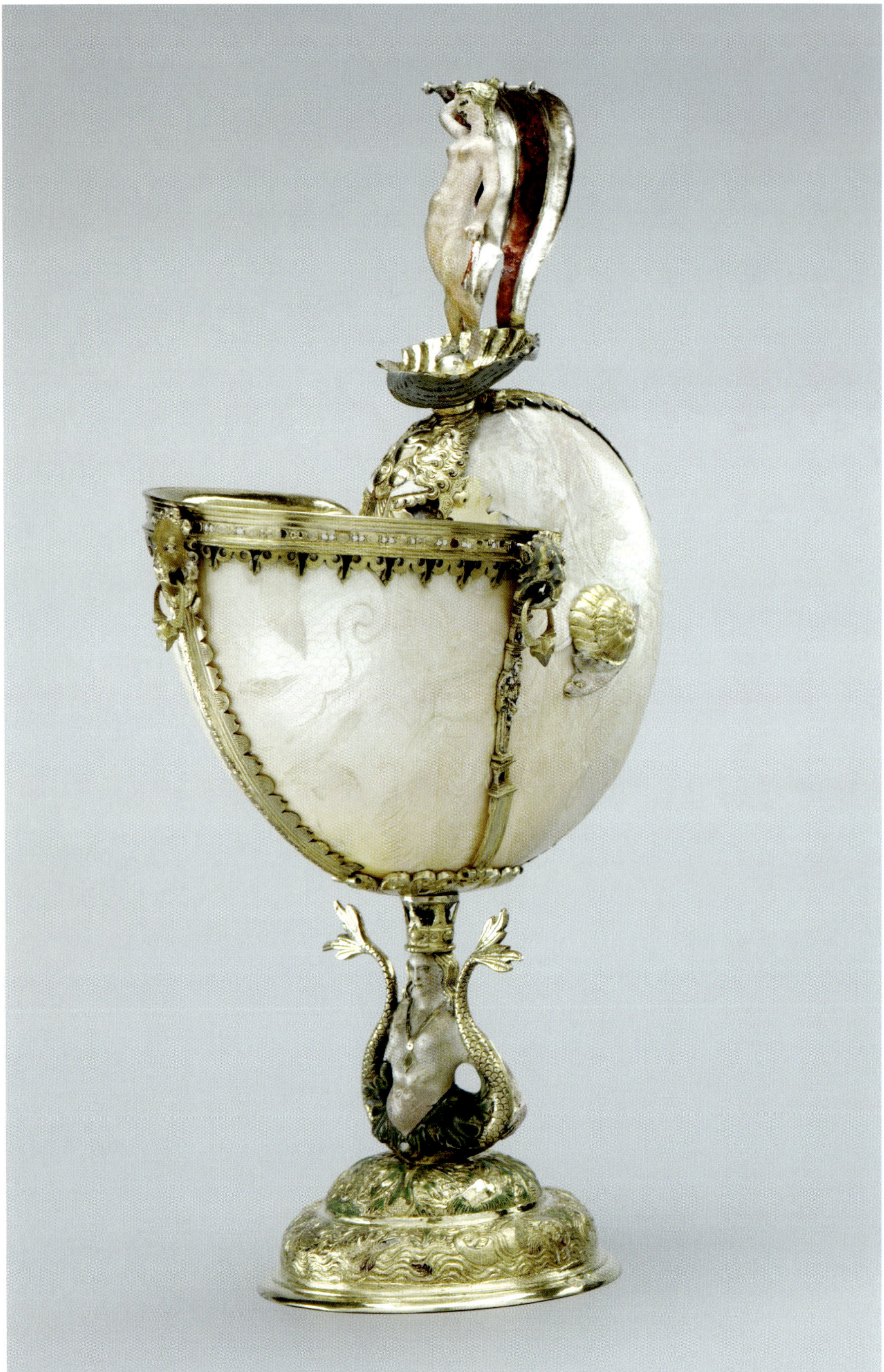

Anna Grasskamp

# Chapter 2
# Shells, Bodies, and the Collector's Cabinet

The 1596 inventory of the collection by Archduke Ferdinand II at Ambras Castle lists among its treasures "a beautiful Indian snail shell with gilded mounts," a description that refers in all likelihood to a cup still extant today (fig. 21).[1] The cup in question, created by the goldsmith Marx Kornblum (active 1570–91), contains at its core a nautilus shell that has been stripped of its natural brown-white surface, exposing layers of mother-of-pearl incised with curved lines, images of foliage, and long-tailed birds. Carefully crafted mounts of gilded silver frame the shell above and below. The lower pedestal takes the form of a mermaid balancing the shell upon her crown. At the shell's summit, a nude figure with the combined attributes of Fortuna and Venus, the goddesses of good fortune and love, stands triumphant. Two gilded silver snails were originally affixed on either side of the cup, one of which has survived in its original position.[2]

**FIG. 21.** Marx Kornblum, *Cup*, c. 1580–90. Nautilus pompilius, gilded and enameled silver, 27.8 cm × 8.5 cm. Kunsthistorisches Museum Wien, Kunstkammer [1063].

When the sixteenth-century German art expert Gabriel Kaltemarckt counted "curious items from home or abroad" among the standard objects in early modern courtly collections, he could easily have had a vessel like Kornblum's cup in mind.[3] Specifically, he referred to "utensils used for drinking or eating which nature or art has shaped or made" out of materials derived from "above the ground, from within the ground or from the waters and the sea." Kaltemarckt wrote that to members of the elite "such things are regularly obtainable in large quantities in Germany and Italy."[4] Indeed, goldsmiths regularly transformed natural items with non-European provenance, such as coconuts and ostrich eggs, into "utensils used for drinking." Vessels made out of Asian shells were displayed in the *Kunstkammern* of German

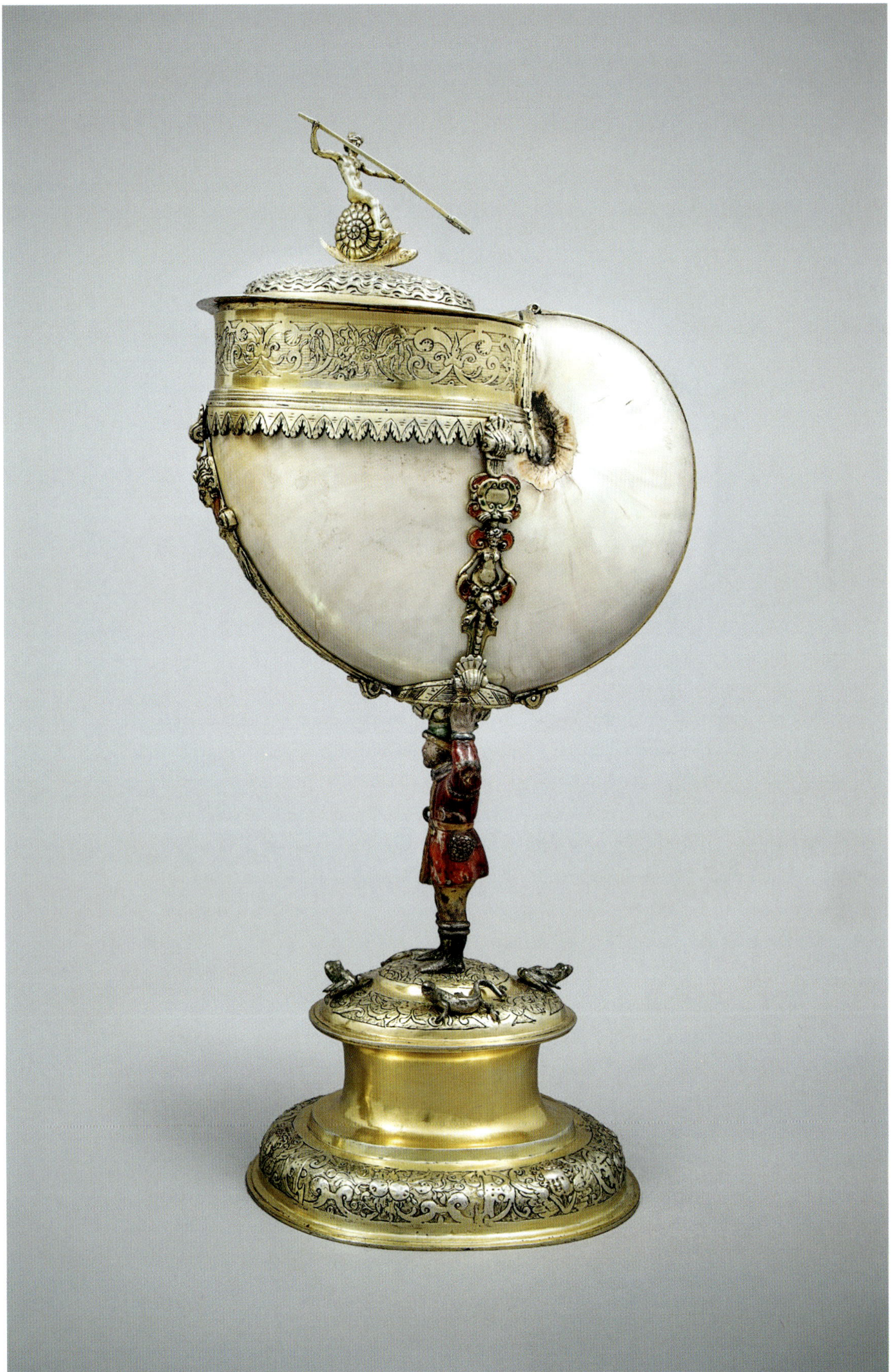

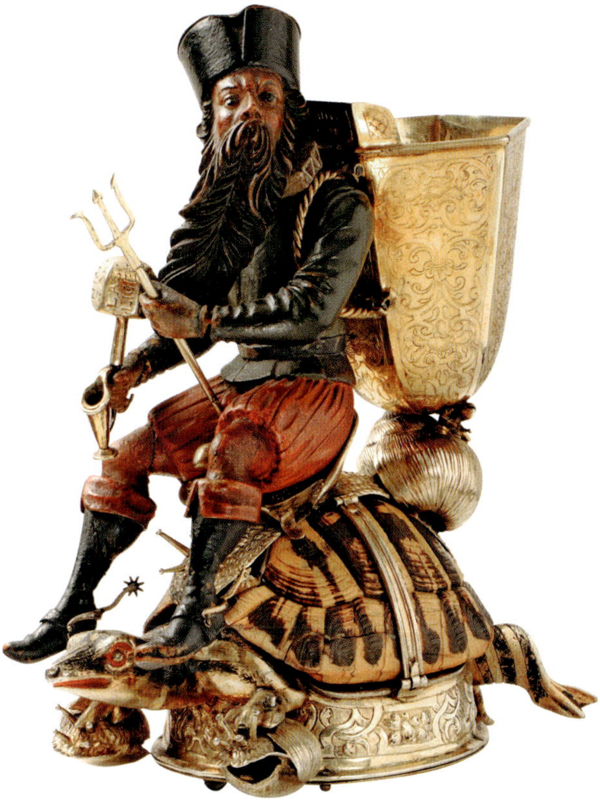

**FIG. 22.** Elias Geyer, *Nautilus Cup supported by a Vintner*, 1606–8, one of a pair. Nautilus pompilius, gilded and enameled silver, 36.8 cm. Staatliche Kunstsammlungen, Grünes Gewölbe, Dresden [III 160].

**FIG. 23.** Leodegar Grimaldo, *Drinking Automata with Neptune as Vintner*, 1626. Wood, painted, gilt silver, iron deck, 22.8 × 10.4 × 20 cm. Museum für Kunst und Gewerbe, Hamburg [2006.57].

princes, the *studioli* of Italian scholars, and the *konstkamers* of elite Netherlandish collectors, and appeared in church treasuries and among the possessions of prosperous merchants.[5] Nautilus cups such as Kornblum's held a prominent place in early modern court collections.[6]

Kornblum's shell cup is not exceptional among early modern vessels in placing the sculptural representation of a snail in dialogue with the nautilus. Another example made between 1606 and 1608 by the goldsmith Elias Geyer formed part of the collection at the court of Saxony, to which Gabriel Kaltemarckt had offered his advice (fig. 22). The base of this cup, which is one of a pair, is adorned with sculptures of frogs, lizards, and a vintner; the sides of the vessel feature bare-breasted female caryatids; and its lid consists of a man with a trident riding a giant snail across the waves. The pairing of vintner and shell was common not only on cups made of actual shells, but also on those adorned with human-made replicas. For example, a drinking automaton created by Leodegar Grimaldo in 1626 features the shells of three snails sculpted in metal alongside a representation of a vintner equipped with a trident and a pannier (fig. 23). This ensemble juxtaposes local snails called "vineyard snails" (*Weinbergschnecken*), a type that a vintner would encounter in his daily work, with the far rarer terrapin shell that forms the vintner's mount. The play between real and represented nature, and between the scale of animal and human figures lends both vessels their rare appeal.

Further, the tridents in both cups evoke the iconography of Neptune, god of the sea, and associate the surrounding shells with the maritime realm.

All three of these cups belong to a category of objects now known as "joke cups" (*Scherzbecher*), made to impress and entertain the participants in lavish dining and drinking sessions held at early modern courts. Grimaldo's automaton, for instance, was designed to move across the dining table and to stop in front of a guest, who was then compelled to drink its contents. The shell cups by Geyer and Kornblum were objects of both wit and erudition. The knowledge required to understand their sophisticated iconography was in playful tension with the drunkenness that resulted from the games in which they were employed. As illustrated by countless period representations of intoxicated men and women of different strata of society, the potential of socially transgressive behavior triggered by drinking included the unleashing of erotic advances and sexual desires.[7] In other words, it allowed Venus, the goddess of love, to make her appearance at the dining table.

Shell cups and shells themselves were evocative vessels, even beyond the dining table, as objects that appealed to the senses. Three different levels of correspondence between shells and bodies impacted their reception among sixteenth- and seventeenth-century collectors: first, perceived resemblances between the shapes of shells and anatomical features of the human body; second, material correspondences between the surfaces of shells and skins; and third, etymological equivalences established between conchs and body parts. This essay will explore all three categories across paintings, textual sources, and, of course, the shell vessels themselves.

To understand the polyvalence of shell cups within this context also offers a new angle of approach on the *Kunstkammer* itself. Past scholarship has highlighted the *Kunstkammer*'s uses as a symbolic space, a microcosm, and a tool for the representation of power.[8] Sixteenth-century collections have also been described as repositories of "practical knowledge" and as spaces for "practical and technical research . . . situated at the center of a complex of workshops and laboratories in which the practical work of court and state was undertaken."[9] The notion that the *Kunstkammer* "stimulated work as well as wonder" is crucial to understanding the encyclopedic dimensions and taxonomies of early modern collecting.[10] Yet neglected in all these accounts are the erotic dimensions of the encounters between object and collector that unfolded in these spaces and deserve scholarly attention in their own right. As this essay argues, to fully understand the early modern engagement with shell artifacts, it is necessary to look beyond the effects that the *Kunstkammer* had in terms of "work" and "wonder" and to acknowledge its potential to function as a site of sexual knowledge and erotic fantasy.

## SHELLS AS BODY PARTS

In sixteenth-century Nuremberg, goldsmiths such as Wenzel Jamnitzer (1507/8–85) and Nikolaus Schmidt (1550/55–1609) artfully engaged the body of the collector with objects that were made to represent human body parts and integrated the shells of Asian sea snails and other kinds of mollusks. Through the drawing and crafting of vessels in which shells (and the spirals contained in their outer and inner shapes) represented the "ideal" proportions of female breasts, the Nuremberg artisans negotiated erotic desires and fears.[11] This is particularly well illustrated by the "breasted vessel" in an ornament print by Erasmus Hornick (1524/27–83), which shows a plump, idealized female breast and an emaciated, sagging one attached to a snail shell's spiral on the left half of the vessel (fig. 24). The suggestion of equivalence between shell and bosom, however, was exclusive neither to the Nuremberg circle nor to the medium of print.

One example is the early seventeenth-century painting *Neptune, Amphitrite and Cupid with a Nautilus Shell* by Jacques de Gheyn II (1565–1629) (fig. 25). Cupid's finger penetrates a nautilus conch, forming an explicit reference to sexual intercourse.[12] As in the "breasted vessels" of Nuremberg, the turban shell positioned in front of Amphitrite echoes the shape of her own breasts, while representing an idealized female breast shape found in nature. The other shells in the painting invite a range of further sexualized interpretations: Amphitrite's vulva is represented by an enormous Asian conch shell that matches her skin tone, while an elongated snail shell takes the place of Neptune's genitals. The pictorial use of inanimate objects to evoke anatomical features and sexual practices was common in sixteenth-century Netherlandish painting, which experimented with double meanings in both visual and verbal language.[13] These anthropomorphic substitutions disturbed the distinction between the inanimate bodies of objects and the bodies of living entities, resulting in a playful "confusion of categories" that would normally separate one from the other.[14] The nautilus shell in de Gheyn's painting exposes the hidden inner parts of a woman; in representing Amphitrite's and Neptune's genitals as larger-than-life, the image demonstrates the potential of shells to evoke the ways that bodies and their parts could be objectified, dissected, and fetishized.

The capacity of a shell to function simultaneously as a rare collectible and a prompt for anthropomorphic substitution is amply on display in a painting from the workshop of the Antwerp artist Frans Francken II (1581–1642): *Ulysses Recognizes Achilles Dressed as a Woman among the Daughters of Lycomedes* (fig. 26).[15] The painting depicts a mythological scene related to the Trojan War: Ulysses and Diomedes visit the family of Lycomedes of Skyros, where Achilles has been living, disguised as a woman.[16] In an effort to uncover Achilles's identity, Ulysses plays a trick on his hosts. Dressed as a merchant, he offers precious gifts to his hosts that include a sword and a shield. When he makes his companion imitate the sound of a war trumpet, it causes Achilles, the proud warrior, to grab the weapons and reveal his identity as a man. In Francken's painting, the white-bearded man behind Achilles appears to lift the latter's skirt in an attempt to find out whether he is one of Lycomedes's daughters or the hero Achilles in disguise. The gesture of the man's right arm echoes that of the woman in the foreground who holds her garments with one hand while

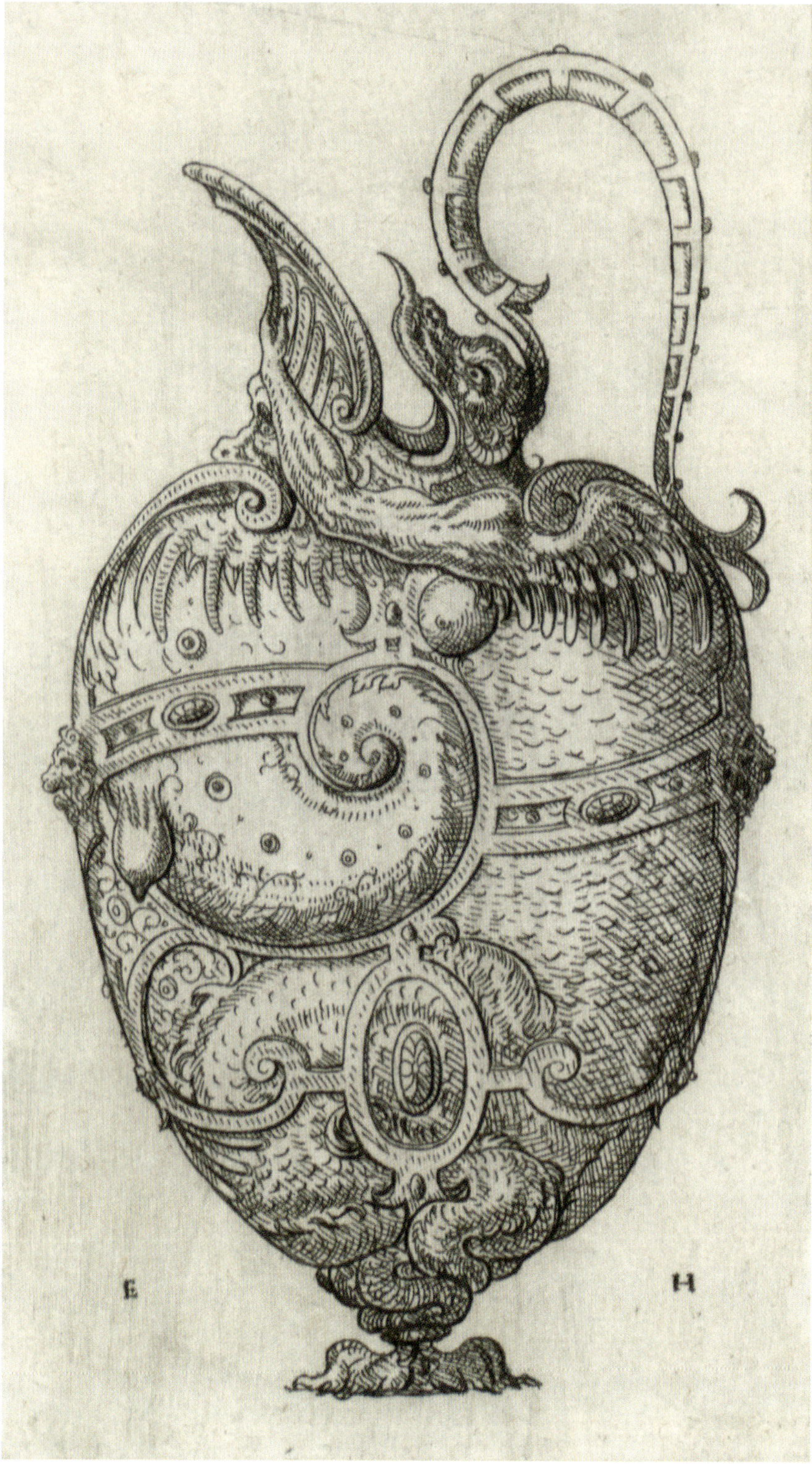
E
H

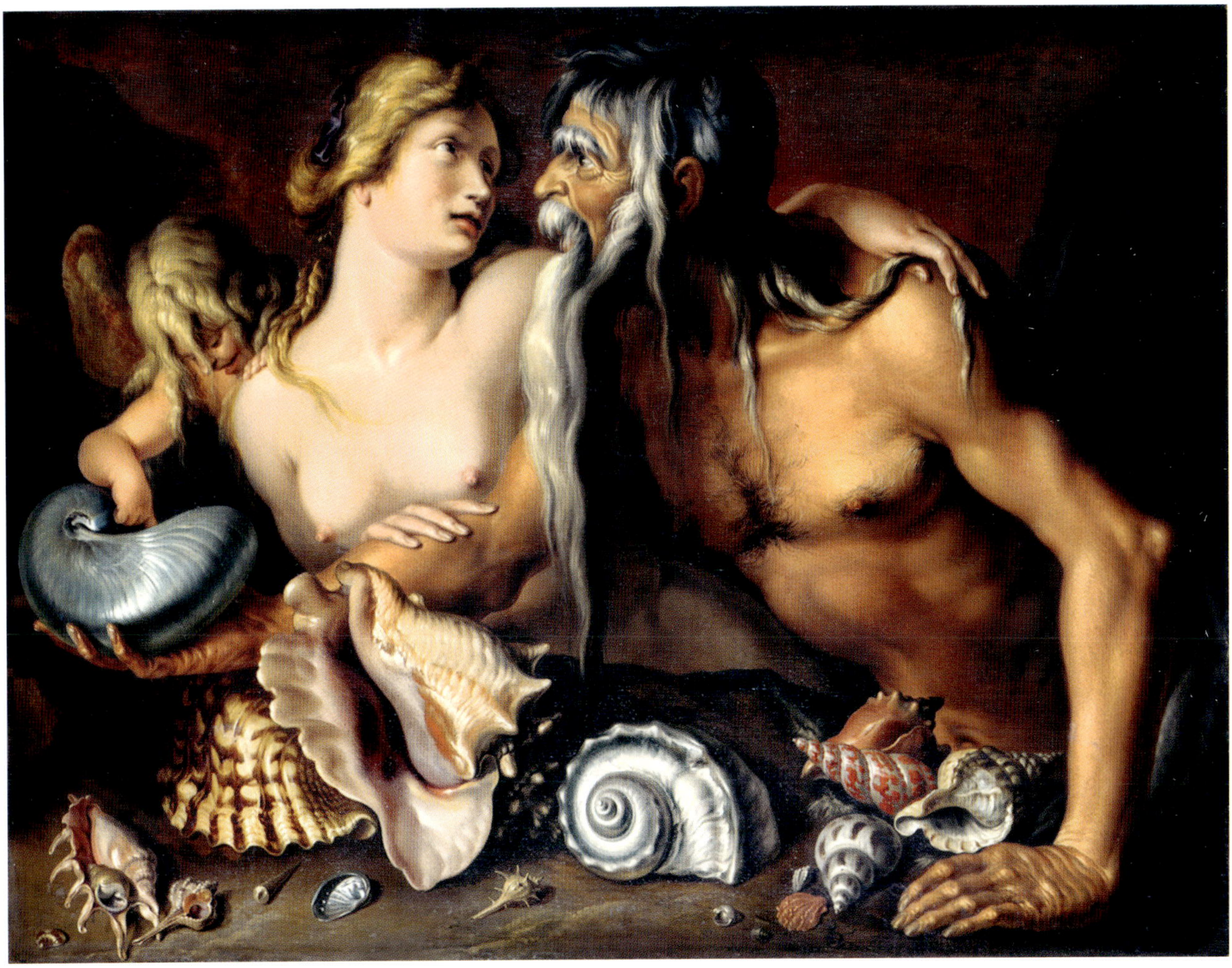

**FIG. 24.** Erasmus Hornick, ewer, 1565. Engraving, 14.5 × 8.5 cm. Staatliche Museen zu Berlin, Kunstbibliothek [OS 907,4].

**FIG. 25.** Jacques de Gheyn II, *Neptune, Amphitrite and Cupid with a Nautilus Shell*, early seventeenth century. Oil on canvas, 103.5 × 137 cm. Wallraf-Richartz-Museum, Cologne [WRM 1792].

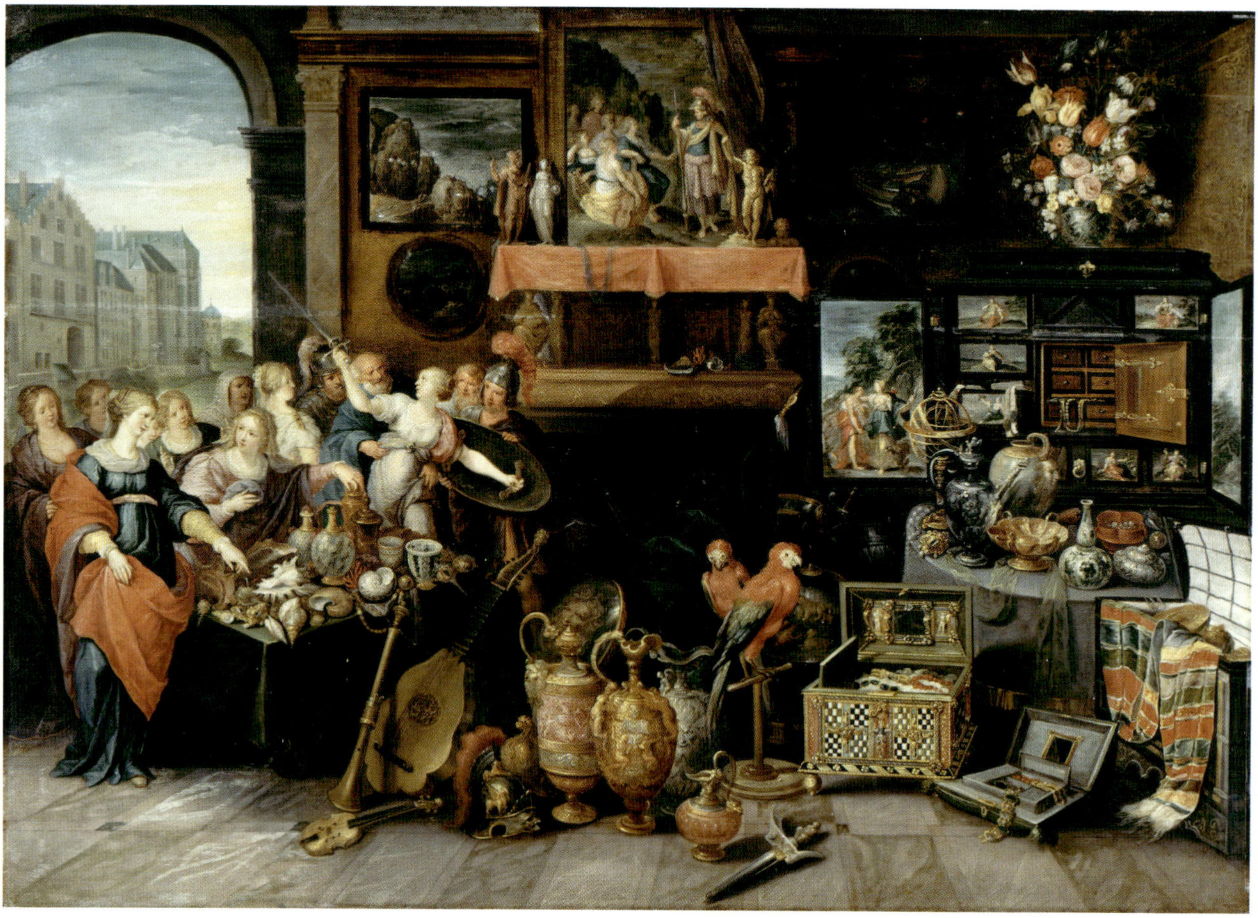

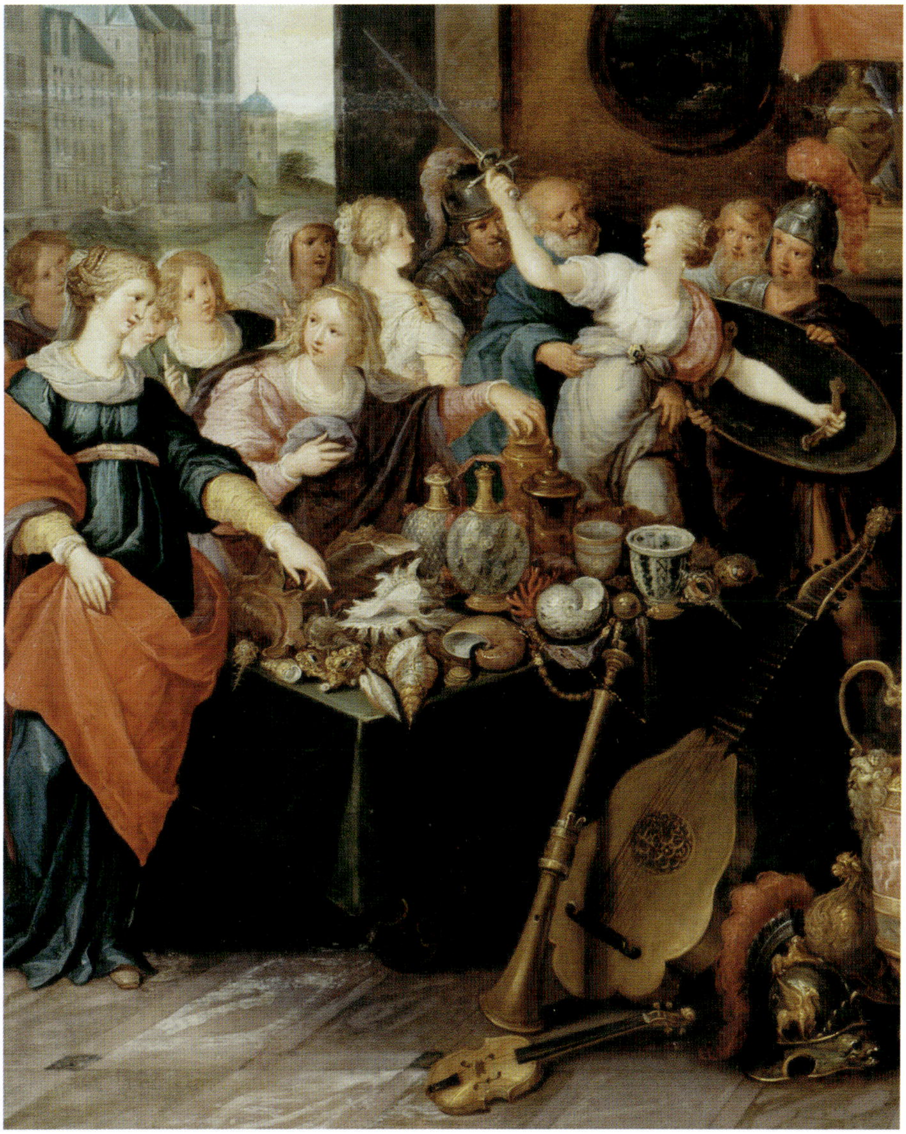

**FIG. 26.** Frans Francken II (workshop), *Ulysses Recognizes Achilles Dressed as a Woman among the Daughters of Lycomedes*, 1620. Oil on canvas, 74 × 105 cm. Musée du Louvre, Paris [R.F. 1535].

**FIG. 27.** Detail of fig. 26.

pointing at a pile of shells with the other (fig. 27). Among those shells we recognize two sorts that, as we have seen in de Gheyn's painting, not only evoke parts of the human body, but can explicitly stand in for male and female genitals. Francken's painting makes explicit the analogy between private parts and shells. The lifting of Achilles's skirt parallels the gesture of the woman further in the foreground who points her finger at a bounty of shells on the table before her, as if to reveal through them the true sex of the hero in female disguise. Achilles also happens to be this woman's lover, as we are presumably looking at Deidamia, the daughter of Lycomedes, who had become intimate with Achilles long before Ulysses's arrival.

The shells in the image represent different levels of the story: the so-called Triton's trumpet (*Charonia tritonis*) displayed on the table refers to Achilles's member but also echoes the brass instrument positioned by its side, which in turn recalls the war trumpet so central to the narrative. The shells are surrounded by ceramic bowls, metal containers, bottles, and branches of red coral. These precious and beautiful rarities laid out on the dark green tablecloth are among the wares Ulysses brought with him to get the attention of Lycomedes's daughters, characterized in Ovid's *Metamorphoses* and the *Achilleid* as "things [that] women buy" and "gifts that should attract maidens' eyes."[17] The painting has been read as animating three different types of desires: Deidamia's infatuation with Achilles, the craving for rarities among Lycomedes's daughters, and the early modern connoisseur's love of art.[18] We can also add lust to the panorama of represented desires, as the shells that so appeal to Lycomedes's daughters present us with artfully rendered objects of female desire in more than one sense: as early modern versions of the precious and beautiful goods that mythological women allegedly desired, and as substitutes for desired parts of the male body. The painting allows for different modes of looking. On the one hand, the narrative clues within the image tell the tale of Achilles's discovery; on the other hand, its details enable the beholder to enjoy the visual spectacle of the early modern rarities on display.[19] A third mode constitutes the "confusion of categories" caused by anthropomorphic substitutions, as described above: the viewer of the image looks at shells as if they were body parts and beholds objects as if their surfaces were skins. The beholder discovers the key to the narrative—the hero-in-disguise—by exploring object surfaces that invite not only our gaze but also touch. This additional reading of the painting allows us to build upon previous understandings of the image as lionizing collecting and expressing early modern anxieties about the commodification of art. In a larger sense, the painting emerges as a reflection on the relationships among visual attraction, sensual stimulation, and sexual objectification.

## CONCHS AND CERAMICS

*Ulysses Recognizes Achilles* also introduces another pairing found in early modern collections and in the paintings that represent them: the juxtaposition of Asian shells with Asian porcelain vessels. In Francken's painting, a typical example of contemporary Chinese kraakware porcelain sits next to a nautilus shell on the table around which Lycomedes's daughters gather. A porcelain bottle with an intricate blue-and-white decoration, also of Chinese provenance, appears next to a nautilus shell on the table at the far right. This association between Asian ceramics and Asian shells is unique neither to the *Ulysses* painting nor to the other similar works produced by Francken and his workshop.[20] It speaks to an interest in the sensual materiality of both shells and porcelain recurrent across the early modern period at large.

Take a depiction of a German *Kunstkammer* from 1668 (fig. 28). This representation of the collection kept by the wholesaler and ironmonger family Dimpfel from Regensburg offers insights into the associations evoked by the assemblage of diverse types of objects within a single space. On the table at the center of the room, a small sculpture of the Crucifixion is placed directly beside a cluster of coral, its branches extending in parallel to Christ's arms. This pairing of human-made object and natural specimen links the blood of the crucified and his distorted body to the color and shape of red coral fragments.[21] The proximity of a kraak bowl to two shells at the far end of the table on the left likewise suggests a similarly meaningful connection between artificial and natural media.

The early modern pairing of porcelain and shell was based on a number of perceived connections. A first link between shells and ceramics is evident in the history of terminology used to describe them. The 1298 account of *The Travels of Marco Polo* first employed the word *porcellane* to refer not only to shells but also to porcelain.[22] The word also bore an implicit relation to the female body: the cowrie shell was referred to as *porcellana* in Italian because of its resemblance to a vulva (colloquially, *porcello*).[23] In the sixteenth century, the term "porcelain" was used in local variations across Europe including *Pourzelanae* or *Porzelana* in German, and *purselyne* or *purselyn* in English. Before diving into a lengthy discussion on the relations between *Porcelaine* and the "myrrhine vessels" of the ancients, French naturalist Pierre Belon (1517?–64) explained that the dual use of the term *Porcelaine* to denote shells as well as ancient vessels was no longer "modern" by 1551, a clear indication that in early modern French the term still carried connotations of both shells *and* ceramics.[24]

In addition to etymology, a second connection was provided by sixteenth-century scholars such as Julius Scaliger (1484–1558), who maintained that the Chinese produced porcelain from ground-up shells.[25] This conclusion was probably based on the qualities of whiteness and transparency common to both shells and porcelain—two of the many characteristics that made them so attractive to European collectors. Early modern European attempts to copy the materiality of chinaware began in Florence's Medici-sponsored workshops and continued at Delft, but remained unsuccessful until the early eighteenth century, when European hard-paste porcelain was first created at the court of Dresden. Part of the challenge lay in figuring out the correct ingredients for producing porcelain. Shells were not part of the recipe, despite the assertions made by some of Scaliger's contemporaries.[26]

Third, shells and porcelain objects were imported to Europe together on Portuguese, Spanish, and Dutch vessels, and often remained together once they entered collections in northern European cities such as Antwerp or Regensburg. According to the 1619 inventory of the Dresden *Kunstkammer*, one of the largest and most comprehensive early modern collections in Europe, two samples of polished Asian nautilus or turban shells (*Perlenmutter schneckenhäuser*) were stored and presented alongside clusters of Chinese porcelain vessels.[27]

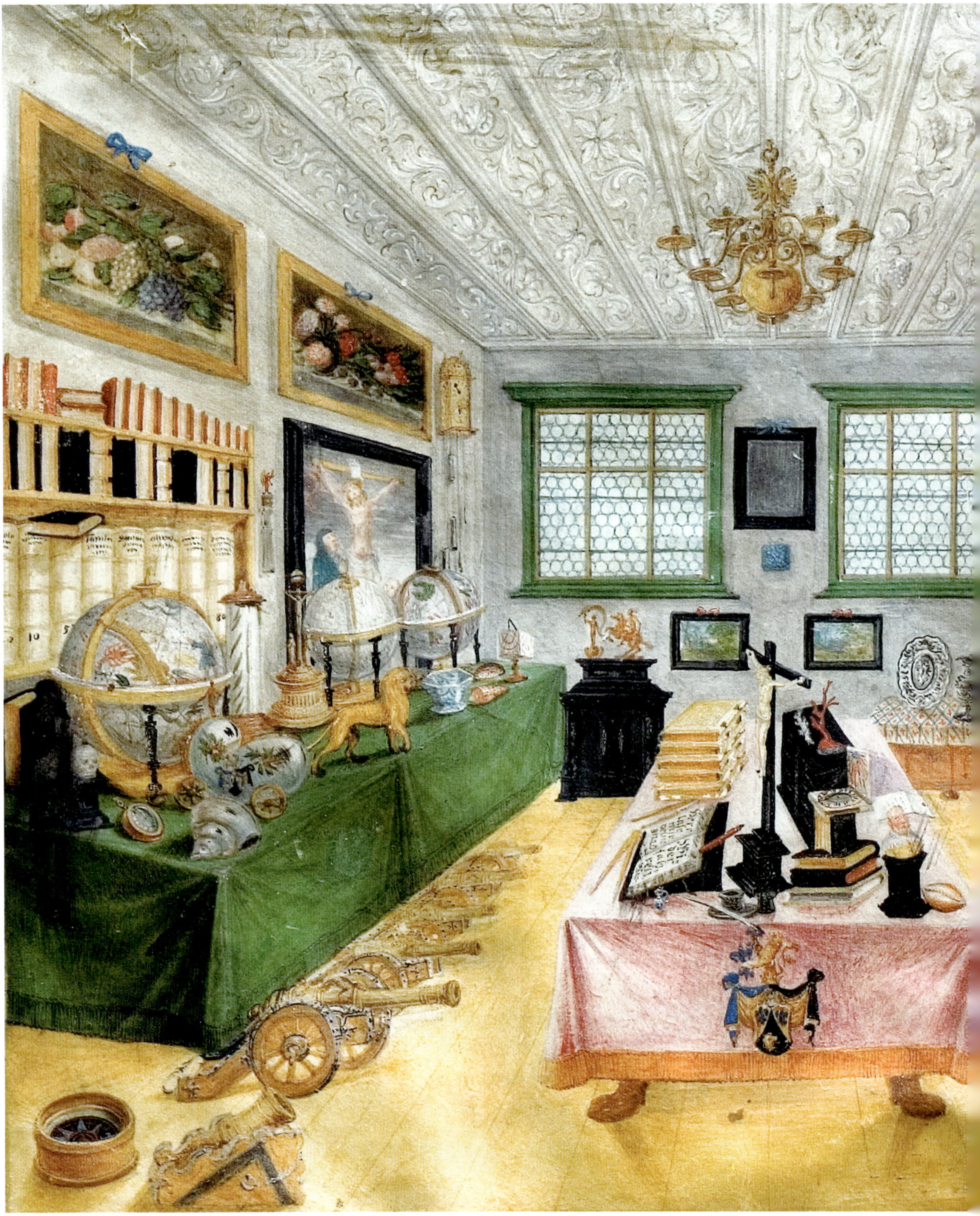

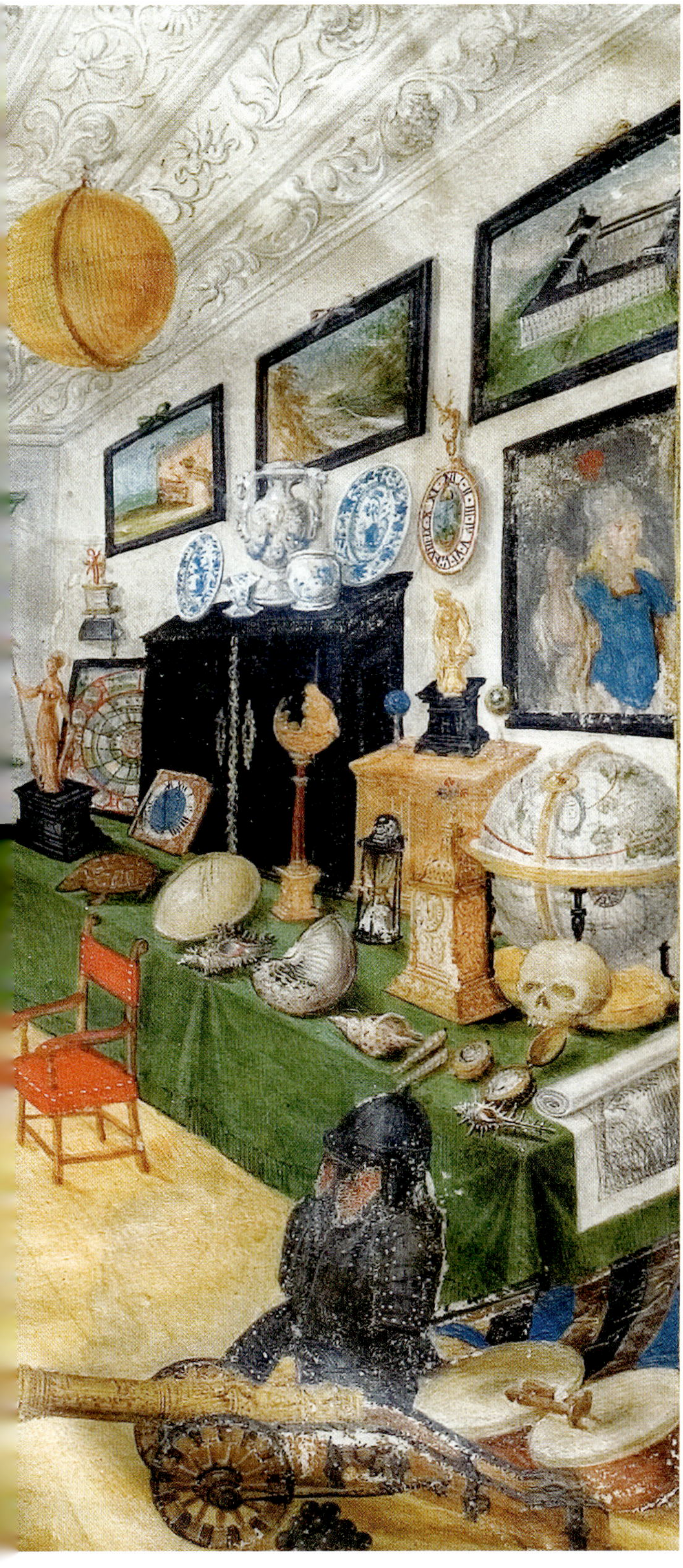

FIG. 28. Joseph Arnold, *The Wholesaler and Ironmonger Dimpfel Family Curiosity Cabinet*, gouache heightened with punched gold on parchment, 1668. 14.9 cm × 19.1 cm. Ulmer Museum, [1952.2611].

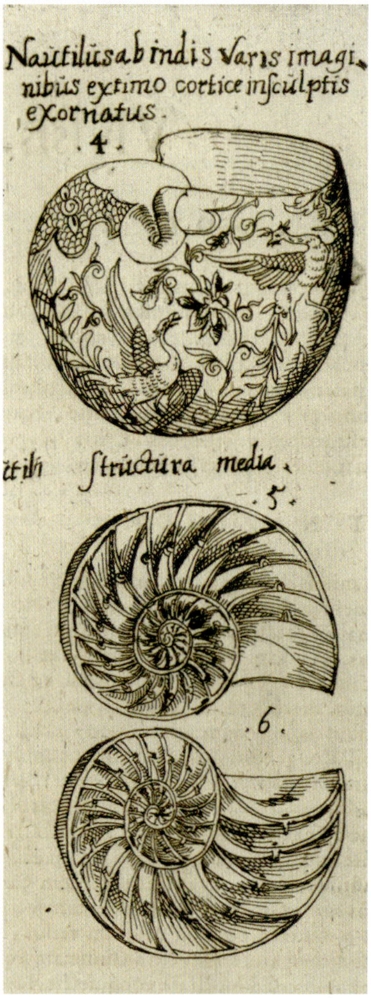


**FIG. 29.** Ulisse Aldrovandi, *De reliquis animalibus . . .*, 1618, 95 (detail). Niedersächsische Staats- und Universitätsbibliothek Göttingen [2 ZOOL V, 77 (1)].

The shells had reached the Dresden court in 1590 as part of a large donation by the Duke of Medici, a gift that also contained numerous porcelain items of Chinese provenance. Shells and ceramics remained alongside one another upon their integration into the *Kunstkammer*. The collection inventory lists them under the common header "Italian vessels for drinking or other purposes that the Duke of Florence has given [to the court of Dresden] in 1590."[28] The description emphasizes that these collectibles had a shared "social life" as gifts before entering the collection.[29] As the Dresden case illustrates, shells and porcelain were sometimes kept together not only—or even primarily—because of perceived material and visual similarities, but also because of their shared provenance.

A fourth link between conchs and ceramics lay in the approach to their decoration. *Ulysses Recognizes Achilles* and the representation of the Regensburg *Kunstkammer* both show shells and ceramics as patterned objects. In the Antwerp painting, an unpolished nautilus is exhibited next to a polished one; the Regensburg *Kunstkammer* features a half-polished nautilus and another that was painted and transformed into a wheeled vehicle. Both representations contrast natural specimens with modified ones, highlighting the shell's potential to serve, when cleaned and polished, as canvas for the inventions of human craftsmanship.

Many sixteenth- and early seventeenth-century collections also contained shells that had been decorated in Asia. The earliest extant Renaissance examples were part of the Medici collections, and are somewhat unusual owing to additional coats of coloring that were presumably applied in Europe.[30] More typical are shells that were incised in Asia with images of birds, flowers, or individual figurative scenes, left uncolored. Even the Italian naturalist Ulisse Aldrovandi, in his 1606 encyclopedia of animal specimens, included a woodcut of a nautilus shell "from the Indies" (*ab indis*) with "engraved" (*insculptis*) ornamentation alongside an image of the object's natural spiraled interior (fig. 29).[31] The decoration on the nautilus depicted in Aldrovandi's treatise closely resembles that on the extant carved shell in the Kornblum cup introduced earlier (see fig. 21). Both show scale- or wave-like line patterns, birds with long, elaborate tail feathers, and vegetal motifs such as leaves, stalks, and peony-like flowers. These motifs were typical, as other examples and images of similarly decorated specimens reveal.[32] In *Ulysses Recognizes Achilles*, the shells' polished surfaces, in particular the turban shell next to the porcelain bottle on the round table to the right, do not feature recognizable motifs engraved on their surfaces. Nevertheless, the lines and highlights indicate the convex and concave surfaces of carved mother-of-pearl.

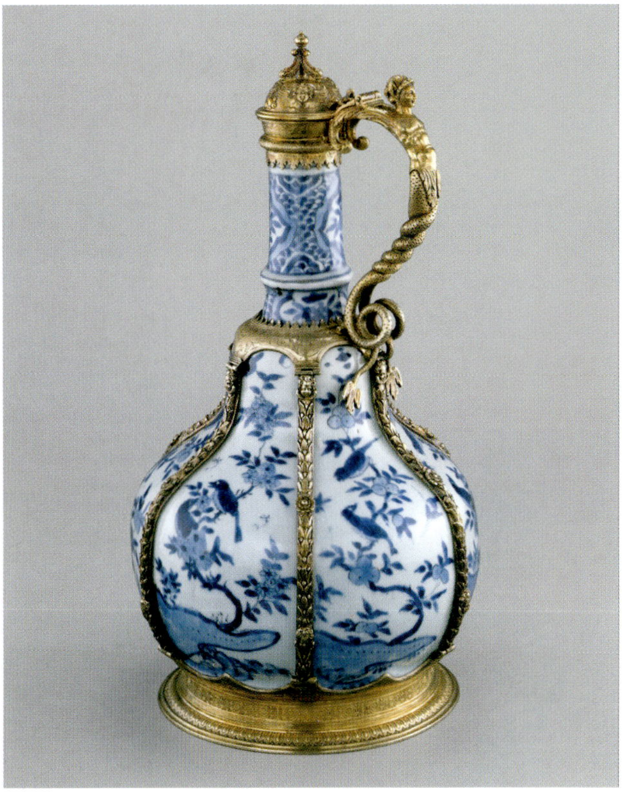

**FIG. 30.** Anonymous, *Ewer from Burghley House, Lincolnshire*. Chinese porcelain 1573–c. 1585, British mounts c. 1585. Hard-paste porcelain, gilded silver, 34.6 cm. The Metropolitan Museum of Art, New York, Rogers Fund, 1944 [AN 44.14.2].

By the sixteenth and seventeenth centuries, Asian craftsmanship was admired and imitated in Europe, as demonstrated by efforts to produce porcelain and by the European interest in conch-carving techniques.[33] Dutch descriptions of shells as "exquisitely polished by the Indians" reveal that during later periods this kind of admiration also had colonial undertones; already by the sixteenth century, these objects seem to have registered on one level as trophies of European conquest.[34] Motifs such as dragons and birds that were carved into shells like the one that Aldrovandi recorded also appear on porcelain.[35] Both porcelain and Asian-carved shells were mounted in precious metals, which simultaneously framed and obscured the complex imagery on their surfaces. Further enforcing commonalities between the conceptualizations of shells and porcelain, the mounts applied to both materials shared a common function as marginal by-works—in Greek, *parerga*. The application of these by-works of sculpted rims, feet, and handles—to natural objects like shells and imported goods like porcelain—Europeanized the haptic engagement with foreign materiality.[36] In other words, these additions amounted to an act of cultural appropriation.

In the cases of early modern porcelain vessels and shell cups, the mounts often added a panorama of maritime motifs. These included metallic sculptural renderings of fish-tailed men and women, dolphins, and other aquatic creatures. The mounts applied to a Chinese porcelain bottle in 1585, for instance, feature three tiny dolphins on the vessel's lid and a somewhat androgynous fish-tailed torso on its handle (fig. 30). Likewise, the mounts on

Kornblum's shell cup combine fish-tailed mythological creatures and sea animals with an oceanscape engraved on its gilt silver foot. One would visually and physically grasp these vessels by their metallic frames. At the same time, these framing elements would not have hindered collectors from directly touching the smooth surfaces of ceramics and shells as well. If anything, such frameworks may have even encouraged a comparison of the haptic experiences of touching metal versus ceramics and shells, and further demonstrated the contrasts between these various materials.

By inviting both the gaze and the touch of the hand, ceramic as well as shell vessels offered seductive surfaces attractive to behold and possess. As part of what Jonathan Hay has called the complex "surfacescapes" of early modern collecting, the cups engage the human capacity for erotic response.[37] "Thinking materially" with the body of the early modern beholder, the artifacts' "sensuous surfaces" embody a powerful affective potential in addition to conveying a plethora of metaphorical meanings.[38] This terminology, developed by Hay in relation to the "erotic economy of decorations" of Ming and Qing dynasty artifacts, requires some qualification when applied to European *Kunstkammer* objects. The unequal trade and colonial relationships between Renaissance Europe and extra-European empires meant that the sensual topography of those objects differed in sociopolitical implication from that of the Chinese emperors' and literati collections with which Hay is concerned. This becomes evident, for example, through the politics of display and the application of mounts that Europeanize the visual and haptic encounter with objects derived from extra-European parts of the world, while in the Chinese tradition the application of metal frames carried different implications.[39] The sensual engagement with foreign surfaces in Europe, unlike in China, also involved the objectification and fetishization of naked "foreign" bodies.[40] As we will see, this tendency toward objectification has implications for *Kunstkammer* encounters with shells in particular.

While a celebration of nature's generative powers in the creation of remarkable things like Asian shells underlay the practice of *Kunstkammer* collecting, the concept of human fertility was seen as akin to nature's reproductive energies. Both nature's urge to create and humans' wish to procreate were seen as intrinsically related and revealing a shared impulse to generate. Celebrating the manifold results of generation through the act of collecting did not necessarily entail the use of sexually explicit visual and material culture in the vein of the previously discussed anthropomorphic substitutions. But even when it was not explicitly sexual, collecting did engage the senses, especially vision and touch. Playing out the full potential of the *Kunstkammer* as a space that celebrates generation and offers multiple possibilities for haptic encounter, early modern goldsmiths explicitly addressed the erotic implications of early modern collectors' engagements with surfaces in the mounting of shells and porcelain.

## SURFACES AND SKINS

The image of Venus sailing in a shell boat, like the one atop Kornblum's cup, is a prominent motif in sixteenth-century Netherlandish and German print culture. Fortuna, the Roman goddess of fortune, was often depicted in a similar way, standing or flying on an orb. The goddess figure on Kornblum's cup appears to be a combination of the two, positioned inside the shell, like Venus, but standing atop a pearl, as Fortuna would. Another visual connotation of this figure is suggested by period literature such as Reginald Scot's 1584 treatise *The discoverie of witchcraft*, which mentions that witches were believed to "saile in an egge shell, a cockle or muscle shell, through and under the tempestuous seas."[41] Witches were shown riding goats in engravings from this period.[42] Riding in this context suggested sexual stimulation but was also a reference to an activity restricted to men, who regularly rode horses while women would do so rarely if ever (and in a different position from that assumed by men). Likewise, sailing was an activity to which women had little to no access. As a goddess, Venus-Fortuna transcended social and visual conventions, and in its German visual and material context, the sculpture carried overtones of transgressive female behavior related to witchcraft, including the appropriation of male practices, spaces, and desires.

In addition to functioning as a sailing vessel, the shell that separates the two figures in the cup's mount is an indicator of geographic space: its surface is engraved with pictorial references to the ocean in the form of lines suggesting scales or waves, much like early modern maps that showed fish-tailed creatures against a backdrop of wavy lines indicating extra-European waters. The golden snail attached to one side of the nautilus swims in these waters, like the mermaid herself. Although Ulisse Aldrovandi asserts (based on Aristotle) that nautilus shells were produced by animals closely related to squids, nautilus and turban shells were associated with snail-like creatures that inhabited foreign oceans.[43] Snails carried strong sexual connotations. Their shells were likened to female genitalia in German encyclopedic writings of the period, in which "the water of the seas of this world" symbolizes "the bitter waters of mundane lust" and sexual ecstasy.[44] By reminding viewers that the nautilus shell is the product of a slimy mollusk, the small detail of the gold snail makes explicit the sexual symbolism that underlies the entire cup.

Some European Renaissance artists depicted foreign oceans and their shores, where pearls, shells, and coral branches could allegedly be harvested in abundance, as erotic paradises. Take Jacopo Zucchi's *Coral Fishers*, created around 1560 for the studiolo of Francesco I de' Medici in Florence, in which fair-skinned mythological goddesses mingle with men and women of color (fig. 31). Alongside naked human figures engaged in retrieving and handling pearls, shells, and coral fragments, Zucchi includes a guenon, a monkey of African origin, who points to the vanity of their pursuits. The painting places the practices of coral, pearl, and shell fishing within a realm where foreign treasures are gathered by foreign bodies: both

**FIG. 31.** Jacopo Zucchi,
*The Treasures of the Sea (Coral Fishers)*, ca. 1560.
Oil on copper, 52 cm × 42 cm.
Galleria Borghese, Rome.

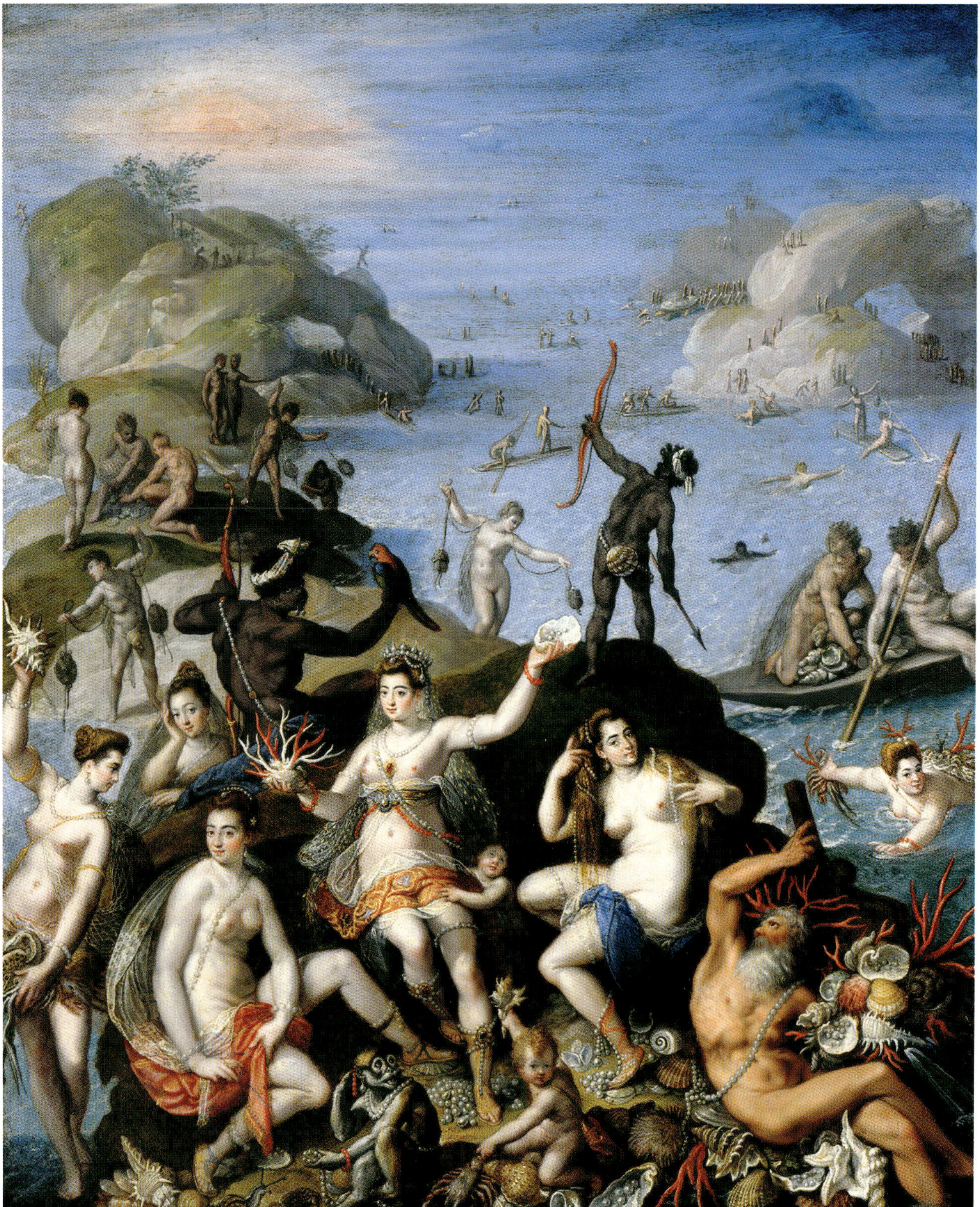

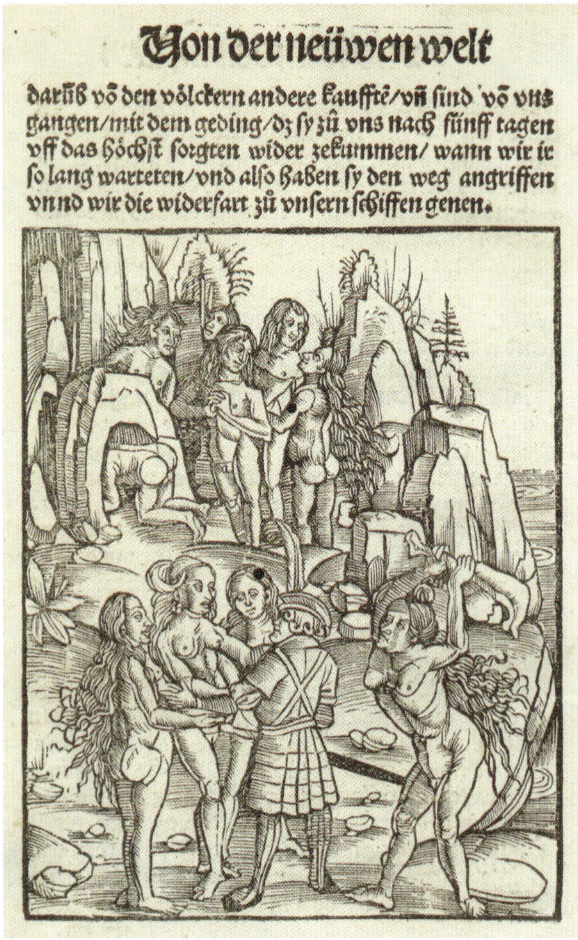
Von der neüwen welt

darūb vō den vōlckern andere kauffte/vñ sind vō vns gangen/mit dem geding/dz sy zů vns nach fünff tagen vff das höchst sorgten wider zekummen/ wann wir ir so lang warteten/vnd also haben sy den weg angriffen vnnd wir die widerfart zů vnsern schiffen genen.

**FIG. 32.** Anonymous, illustration from Amerigo Vespucci, *Diß Büchlin saget, wie die zwen durchlüchtigsten Herren Her Fernandus K. Zu Castilien und Herr Emanuel K. zu Portugal haben das weyte Mör ersuchet und funden vil Insulen unnd ein nüwe Welt . . . vormals unbekant*, 1509, 24b. Woodcut. Universitätsbibliothek Freiburg [i. Br., J 4672,m*].

mythological figures and men of color alike. Two African men stand out as they are both equipped with bows that identify them as hunters of animals rather than of *Kunstkammer* collectibles. The posture in which one of them holds a parrot is reminiscent of early modern representations of huntsmen with their falcons. The other bowman directs his arrow into the waters. Surrounded by more than fifty figures who are engaged in the exploitation of the sea, the man's arrow functions as a pointer hinting at the treasures to be found in the deep. Through the figures of the two bowmen, the activities of coral fishing, pearl diving, and shell collecting are represented as the aquatic equivalent of overland hunting practices. The painting's foreground displays the prey, inviting viewers to engage in a treasure hunt of their own: rare shells alongside pearls, coral fragments, and extra-European animals. Within this allegory of desire and possession, both shells and skin color are closely entangled.[45]

In addition to paintings in *Kunstkammer* and Italian *studiolo* collections that blurred the boundary between mythological realms and extra-European spaces, Renaissance illustrations in travelogues visualized eyewitness accounts describing inhabitants of "the Indies"—where nautilus conchs and porcelain came from—as scarcely covered or entirely naked (fig. 32). The nakedness of the "Indians" made them comparable to mermaids and mermen, believed by European collectors to populate the oceans that connected local seas to the rest of the world. Sixteenth-century texts on extra-European spaces and their inhabitants use the rhetoric of sexualization and commodification in their descriptions of foreign people, and contain an abundance of narratives featuring "sexually overactive" non-European men and women.[46] The woodcut mentioned above served to illustrate a (fictive) travel report on the New World in letters attributed to Amerigo Vespucci (1454–1512).[47] It depicts an imagined encounter between "Indian" women and a European explorer, the latter distinguished by his clothing and feathered hat. As three "Indian" women use their seductive agency as a weapon of distraction, a fourth approaches from behind, ready to bash in the foreigner's head.

These themes of sexual seduction and hidden danger also appear on Kornblum's cup. Venus-Fortuna's flawless beauty stands in contrast to the image of the mermaid at the foot of the cup. While the nudity of the goddess above is portrayed as "innocent" and "natural," the mermaid has seductively embellished her body with jewelry. A sparkling necklace draws the viewer's gaze to her breasts. Both Venus-Fortuna and the mermaid belong to the sea, but they occupy different domains. One sails aboard a Mediterranean cockle or scallop shell, while the other reigns over an underwater empire that extends to non-European spaces. The mermaid bears some resemblance to the mythological Nereid Galatea, whose name can be translated as "milk white," and who is referred to as "the milk white one" or "the one who has milk-white skin."[48] Precious objects that were whitish in color were often compared to the skin of seductive women, which, according to early modern beauty standards, was meant to be as pale as possible.[49] Even today, pearl and shell extracts are mixed into certain cosmetic products, and "porcelain skin" continues to be a laudatory expression. In fact, Ovid's *Metamorphoses*, a major source of inspiration for *Kunstkammer* artifacts, frequently compares skin to the surfaces of objects. Ovid described Galatea as "more radiant than crystal, smoother than shells," but also as "wilder than an untamed heifer, . . . more truculent than a pregnant bear, . . . crueller than a trodden snake."[50] In serving as inspiration for a number of sixteenth-century *Kunstkammer* artifacts, Galatea offered one of many examples of females who hid their power beneath an erotically charged "radiant" and "smooth" appearance.[51] As seductive as Galatea, but also as dangerous as she is, the "Indian" women in the illustration above, as well as the mermaid queen on Kornblum's cup, likewise hide their "wild" nature behind an attractive appearance.

While the connection between Kornblum's Venus-Fortuna and women who sail atop shells (and qualify as witches) is implicit, the mermaid on the cup's foot more explicitly evokes the "dangerous" seductiveness of Galatea's ambiguous nature, which comprises both outer beauty and inner immorality, and oscillates between human and animal natures. It is no coincidence that Venus-Fortuna is separated from the nautilus by a European shell, while the Galatea-like mermaid sits beneath an Asian shell associated with enormous slugs. In the rapidly expanding seafaring kingdom of England, but arguably also in early modern Europe at large, the mermaid functioned as a symbol of the unfamiliar, the enticement and the danger of the European interest in and "desire for the 'strange' and 'exotic' body."[52] Accordingly, Kornblum's cup contrasts a local shell's symmetrical proportions and the ideal beauty of a porcelain-skinned woman above with the "dangerously" seductive foreign bodies of an Asian shell and a mermaid below. The vessel rendered the skin of an elusive queen and the surface of an item from her maritime kingdom touchable, bringing exoticized foreign bodies and objects within reach of the armchair-traveling *Kunstkammer* collector.

## CONCLUSION: SHELLS AS BODY PARTS

In the first half of the twentieth century, American photographer Edward Weston (1886–1958) experimented with a nautilus and other kinds of shells. He took pictures of single shells or combined them (fig. 33). Upon learning that this set of photographs was described as "perverse," "erotic," and "sexual" by those who first saw them, he was baffled. "Why were all these persons so profoundly affected on the physical side?" he wondered in his diary. "For I can say with absolute honesty that not once while working with shells did I have any physical reaction to them: nor did I try to record erotic symbolism."[53]

As the response to Weston's work shows, associations between the surfaces of shells and human skin, nautilus conchs and body parts are not restricted to early modern Europe. They have in fact appeared in a variety of cultures throughout history.[54] What makes the early modern European experience of shells and sexuality fundamentally different from the twentieth-century American one is the fact that shells arrived in Europe as rare, treasured goods from Africa and Asia, regions that very few Europeans had access to or comprehensive knowledge of during the sixteenth and seventeenth centuries. Asia, where many of the most desired specimens were acquired, could by then be reached by crossing oceans that early modern maps depict as filled with extraordinary creatures, sea monsters, and mermaids, who appear prominently on the elaborate mounts that northern European goldsmiths added to some shells. A neatly woven web of symbolic meanings and metaphorical implications tied nudity and foreignness to shells, and informed the decorative framing that European craftsmen added to Asian shells in early modern *Kunstkammer* collecting.

In the early modern period, the oceanic world was a site loaded with sexual connotations, imagined to be inhabited by lascivious mermaids and wanton seamen as exemplified by the painting *Neptune, Amphitrite and Cupid with a Nautilus Shell*. Likewise, the *Kunstkammer* was an eroticized space. Shells stood squarely at the intersection of material collecting and visual fantasies of oceans; they were etymologically loaded with references to

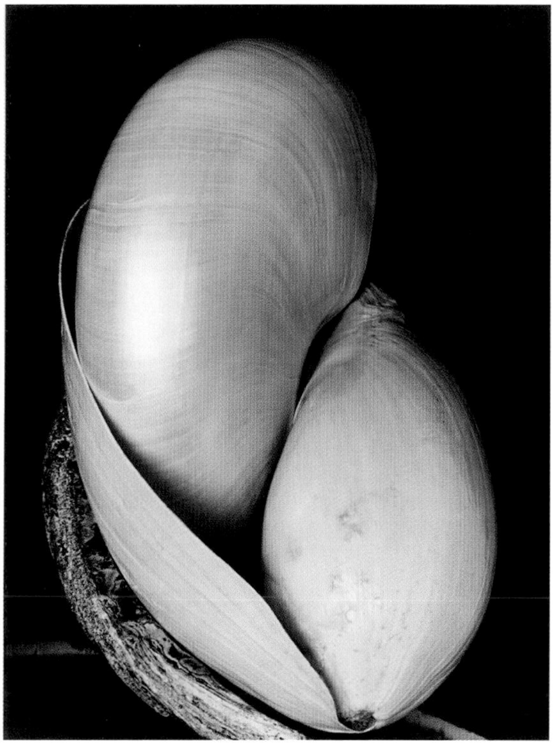

**FIG. 33.** Edward Weston, *Shells*, 1927, photograph, 22.86 cm × 17.78 cm. Sheldon Museum of Art, University of Nebraska-Lincoln, Anna R. and Frank M. Hall Charitable Trust [H-513.1958].

the female body, to which their shapes, colors, and surfaces bore a striking resemblance in the eye of early modern collectors. While the formal resemblance between shells and genitals has remained a visual trope throughout the centuries, the intricate web of early modern references between Asian shells and Chinese porcelain has lost its significance. Both shells and porcelain were considered "foreign" and associated with the sea. Much Renaissance imagery depicted inhabitants of "the Indies," where shells and porcelain originated, as partially or entirely naked. This nakedness made them comparable in the European imagination to mermaids and mermen, who were believed to populate the spaces between Europe and the Indies, as well as to mythological creatures like the seductive, dangerous Galatea. As Jill Burke has argued, we cannot fully understand the emergence and importance of the European nude without rethinking it in relation to depictions of the *non*-European nude.[55] This chapter makes a smaller claim, namely, that we cannot fully understand the meaning of foreign shells in Europe without taking into account fantasies of foreign bodies, such as those of mythological figures and women of extra-European origin. The fantasies of seduction and danger evident in objects and images that frame Asian shells in the *Kunstkammer* illustrate that some aspects of European conchophilia, a phenomenon that this book argues was of high cultural and artistic impact during the early modern period, amounted to cultural essentialization and material fetishization.

# Part II

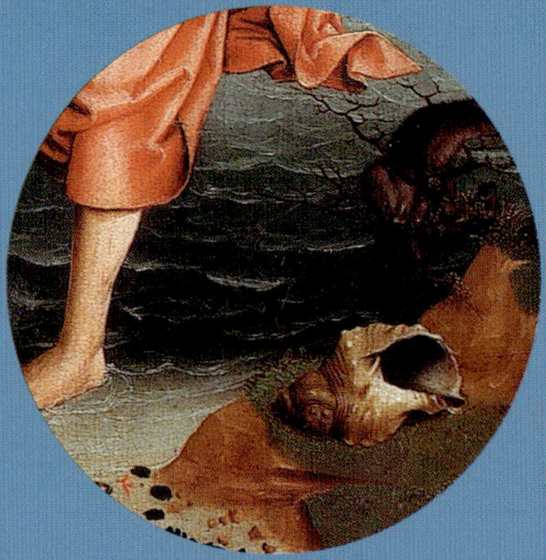

# MICRO-WORLDS OF THOUGHT

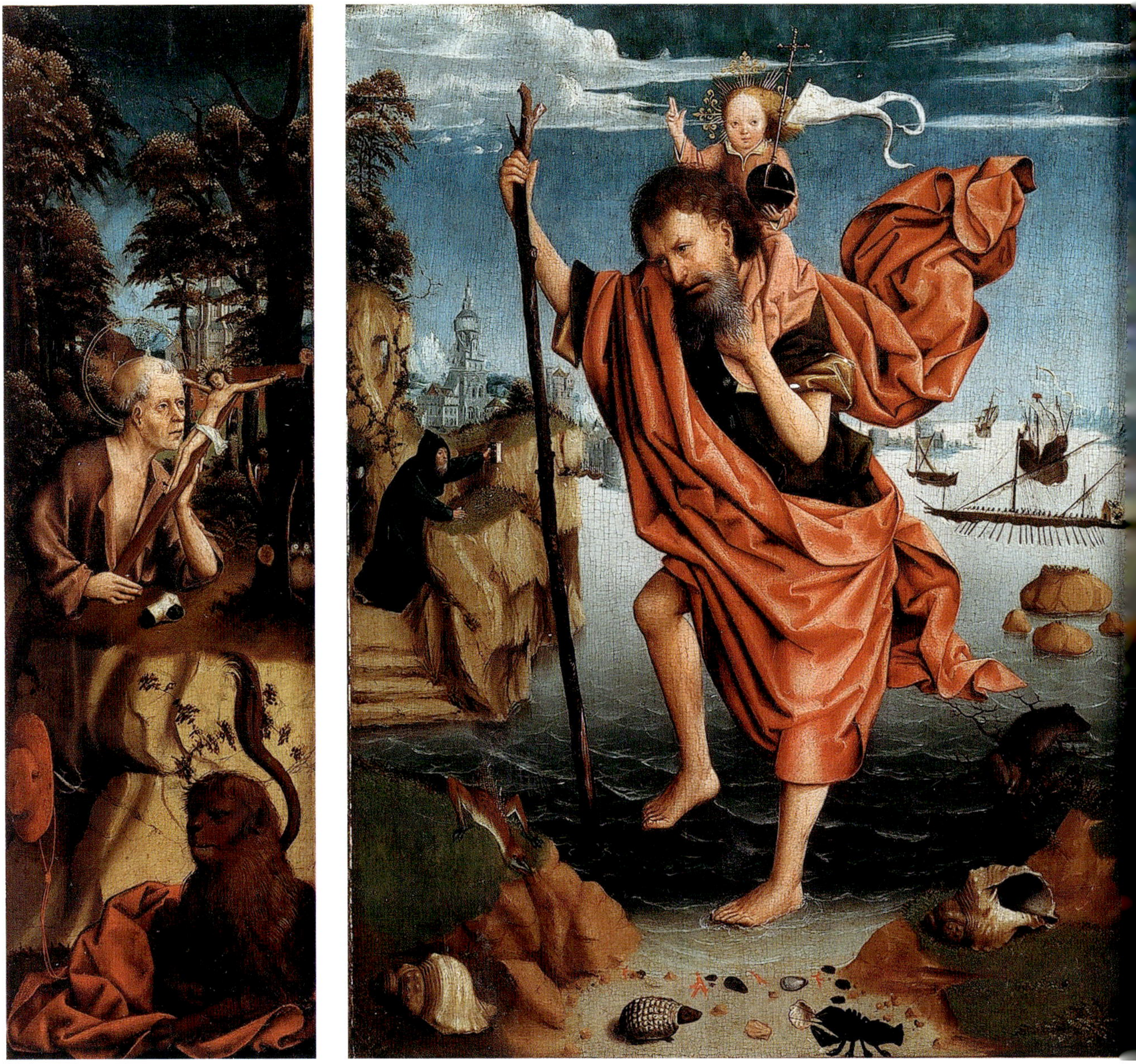

**FIG. 34.** Master of the Holy Figures. *Triptych of St. Christopher, St. Jerome, and St. Anthony*, c. 1510–20. Oil on panel, 57.5 × 93.4 cm. Museum Mayer van den Bergh, Antwerp [MMB .0024].

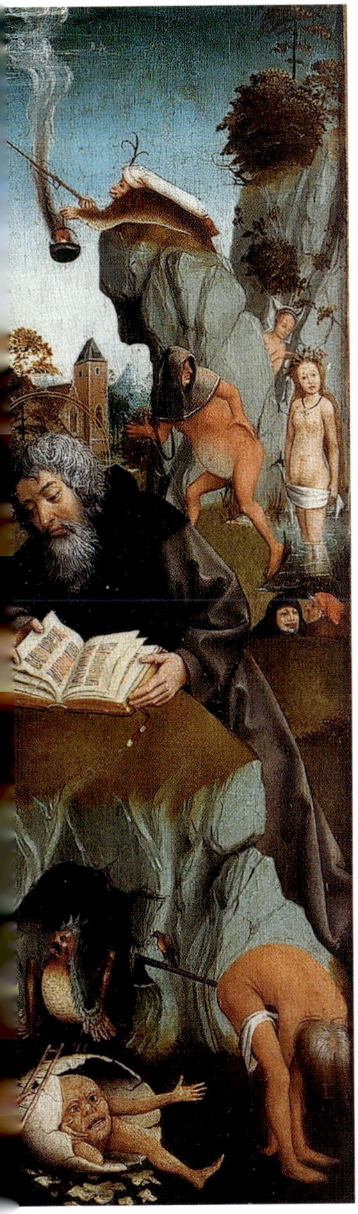

Marisa Anne Bass

# Chapter 3
# Shell Life, or the Unstill Life of Shells

Shells begin as movable dwellings for the mollusks who create them, but where a shell ends up is far from predetermined. They may go on to have many lives once their makers have left them behind. Whether subsumed into the geological landscape, reoccupied by a hermit crab, or repurposed by humankind, shells are more than exoskeletal remains of life past; they are persistently animable things. To admire a shell's variegated surface, to feel its sculptural form in one's hand, or to hear the echo of the sea in the architecture of its inner chamber, is to engage with a curious object that is also an uncanny subject. This chapter explores how early modern artists employed the pictorial realm to think through shells and their resonant histories, and it seeks to understand the encounters with shells that these pictures stage. I take as my starting point a simple premise: the life of a shell in a painting is never a *still* life.

An anonymous Netherlandish triptych from the first quarter of the sixteenth century offers a first case in point (fig. 34).[1] Dominating the triptych's central panel is the giant St. Christopher, who has just traversed a river with the infant Christ upon his shoulders. He is about to learn that his passenger is the true ruler of the world, one more powerful than either the king or the devil by whom he was previously employed. His spiritual conversion is imminent.

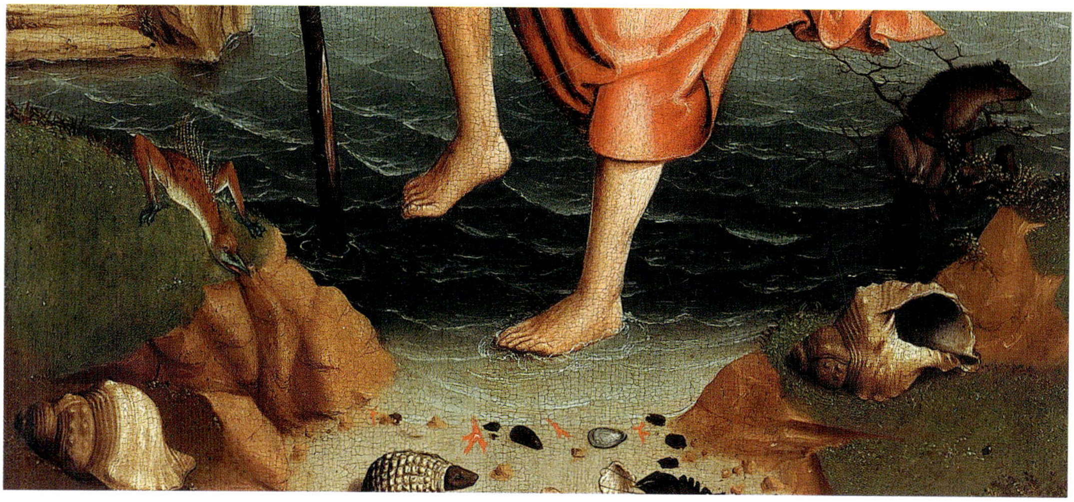

**FIG. 35.** Detail of fig. 34.

Yet the shoreline before him teems with menaces, and his next step may prove a perilous one (fig. 35). On that strip of sand, a turtle and a lobster face off amidst scattered coral, stones, and little bivalves, while a still more unsettling congregation occupies the embankments to either side: a lurching bird-like monster, a surprisingly well-endowed beaver, and two sizable trumpet shells—one downturned and the other positioned to reveal the shadowy depths of its interior.[2]

This collection of creaturely specimens elides nature and fantasy in a manner that recalls the works of Hieronymus Bosch, a Renaissance artist who knew well how to transform a thing observed from life into an invention of an otherworldly kind.[3] Enthralled by his ingenuity, Bosch's Netherlandish followers did their best to keep up. In Alart du Hameel's Boschian engraving from the end of the fifteenth century, a panoply of somersaulting demons, drunken dwarves, and warring mermen embody the battles that Christopher must wage—and the dangers that he must overcome—on the arduous path to faith (fig. 36).[4] But so too does the lobster snapping at the saint's muscular calf. Depicted larger than life and in such frenetic surrounds, the crustacean resides within the image at once as a familiar and a monstrous thing: a reminder that even the real inhabitants of the watery deep were perturbing enough to rouse the imagination. Within the realm of the anonymous triptych, the two trumpet shells play an equally ambiguous role. The upturned shell in the lower right is eerily echoed by the eggshell depicted in the nearby wing, from which a fleshy monster has just broken free. The shells lie still for now, though perhaps not for long, and their restless stillness is in part what makes them so arresting.

At the same time, the shells stand out among the specimens within the triptych because they are represented with the greatest truth to nature, even more so than the lobster in du

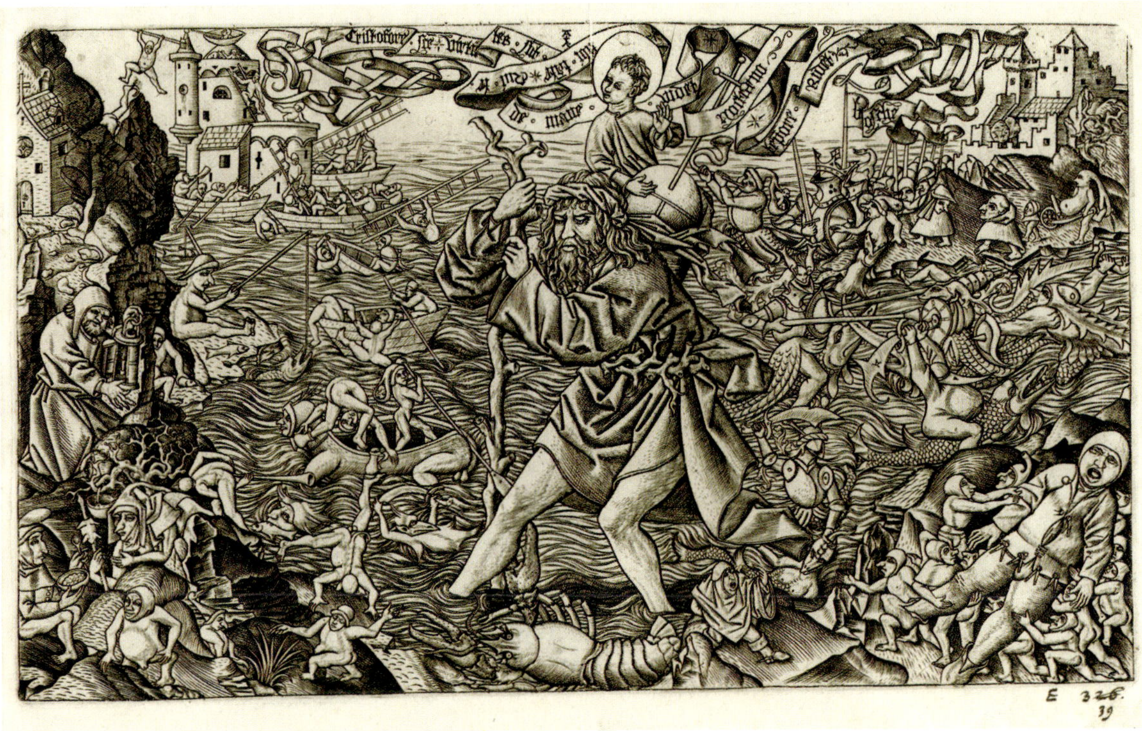

**FIG. 36.** Alart du Hameel, *St. Christopher*, c. 1478–1506. Engraving, 19.9 × 33.6 cm. British Museum, London [1910,0409.6].

Hameel's engraving. So carefully delineated are their forms that both can be identified as depictions of *Charonia lampas*, a species of large trumpet nonnative to the Low Countries. By representing the same type from two different angles, the artist amplifies the shells' dimensionality and presence, as if inviting us to imagine turning them over in our hands, or lifting one to our lips and letting loose its droning sound. The doubling also signals that the shells, however singular they appear within the space of the picture, exist as multiples in the outside world. Modern archeological excavation of a fifteenth-century site along the coast of Belgium uncovered a cache of shells from this very species, which were most likely imported to the region from the West African coast and via trade with Portugal.[5] The find attests that well before the seventeenth century, when the Dutch East and West India Companies were navigating the globe, shells were already among the cargo of ships like those present in the painting's background riverscape. The two trumpet shells might thus be taken to embody both the perils and the prospects of seeking distant shores, and to frame Christopher's storied journey of conversion as a simultaneous journey of discovery. Their exoskeletal forms—as ingenious as any monster of human imagining—are curiosities of nature that have been carried across the seas and repossessed within the space of a European painting to become curiosities in yet another sense: objects that provoke in us a desire to understand them.

So what are these shells really doing on the shoreline of this triptych's central panel, within an image meant first and foremost to serve in the contemplation of Christopher and his saintly companions? Too vibrant to be treated as mere specimens, too palpable to be dismissed as mere fantasies, they might seem to embody the paradox associated with the genre of painting called "still life" or "dead nature" (*natura morta, nature morte*, etc.), which came to be known as such only in the mid-seventeenth century. What we would now refer to as still life—namely, tableaus of flowers, foodstuffs, and other objects both natural and human-made—are often found within Renaissance paintings like our triptych that were created prior to the genre's full-fledged emergence, most often in the margins of figural compositions such as depictions of saints, portraits, and market scenes.[6]

To take the shells in the triptych as examples of proto–still life is to take seriously the contradictions of the genre's name. Seventeenth-century writers like the art theorist Joachim von Sandrart employed "still life" not only in reference to the representation of subjects both living and dead but also as a way to describe how artists *still* the life of those subjects with the pictorial realm.[7] In other words, the term "still life" refers less to the subjects themselves than to what their representation achieves: the demonstration of painting's ability to hold nature in an artificial state of suspension by endowing its subjects with a restive vitality at once seductive and strange. If the attention lavished on the two shells at Christopher's feet seems to give them semiautonomous status within the composition, one might argue that they are so placed to make this very point. It is as if they are precociously poised to be picked up and carried off from one pictorial world into another.

That said, the painting's anonymous inventor was surely oblivious to presaging an as-yet-nonexistent genre. Instead, the triptych is among many devotional images produced in the sixteenth-century Netherlands that pushed the boundaries of its *own* genre beyond the iconic, which challenged viewers to think through the ways in which divine, natural, and human histories are all inextricably entangled.[8] The trumpet shells participate in this call to contemplative action, compelling us to make sense of their simultaneous status as eerie life-forms, exotic specimens, and signifiers of the parallel between the saint's spiritual journey and the viewer's own. They are integral to the design and experience of the triptych as a whole. Rather than treating the two shells as part of a precocious "still life"—rather than naturalizing them within a now-accepted pictorial category—I want to consider instead what happens when we take these uncanny specimens for what they are: unruly examples of "shell life" in the history of early modern painting.

By thinking with the concept of "shell life," my aim in this chapter is not to resolve the complexities of the individual works that I discuss below but instead to draw out some of the potential associations that shells inspire within and across them. Although several of the works in question have been designated shell still-life paintings in past scholarship, attempts to interpret them within that genre do not address the peculiarities of shells in the realm of two-dimensional representation as objects that resist binary categories—above all, as things neither wholly still nor altogether alive. Regardless of whether they are lingering on the sidelines or appear at the center of a composition, shells demand to be understood as more than another example of a still-life subject. From the shoreline of one picture to another, from motion to stillness and back again, shell life is by nature unconfined.

## SHELL LIFE WITHIN STILL LIFE

Every shell implies movement, even when at rest. A vacated shell of a gastropod or bivalve was once a thing put in motion by the creature who lived within it, rode the waves, burrowed into sediment, or scoured the ocean floor. At the same time, a shell's very contours mark out the stages of its growth and the annals of its making.[9] Still-life painting, in the constant dialectic that it poses between stillness and enlivenment, might be said to meet its match whenever a shell enters its domain. In one sense, a shell represented on a table appears wholly divorced from its natural environment, the context of its original function, or the history of its creation. Yet painters of still life were far from oblivious to shells as subjects with multiple lives.

Over the course of the seventeenth century, as still life became a productive arena of specialization, several early modern artists made their names as painters of flowers or sumptuous banquets. Within both of these subgenres, shells often make appearances. In banquet scenes, shells turn up in the form of elaborate drinking vessels such as nautilus cups, their natural forms made lustrous and ornate by means of human artifice (see fig. 1).[10] In flower still-life paintings, shells are often found scattered in the foreground, competing with the blooms arranged beside them in the diversity of their colors and allure of their forms (see fig. 13, fig. 79). Like their costly floral counterparts, exotic shells became increasingly valued commodities and collectibles during this period. The impulse to represent them undoubtedly stemmed from their desirability as rare and intimate possessions, as well as from the growing fervor for their study among naturalists and nature-loving amateurs.[11] At the same time, market forces and natural-historical inquiry alone cannot account for what shells actually do within these paintings, let alone for how they differ—both as curiosities and as collectibles—from the other objects depicted around them.

An ingenious painting by the Flemish artist Clara Peeters, signed and dated 1612, already suggests that the early innovators of the still-life genre were awake to the unique potential of shells as subjects (fig. 37).[12] Peeters presents us with a medley of surfaces and shapes: an earthenware vase holding a delicate bouquet, two glistening gilt goblets, stacks of coin, a linked chain, a shallow porcelain bowl with an almost opalescent sheen, and a cluster of four shells.[13] Displayed so as to reveal their apertures, these small univalves together rival the craftsmanship and variety of all the other objects in the composition. They are matte and shiny, colored and textured, rounded and filigreed—masterful demonstrations of nature's handiwork.

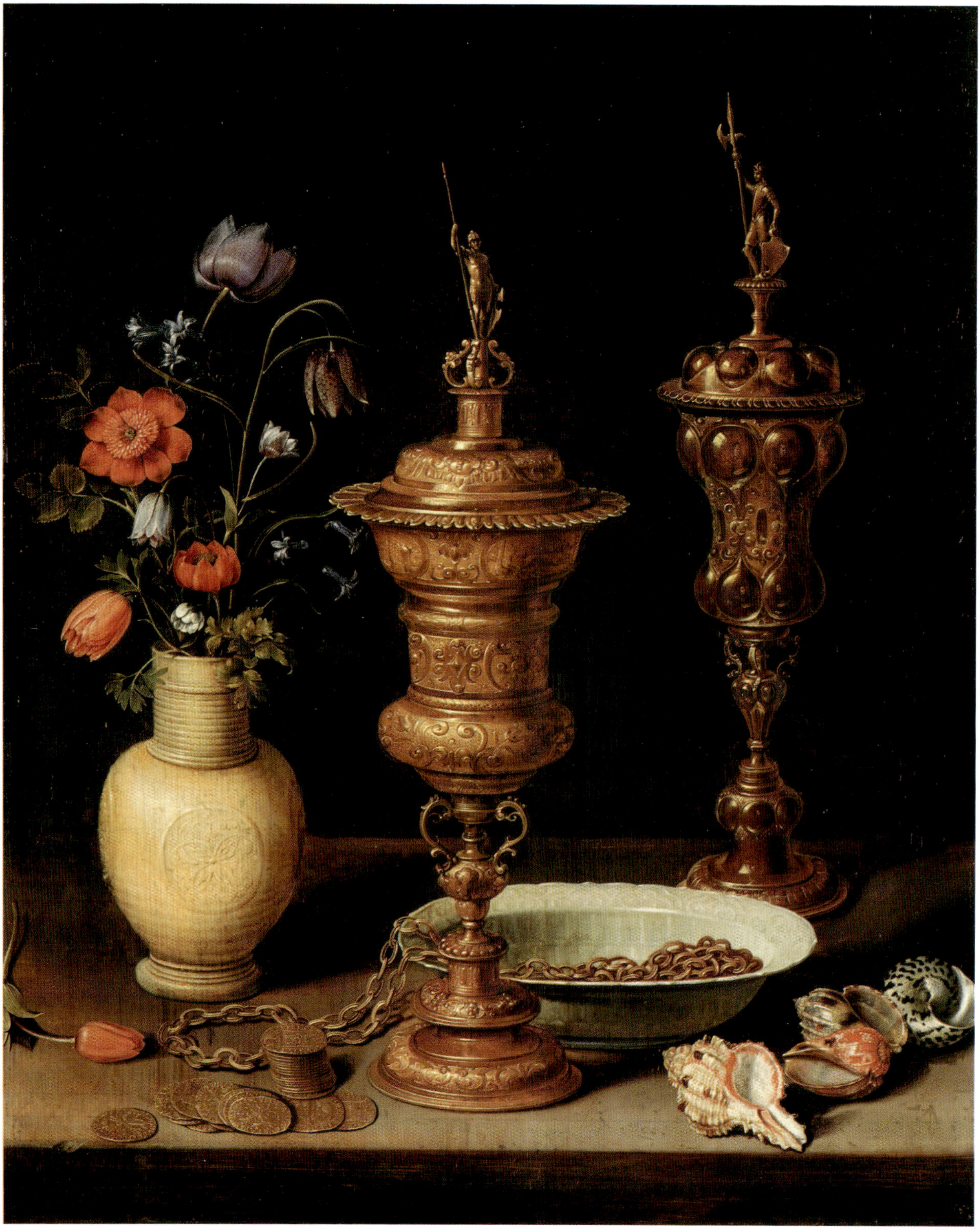

By aligning the gastropods with the coins on the foreground ledge, Peeters signals that these nature-made things are equal in value to human-made currency. Both might be picked up, passed between hands, and circulated beyond the space of the painting. More provocatively still, Peeters positions the shells along a vertical axis with the convexities of the rightmost goblet, in which we see her likeness reflected many times over. The discovery of Peeters's self-portraits heightens our awareness of the tension between what is outside and what is inside the picture, an awareness that extends to the specimens below; in counterpart to the bulbous forms on the goblet, the shells have all been arranged with their concavities facing toward us. It is as if Peeters were asking: Is there anything more difficult to paint than a shell? As she too faces toward us, holding her palette and brush, the singularity of the shells reflects her own singular abilities as an artist.[14]

**FIG. 37.** Clara Peeters. *Still Life with Flowers, Goblets, and Shells*, 1612. Oil on panel, 73 × 62 cm. Staatliche Kunsthalle, Karlsruhe [2222].

Peeter's near contemporary Balthasar van der Ast was among the first still-life painters in the Netherlands to devote special attention to shells, and his compositions reveal an awareness of their peculiar vitality. A still life now in the Rijksmuseum, painted around 1620, might be taken as an exercise in exploring the distinction between the representation of shells and that of other natural subjects within a single compositional frame (fig. 38).[15] Van der Ast's painting abounds in suspense and suspended life. The fruit in the foreground left is yet green, a sign of how recently it was plucked from the tree. The flowers arranged in the porcelain vase remain in full bloom, even despite the spiders that have spun their webs beneath the dark purple fritillaries on the right side of the bouquet. Yet the longer one looks, the more peril one sees: insects everywhere hang, perch, and lurk in wait to devour the bounty on display.[16] Only the three shells on the edge of the table share with the Wan-li vase and plate an immunity to the inevitable fate of the nearby tulips, carnations, and fruits. Like the porcelain itself, they are delicate and intricate yet durable creations, which will remain after the insects have moved on and the surrounding specimens have perished or faded. But still and inanimate they are not. The rustle of the tablecloth beneath the turban shell (*Turbo chrysostomus*) in the foreground directs our eye to the creature emerging from its interior: a hermit crab clawing its way into the world of the living and out of the depths of the picture. Van der Ast shows us, quite literally, that there is something creepy about shells within the not-so-still domain of *natura morta*. The crab emerges from the shell—from the dead space of its interior—to actively disrupt the stilled world that surrounds it.

The shell life within van der Ast's composition refutes any notion that still-life painting is reducible to a moralizing reflection on the vanity of earthly things, even if moral lessons were sometimes attached to the covetous desire for costly flowers and shells in the emblematic literature of the artist's Dutch contemporaries (see fig. 2).[17] The shells in the painting embody resurgence rather than evanescence, and they speak to the playful tactics of the artist himself. Technical examination has even revealed that van der Ast did not plan the turban shell in advance but instead added it after the tablecloth had already been painted.[18]

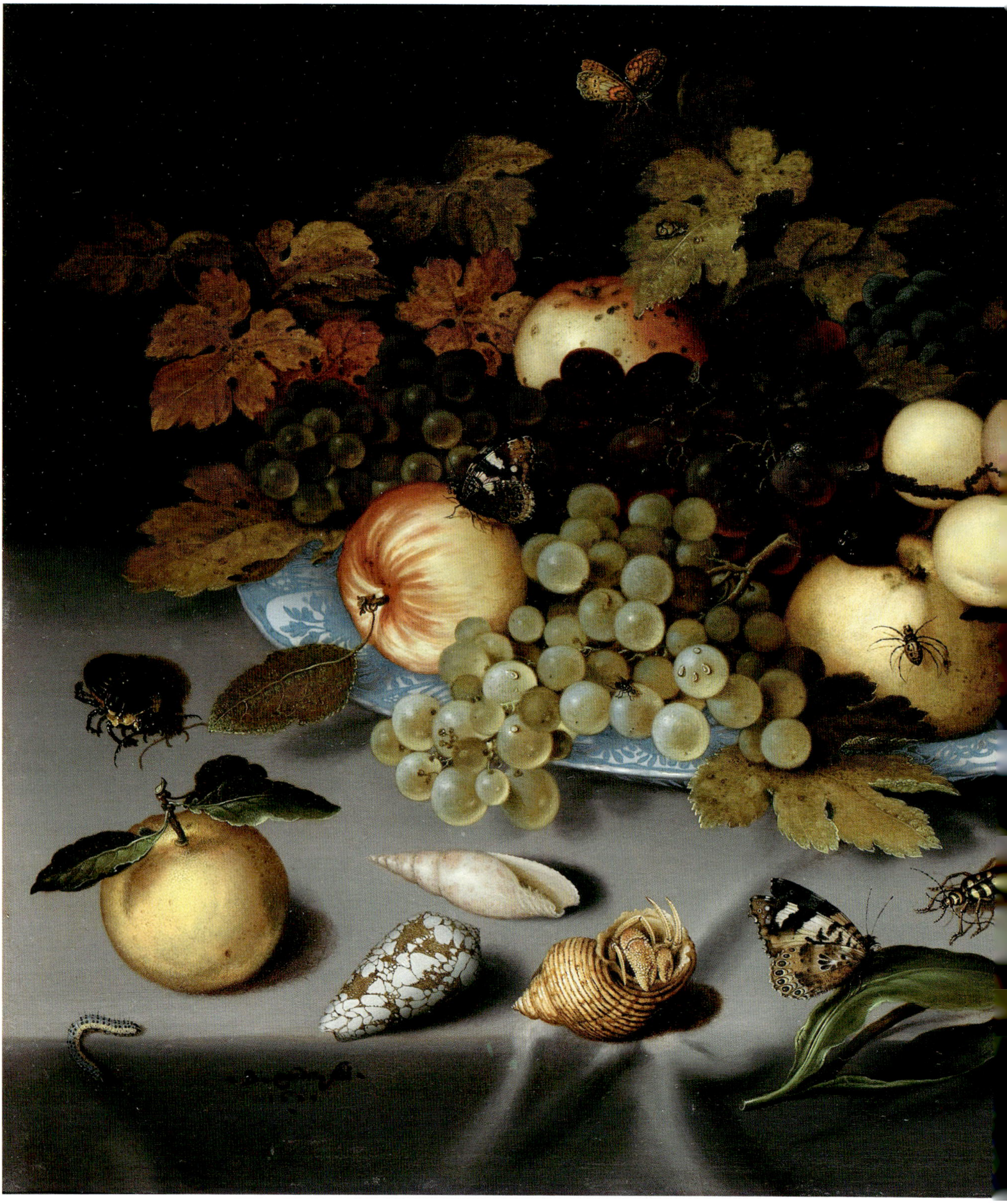

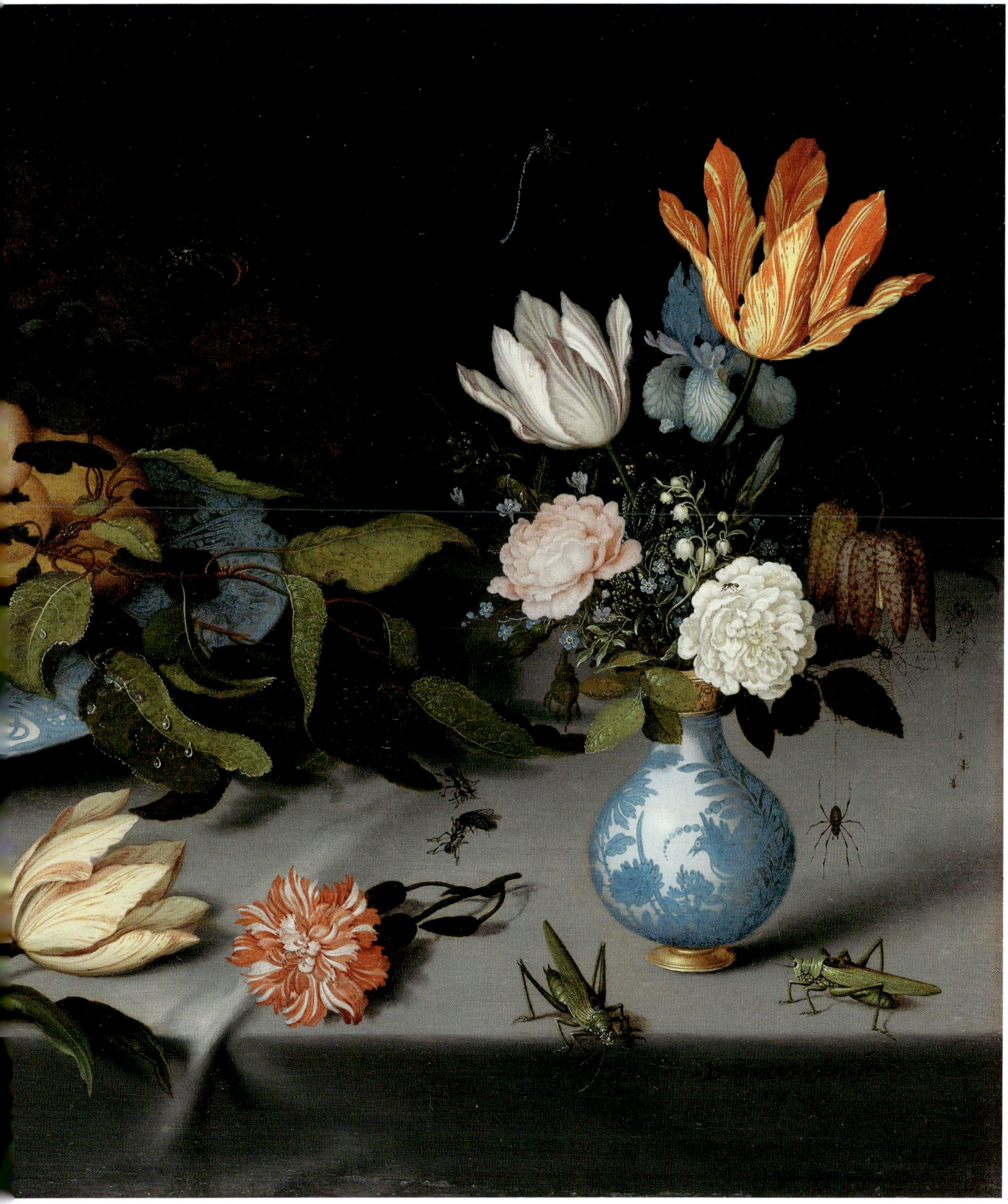

**FIG. 38.** Balthasar van der Ast, *Still Life with Fruit and Flowers*, 1620–21. Oil on panel, 39.2 × 69.8 cm. Rijksmuseum, Amsterdam [SK-A-2152].

**FIG. 39.** Pieter van de Venne, *Still Life with Shells and Insects*, 1656. Oil on panel, 26 × 20.6 cm. Detroit Institute of Arts, Founders Society Purchase, General Membership Fund [39.671].

It is a crucial late addition, one that punctuates the temporal push-and-pull of the entire composition. Van der Ast's repeated return to this motif in later paintings suggests that he was fond of its effect.[19] The turban shell, together with the adjacent magnificent cone (*Conus magnificus*) and episcopal miter (*Mitra mitra*), are all exotic shells from the Indo-Pacific region that complement the other nearby specimens in their shapes, colors, and surfaces. Van der Ast nonetheless goes beyond the presentation of shells as beautiful or valuable things to insist that their forms are inseparable from their function as vessels of creaturely life.

Van der Ast hailed from a family dynasty of still-life painters based in Middelburg, the main port city of the Dutch province of Zeeland, which had close ties to the sea as well as a vibrant community of artists, scholars, and naturalists. Alongside Middelburg's thriving garden culture, which has often been linked to the rise of flower still-life painting in the region, emerged a local circle of enthusiasts for the study and collecting of shells.[20] Among them was Philibert van Borsselen, whose verses titled *The Beach* (first published in 1611) treat his subjects not only as works of divine artifice but also as strange and lively things. Toward the beginning of his poem, van Borsselen urges the painter to rise to the challenge of the diverse colors observable in nature, calling for admiration of the hues found in flowers like the lily, hyacinth, and violet. "But ultimately," he concludes, "all these must concede to the beauty of shells, to which I have never seen a tulip that can compare."[21] Van Borsselen writes that in studying shells, "one sees wild watery pools come to life, here in ebb, there in flow," and describes the sun, moon, stars, and sea as the "master painters" (*meester-schilders*) of their artfully turned forms.[22] He exhorts his readers to recognize their peculiar properties of reflection, luminescence, and geometry. Shells for van Borsselen are not only visible manifestations of Mother Nature's skillful crafting; they are also creations that embody the elemental movements and transformations that are ever ongoing in the natural world.

A curious little picture by Pieter van de Venne, a Middelburg artist of the next generation, stages the especial liveliness of shells even more directly than the paintings of Balthasar van der Ast (fig. 39).[23] Signed and dated to 1656, van de Venne's composition presents eight subjects arranged in artificial symmetry on a plush burgundy surface. The contrast between the dusky background and the luminous conchs in the foreground dramatizes the properties that van Borsselen praised in his poem. Not only do the shells seem to glow in the dark; van de Venne's loose brushstrokes and use of flickering highlights give the impression that the two flesh-colored conchs are stirring restlessly in place. At the same time, the contrast of light and dark almost obscures the black murex shell (*Murex radix*) that lurks between the two conchs, and which is positioned like a nighttime predator waiting to pounce. While the monstrous murex seems to menace us, the insect perching on the conch to the right, and the moth entering from the upper left, pose no danger to the life of the shells upon which they descend. The painting originally had a pendant, on a panel of near equal size, depicting a pile of cut blooms drooping from lack of water and threatened by a parallel butterfly flying down

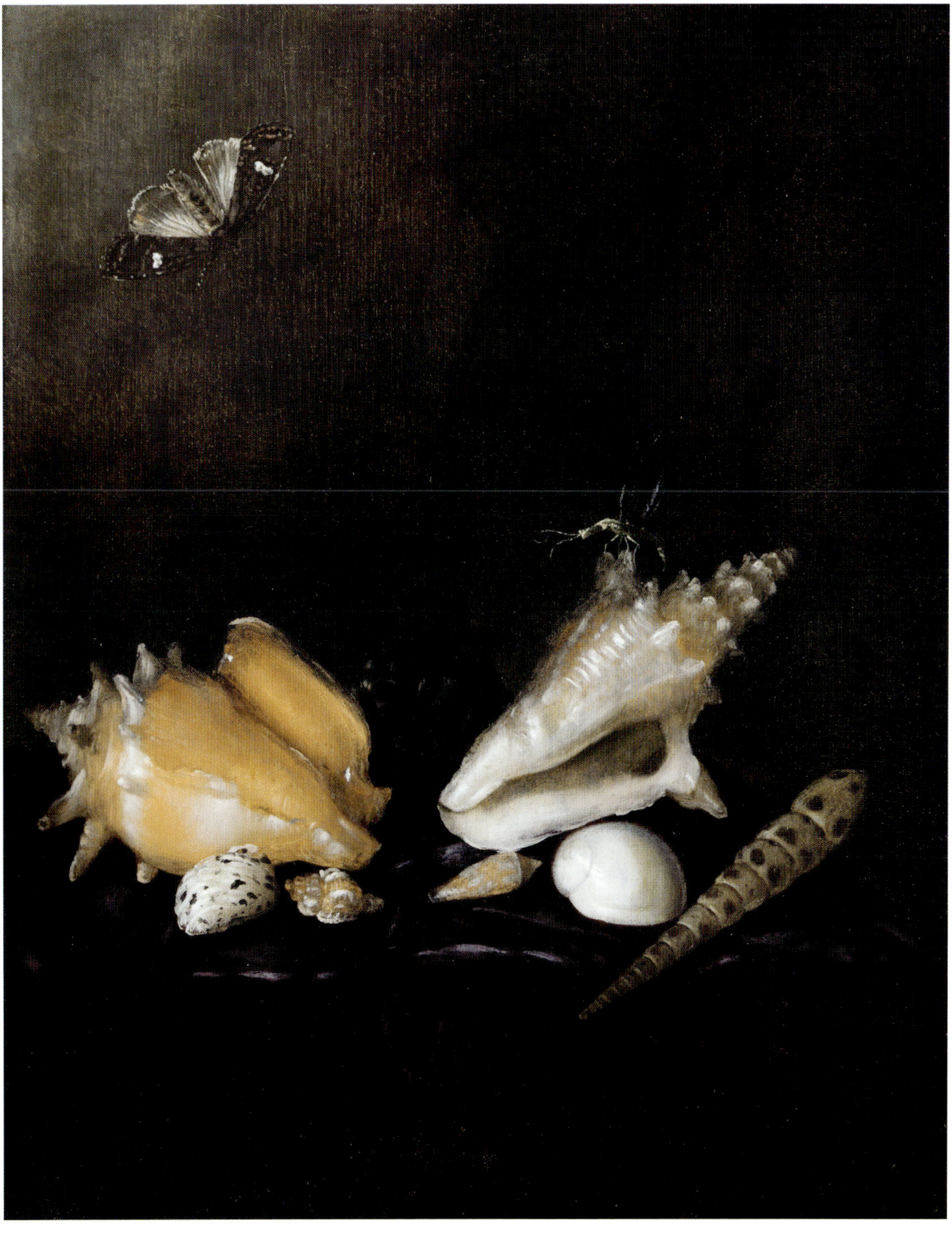

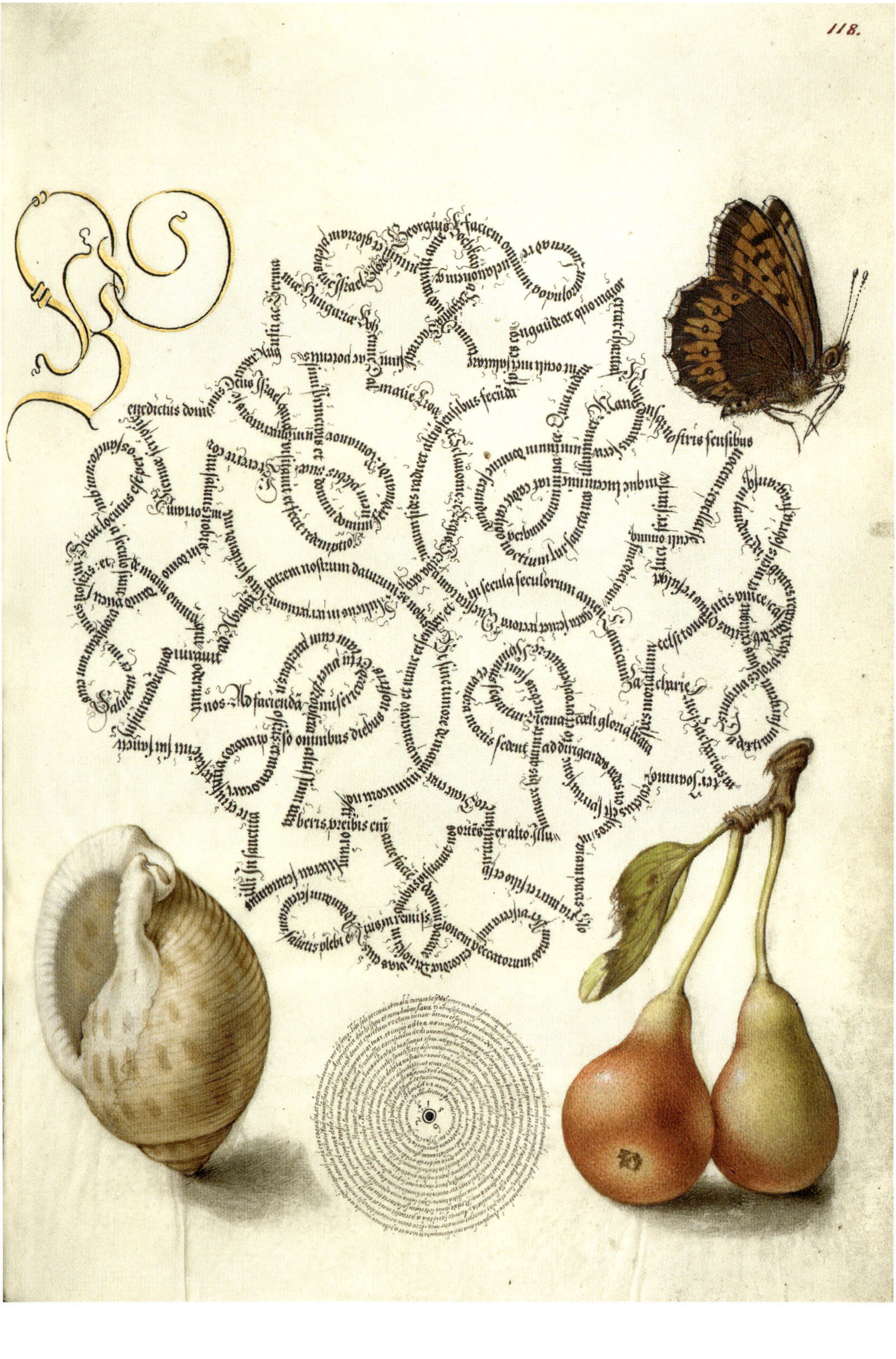

from the upper corner.[24] The juxtaposition would have affirmed that no flower can compare to the eerie luster of shells, which surge with inner life even within the shadows of painting's stilled realm.[25]

FIG. 40. Georg Bocskay and Joris Hoefnagel, *Mira calligraphiae monumenta (fol. 118r)*, 1561–62 and 1591–96. Tempera colors, watercolors, gold and silver, and ink on vellum, 16.6 × 12.4 cm. J. Paul Getty Museum, Los Angeles [Ms. 20].

## CADAVERS AND CAVITIES

Depictions of shell life within any genre of picture—still life or otherwise—are never only surface deep. Pliny the Elder, the great natural historian of antiquity, set the precedent when he mused that shells display to full effect "nature's love of play," and reveled in cataloguing their "many differences of color and shape."[26] Whether flat, long, circular, humped, smooth, or grooved, shells presented Pliny with resemblances to other things, particularly to body parts and objects of human craftsmanship. He describes one shell adorned "with flowing locks or curls," another that recalls "the teeth of a comb," still others "corrugated like tiles or reticulated into lattice-work" or "curved so as to make a trumpet."[27] It was surely no coincidence that the Netherlandish miniaturist Joris Hoefnagel, when commissioned to illuminate a stunning calligraphic manuscript for his patron Emperor Rudolf II in the 1590s, also recognized in shells an opportunity both for formal play and for appreciation of nature's limitless powers of invention.[28] Alongside a circle of minute script at the bottom of one folio, Hoefnagel painted a striated mollusk poised impossibly on its tip like a spinning top, as if inviting us to think into the shell's geometry and to recognize divine nature as the model for the artifices of the human hand (fig. 40).[29]

Thinking into shells like mollusks, the interiors of which we cannot see, is a mode of inquiry that early modern artists like Hoefnagel often encouraged through compositional means. A shell within the realm of representation may register first as appealing surface and form, and, to those in the know, as a specimen with a particular geographical origin and set of distinctive traits. But as the twentieth-century French philosopher Gaston Bachelard wrote in his own ruminations on shells and the poetics of their inner spaces, if we really want to understand inhabitable things, we "must not be subject to the charms of external beauty."[30] However much we admire a shell's surface, so Bachelard suggests, we should appreciate in equal measure the mysteries of its depth.

A painting from 1611 by the Dutch artist Joachim Wtewael reverberates Bachelard's words (fig. 41).[31] An unusual rendition of its subject to say the least, Wtewael's *Perseus and Andromeda* takes a popular myth and renders it strange. There is something absurd in the depiction of the hero Perseus about to slay the creature to whom Andromeda was a promised meal. It seems impossible that the stout Pegasus could fly on such pathetic wings, and the technicolor dragon shooting spouts of water from its nostrils hardly seems a formidable enemy. The pathos of the narrative is expressed instead through the tension and torque of Andromeda's body and the insistent gesticulations of her hands, but even here, Wtewael resists a straightforward approach to history painting. This Andromeda might just as readily

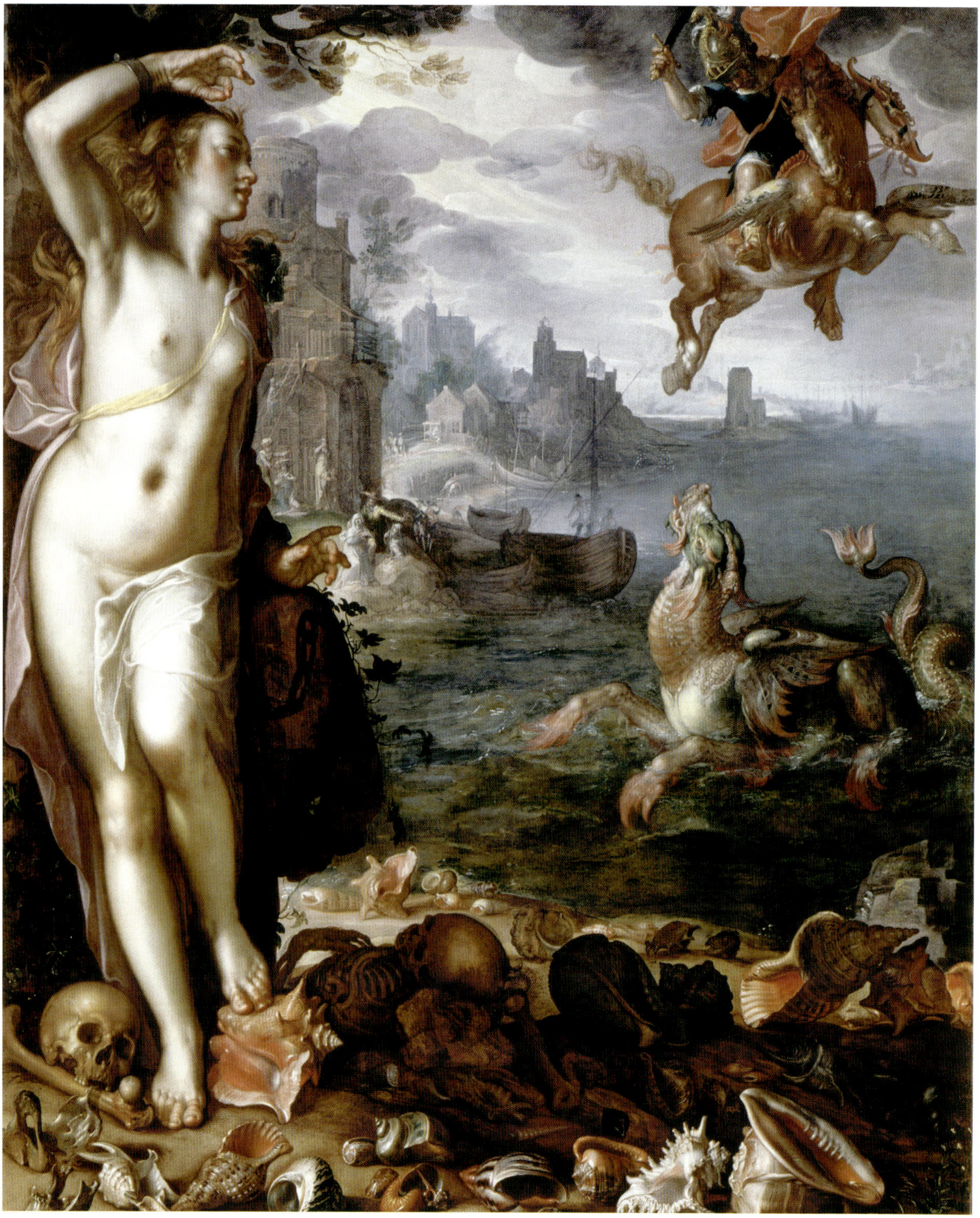

be taken for Aphrodite, that goddess born from the sea who is the very figuration of desire.[32] The positioning of Andromeda's left foot on an impressive *Strombus gigas*, a conch that hailed from Caribbean shores—and a prized collector's item in Wtewael's contemporary milieu—further suggests a connection to the amorous deity.[33] Its glistening pink interior, turned in our direction, exposes the erotic allure of the female body in a state of subjection. Both she and the shell seem to be precious objects for the taking.

**FIG. 41.** Joachim Wtewael, *Perseus and Andromeda*, 1611. Oil on canvas, 180 × 150 cm. Musée du Louvre, Paris [R.F. 1982-51].

Yet the carpet of shells at Andromeda's feet ultimately grounds the atmosphere of the mythological scene in a harsher reality. At the painting's threshold, closest to the level of the viewer, human and molluscan remains seem to be the refuse of the dragon's former conquests. For every shell that echoes the peachy lips of Andromeda's footrest, another lies shrouded in darkness and scattered alongside human bones. The analogy between exoskeleton and corpse is condensed in the juxtaposition of the *Strombus gigas* with the glistening skull beside Andromeda's right ankle, the deep cavities of its nose and eyes parallel to the conch's seductive interior. One may well discern a moral here in the link between death and desire, but the attention that Wtewael lavished on the shells in the foreground is in excess of what is required to convey that message. There seems little doubt that Wtewael studied these specimens from life given the detail and precision of their rendering. It is as if he has taken a conchological collection and released it back into the wild. The shells appear not just dead but far more deadly than the dragon beyond them, and any endeavor to gather them up again would be at the collector's peril.

Just over a decade prior to Wtewael's magnificent painting, Shakespeare put in words an experience similar to the encounter with the artist's necropolis of shells. In the first act of *King Richard III*, Clarence recounts—on the eve of his own murder—a stygian nightmare of shipwreck. Within the dream, Richard's doomed brother envisioned bodies devoured by fishes and condemned to watery graves, their remains taunted beneath the sea by stones and pearls:

> Some lay in dead men's skulls, and in the holes
> Where eyes did once inhabit, there were crept—
> As 'twere in scorn of eyes—reflecting gems
> That wooed the slimy bottom of the deep
> And mocked the dead bones that lay scattered by.[34]

Richard's lines are evocative not just for their macabre imagery of the underwater realm, which Shakespeare employed in other plays as well, but still more for their almost cinematic quality.[35] The first three lines progressively zoom in, from skulls to their cavities to the pearls that have come to occupy their insides, while the last two lines pan out to the murky seabed. A similar movement results from Wtewael's blending of painterly modes. Standing before

**FIG. 42.** Giovanni Battista Recco, *Still Life with Fish and Oysters*. Oil on canvas, 100 × 126 cm. Nationalmuseum, Stockholm [NM 759].

the painting, we might step back to admire the softness of Andromeda's arching body and the vanishing expanse of the landscape. But the impulse to survey those eerie and pearlescent bodies on the shoreline inevitably draws us closer again. The most uncanny word in Clarence's account of his dream is the verb "crept," which transforms the "reflecting gems" from inert objects into beings that glare back at us.[36] The heightened detail that Wtewael employs in his depiction of the shells has the same uncanny effect, endowing those lifeless things with a sense of latent power.[37] The shells plunge us into a spiral of reflection on their haunted depths, threatening to pull our attention away from the mythological drama unfolding around them.

A long literary tradition of associating the depths of the sea with the descent into Hades preceded both Wtewael's painting and Shakespeare's lines.[38] The Neapolitan poets of the early sixteenth century were already writing of the oceanic realm as a sinister site and stimulus for melancholic imaginings. Bernardino Martirano installed a shell-covered grotto at his villa of Leucopetra south of Naples where he invited fellow poets and scholars to gather in conversation, reflecting on the watery landscape that they so often took as inspiration for their writings.[39] Among the most influential verses to emerge from the literary circles of early modern Naples were the *Piscatorial Eclogues* of Jacopo Sannazaro, which were read and imitated across Europe long after his death, not least by the likes of John Milton.[40] The first of Sannazaro's eclogues, a dialogue between two fishermen named Lycidas and Mycon, is a lament on the loss of their mutual beloved Phyllis. Describing the funerary offerings that they have bestowed upon her tomb, Mycon declares to Phyllis's departed shade, "See, we are bringing you moss from the green-blue sea, we are bringing you purple conchs, and coral for which we scoured throughout the deep and scarcely wrenched from rocks in the abyss."[41] Lycidas, who can no longer bear to walk the busy city streets, wanders the shoreline longing to become one with the sea and its tumultuous waves.[42]

The offerings of Lycidas and Mycon, and the doleful mood of Sannazaro's poetry, share much with the shadowy still-life paintings of marine subjects that emerged in Naples during the subsequent century. Giovanni Battista Recco's *Still Life with Fish and Oysters*, signed and dated 1653, presents an odorous haul wrenched from the deep, as if a Neapolitan fisherman had just stepped ashore and unloaded his catch before us (fig. 42).[43] These shells may be comestibles, but they hardly appeal to the appetite. Occupying the upper corner is a tray of bivalves, garnished with mossy sponges, which echo in their colors and sheen the bodies of dead fishes splayed out beneath and around them. In the foreground, a pair of trumpet shells gleam with ocean spray. Their cavities—juxtaposed like two eyeholes in a skull—are shrouded in the same darkness that imbues the painting as a whole. Only on close inspection does the foremost shell reveal a creature lurking inside it, not unlike the hermit crab in the earlier still life of Balthasar van der Ast. The fishes' vacant eyes, along with the open bivalves that look like a pile of eyes and tongues, cast this emergent creature in a far less playful light. Recco's subjects themselves seem to mourn the loss of their underwater realm.

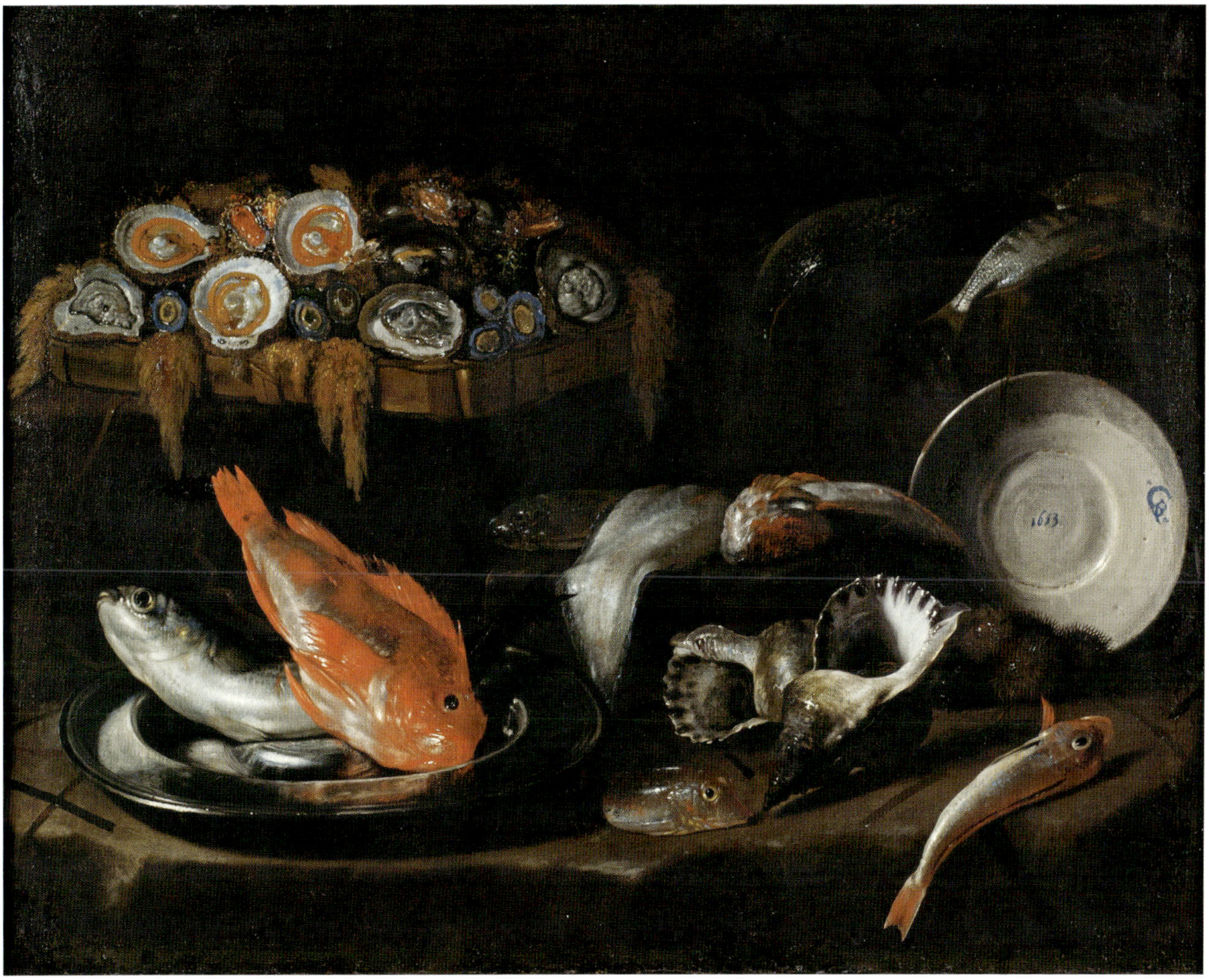

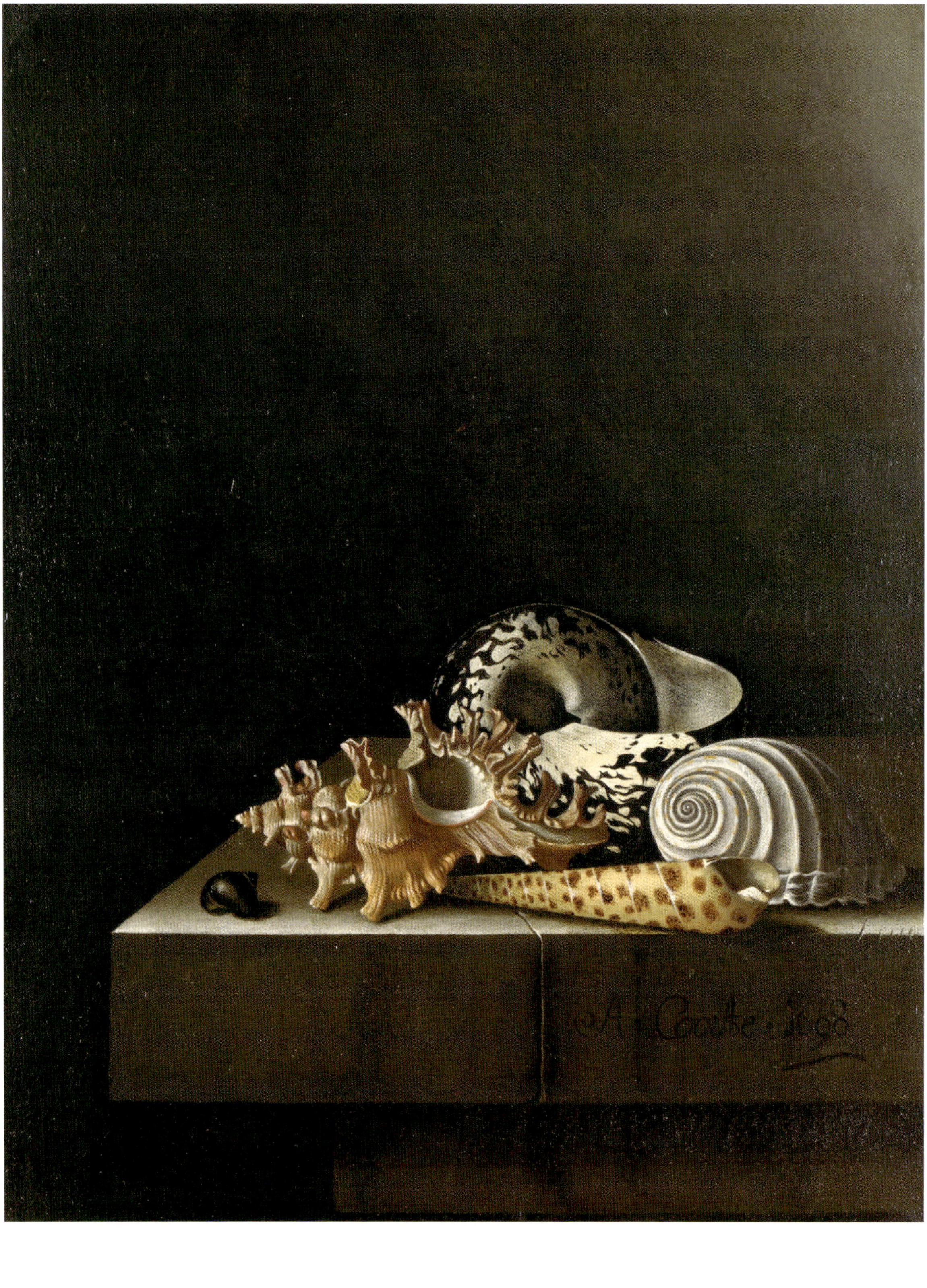

At the end of the seventeenth century, the virtuosic Dutch artist Adriaen Coorte explored the poetics of the vacant shell on a more intimate scale (fig. 43).[44] Five specimens appear on the solemn stone plinth that Coorte so often used to stage his pictures, their presence amplified by the darkness that surrounds them. The painting is so small, and so delicate in its execution on paper pasted to panel, that the picture seems as rarified as its subjects. But there is something haunting about the composition that mingles with the pleasure to be taken in admiring the diversity in size, shape, texture, and color of the shells that Coorte selected for display. With the exception of the small black shell (*Euspira nitida*) set apart on the left, the specimens seem to merge together and form the mask of a creature not quite human. As in Recco's painting, here too the paired hollows of the spiny snail shell (*Chicoreus sauli*) and of the magpie shell (*Cittarium pica*) just behind it join together to form the mask's eyeless sockets, while the lip of the latter suggests a drooping ear and the slender auger shell (*Terebra subulata*) in the foreground a contorted nose. Coorte, as did Wtewael before him, denies the collector's inclination to treat each shell as a single precious thing. The composite head without eyes, once observed, cannot be unseen. Yet like a skull at the slimy bottom of the deep, the shells' inner depths are penetrable only through the workings of the imagination. Coorte presents us the most paradoxical kind of picture, one that simultaneously entices and scorns our desire to see within.

**FIG. 43.** Adriaen Coorte, *Shells on a Stone Plinth*, 1698. Oil on paper, pasted to panel, 29 × 22.5 cm. Rijksmuseum, Amsterdam [SK-C-1761].

## ARTIFACTS OF DEEP TIME

Every evacuated shell is in some sense a fragment of a former life, an architectural remnant that has outlived its architect. This understanding of shells refracts onto the ways that they were treated within early modern collections. Once stored within boxes and drawers, shells must await reactivation by human hands. Part of the pleasure that a collector might derive from taking a shell out of its drawer is the sense of revival that comes with allowing its surface to catch the light, caressing its form, and placing it in conversation with other specimens. Grand edifices on miniature scale, shells offer an invitation to dwell upon natural histories that run parallel to our own.

The collecting and appreciation of shells in the seventeenth century coincided with the antiquarian interest in other precious small-scale objects such as Renaissance medals, ancient coins, gems, and cameos. The English naturalist Robert Hooke turned this affinity into a metaphor when he described a cache of fossilized shells uncovered during an excavation in central London as "the medals, urnes, or monuments of nature," proclaiming their antiquity far greater than any monuments of human making.[45] Comparing Hendrik Goltzius's portrait of the shell collector Jan Govertsz van der Aar from 1603 with a contemporary etched portrait of the Antwerp antiquarian and shell collector Abraham Gorlaeus shows that shells and coins alike were studied in the same way: through touch and sustained

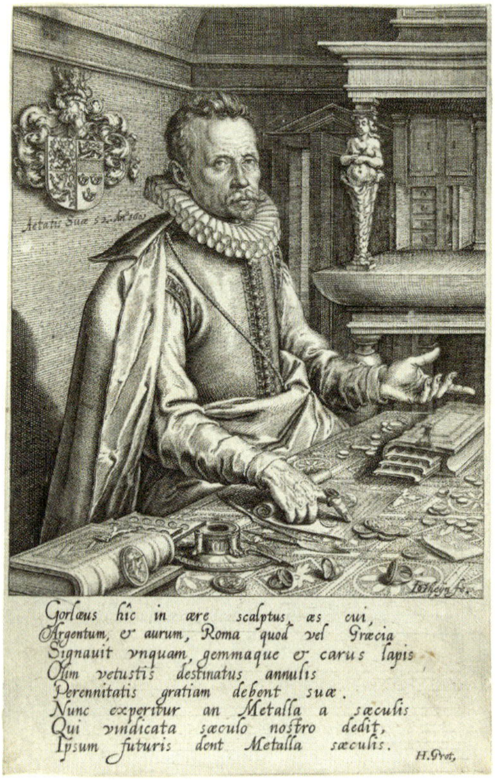


**FIG. 44.** Jacques de Gheyn II, *Portrait of Abraham Gorlaeus at Age 52*, 1601. Engraving, 16.8 × 10.9 cm. Rijksmuseum, Amsterdam [RP-P-OB-9952].

contemplation (see fig. 12; fig. 44).[46] As van der Aar fondles his prized natural artifacts acquired from distant shores, Gorlaeus points to the venerable artifacts from ancient Greece and Rome that he collected on his travels—objects that he professed a perpetual desire "to examine closely and to feel by hand."[47] Cabinets like the one behind Gorlaeus were used in the seventeenth century to house both shells and small antiquities, each drawer opening to reveal a different material history.

This antiquarian bent underlies the representations of shell life by the painter Jacques Linard, who was active in seventeenth-century Paris, and whose position as an official artist of the royal court afforded him access to the elite collections of the period.[48] Shell collecting was ascendant in this milieu at the time, not least through the endeavors of Gaston of Orléans, the eldest brother of King Louis XIII. Gaston amassed an enormous collection of *naturalia* that included over one thousand shells; many were inventoried upon his death in boxes custom-designed to house them.[49] He also assembled a vast cabinet of human-made curiosities—particularly antique coins, medals, and engraved gems—that was no less comprehensive in scope.[50] Gaston's tandem interest in natural history and numismatics had its roots even earlier in sixteenth-century France, from the antiquities collection of King François I to the writings of the naturalist, antiquarian, and adventurer Pierre Belon, who published one of the first illustrated treatises devoted to marine life.[51] Belon's close

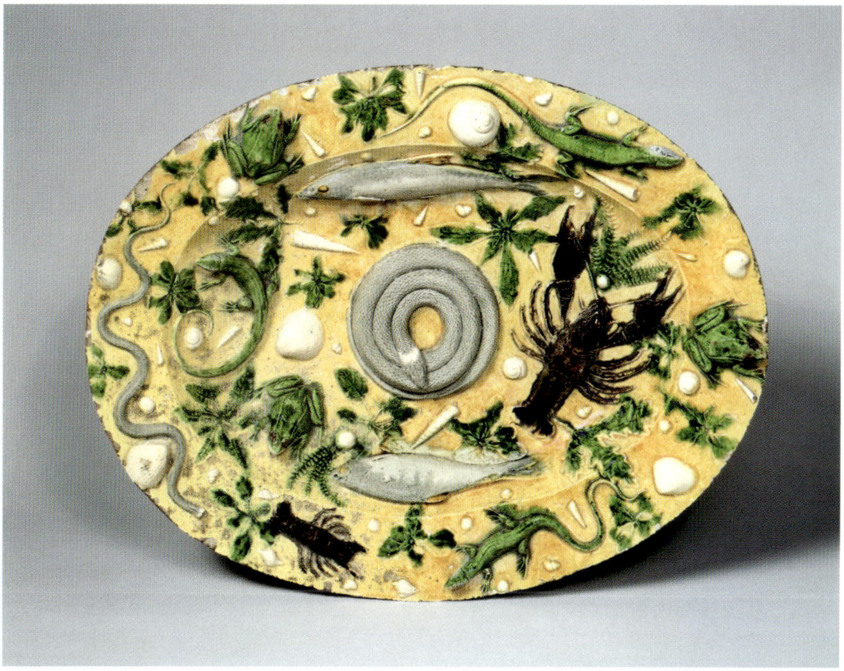

**FIG. 45.** Bernard Palissy (attributed to), *Oval Basin with Life Casts of Reptiles and Sea Creatures*, c. 1550. Lead-glazed earthenware, 47.9 × 36.8 cm. The J. Paul Getty Museum, Los Angeles [88.DE.63].

contemporary and fellow polymath Bernard Palissy produced ceramic plates and vessels adorned with casts of reptiles, fishes, crustaceans, plants, and shells that seem to creep uncannily across their surfaces; he also expounded upon the singular artifice of shells and other natural specimens in his writings (fig. 45).[52] On the subject of molluscan architecture, Palissy is unequivocal. "Do you think," he writes, "that the little concavities and ribs of these shells were made only for ornament and beauty? No, not at all. These are designed to advantage, to augment the strength of these fortresses, as do flying buttresses positioned against a wall as reinforcement."[53] In describing the architecture of shells, Palissy makes analogy to nothing less than the great Gothic cathedrals of his native land and, in so doing, celebrates the union of form and function in their design.

The natural fortresses that Palissy describes are fully on view in Linard's most popular composition, one that was frequently copied, and which exemplifies his witty approach to animating the shell as both collector's object and historical subject (fig. 46).[54] At center is a simple wooden box that appears as a motif in many of Linard's pictures, its facture evident from the visible nail and seams where the strips of wood have joined together. Linard seems to mock the box's function as container by piling shells and a gleaming branch of coral atop its lid, as well as positioning in the foreground numerous specimens far too large to fit within it. On the one hand, the angle of the table pushes the shells forward in space, as if they are defensive structures barring our access to the picture itself. On the other hand, the manner in which the shells are spaced out in neat rows implies the recent presence of a collector who

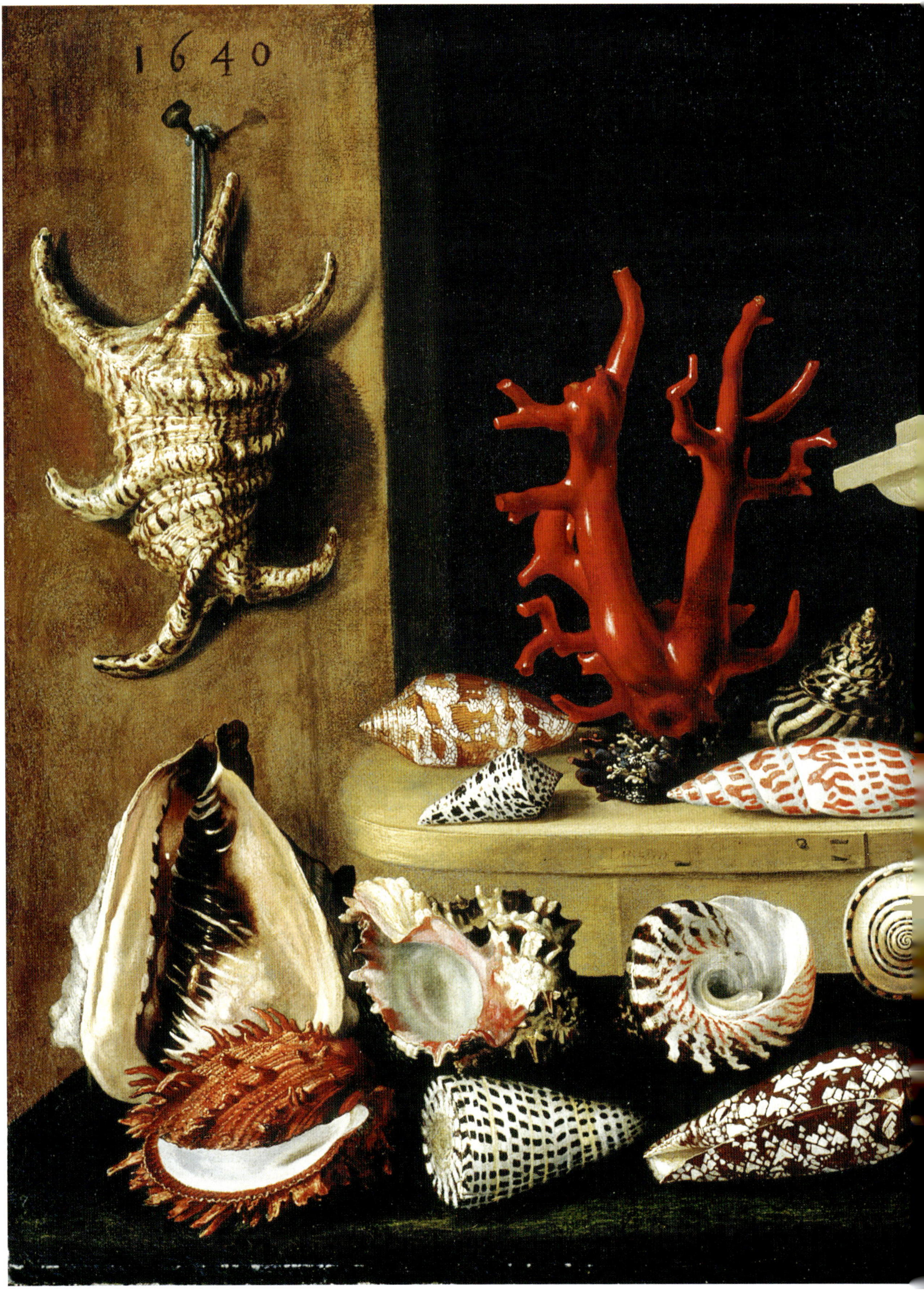
1640

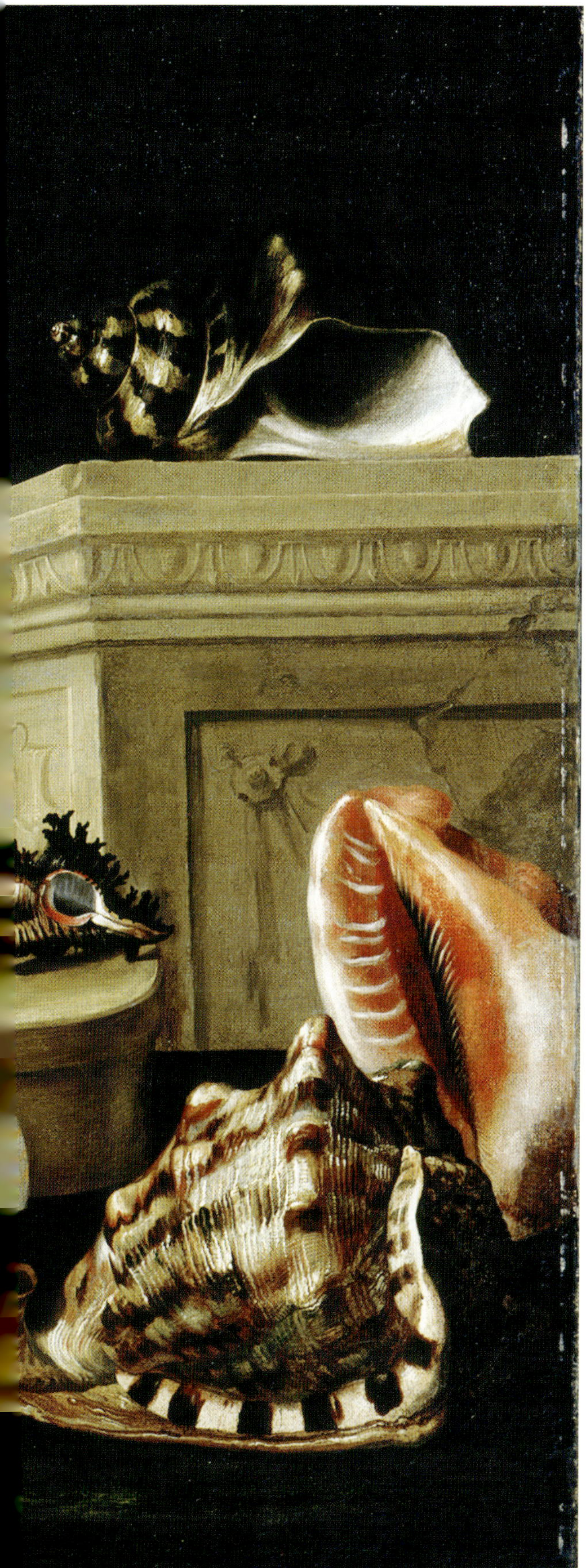

**FIG. 46.** Jacques Linard. *Still Life with Shells and Coral*, 1640. Oil on canvas, 53.5 × 62.2 cm. Musée des Beaux-Arts, Montreal [1999.149].

has taken them out one by one and placed them down upon the table. Against the wall in the background left, a spider conch (*Lambis chiragra chiragra*) hangs from a blue thread attached to a nail just below where Linard inscribed the picture's date. We sense not only the invisible hand of the collector at work but also the hand of the artist who contrived this display. The spider conch embodies the tension of the entire painting, which oscillates between presenting the shells as beautiful objects and enlivening them as lively subjects. Despite being tethered in place—despite the fact that we know the shell no longer has a mollusk within it—the conch still looks as though it might at any moment break free and creep down the wall.

Linard's selection of shells is itself remarkable in scope, representing a diversity of geographical origins that may well have been intentional, and which is evident in the front row alone. The vibrant spiny oyster (*Spondylus princeps*) in the lower left corner hails from the west coast of South America. The striped king helmet (*Cassis tuberosa*) in the opposite corner is found in the eastern Atlantic, while the lettered cone (*Conus litteratus*) and princely cone (*Conus aulicus*) between them are among the venomous snails of the Indo-Pacific. Spanning from east to west, the shells chart a history of travel and conquest, indexing the expansion of global trade across the seventeenth century. In this respect, the painting's composition encourages the recognition of each shell as an individual entity within a larger oceanic kingdom, appealing to the imagined collector who aspires to achieve a comprehensive assemblage of specimens from every corner of the known world.

At the same time, Linard asks the knowing viewer to reflect on the represented shells in both space and time, to think beyond longitude and latitude to the question of longevity. Hovering in the background is an ancient funerary altar with a libation vessel depicted in relief on its receding side and a prominent crack that marks its age. The altar is surmounted by a marbled turban shell (*Turbo marmoratus*) with its shadowed mouth opened toward us; one can almost hear the sounds of ancient rites resonating from within its inner chamber. By placing the shell atop this stone structure, Linard may have been appealing to the antiquarian knowledge of his courtly patrons. Shells in Greek antiquity were often presented as votive offerings to the goddess Aphrodite-Arsinoe—protectress of seafarers—a practice documented not least through a famous epigram by the Greek author Callimachus.[55] The nautilus shell in the poem addresses the reader in the first person, reminiscing about its former life when it rode the open seas with wind in its sails. Now that it has been transformed into a "plaything" (*παίγνιον*) of the gods, the nautilus declares itself lifeless (*ἄπνους*)—literally "without breath" or "without wind"—even as its capacity to speak belies that very claim.[56] The tension between the shell's liveliness and lifelessness is central to the playfulness of the poem as a whole.

Much like the nautilus in Callimachus's epigram, the shells in Linard's painting are effectively lifeless, yet the artist's emphasis on their electric hues, vibrating patterns, and curious structures brings them to life for the delectation of our senses and the stimulation of our minds. As we study the tableau before us, Linard asks us to remember that these wondrous and exotic possessions from faraway places are also the structural remains of an underwater world. Like the buttresses of a Gothic cathedral, the shells uphold the history of their creators. They are survivors of the vagaries of time. While an empty shell might readily be understood to evoke the transience of earthly things, Linard makes the case for the opposite view. The collector, like the antiquarian with his ancient coins and gems, could take up the study of shells as a way of animating the past in the present. Linard's painting is not a mere surrogate collection. It is also an image about the need to keep the strangeness of shells alive, to recover their origins, and to look beyond their beauty to the architecture of their endurance. Shells, so the artist shows us, are monuments of nature indeed.

## WHAT A SHELL KNOWS

In a 1937 essay, the French poet Paul Valéry reflected on what humankind can and cannot perceive about a shell's origins and making. Rather than coming to a definitive conclusion, the essay suggests that such reflections are always open-ended. The shell, writes Valéry, "has *served* me, suggesting by turns what I am, what I know, and what I do not know."[57] To recognize the perpetual mobility of shell life across place, time, and pictorial genre is to heed Valéry's words. In their manifold artifice and complex histories, shells are too restless to be held in place by symbolic meanings or fixed interpretations. They are, in art as in nature, the ultimate curiosities.

An anonymous painting from the mid-seventeenth century affirms the point and offers a fitting place to conclude (fig. 47).[58] This large canvas charts the movement of shells from the sites of their creation in the background to the spaces of their possession in the foreground, a movement punctuated by the lone nautilus shell at foreground center that has already been altered by human intervention. The signature helmet sculpted into its face, and pointing directly toward us, associates the nautilus with the work of the Bellekin family, master shell carvers active in the later seventeenth-century Netherlands (fig. 48). At the same time, the inclusion of the parrot and a cluster of dead exotic birds recalls the advice of the Dutch art theorist Karel van Mander, who proclaimed that "the speaking parrot, birds, and shells offer examples of how to unite colors with one another," and exhorted the young artist to acknowledge Nature herself as the mother of painting.[59] In this light, the carved nautilus—dwarfed by the enormous conchs and glistening turban shells, overwhelmed by the colors and forms scattered around it—seems little more than a trifle. What the painter can depict, what the sculptor can carve, or what the architect can build is always lesser than nature. Still life, in the end, proves a profoundly inadequate category for describing the inherent unstillness that defines the representation of any shell within a picture. In a century of expanding trade, travel, and knowledge, this anonymous artist saw what Valéry would see centuries later: a reminder that however much we might strive to think into shells, there are limits to how deep we can get.

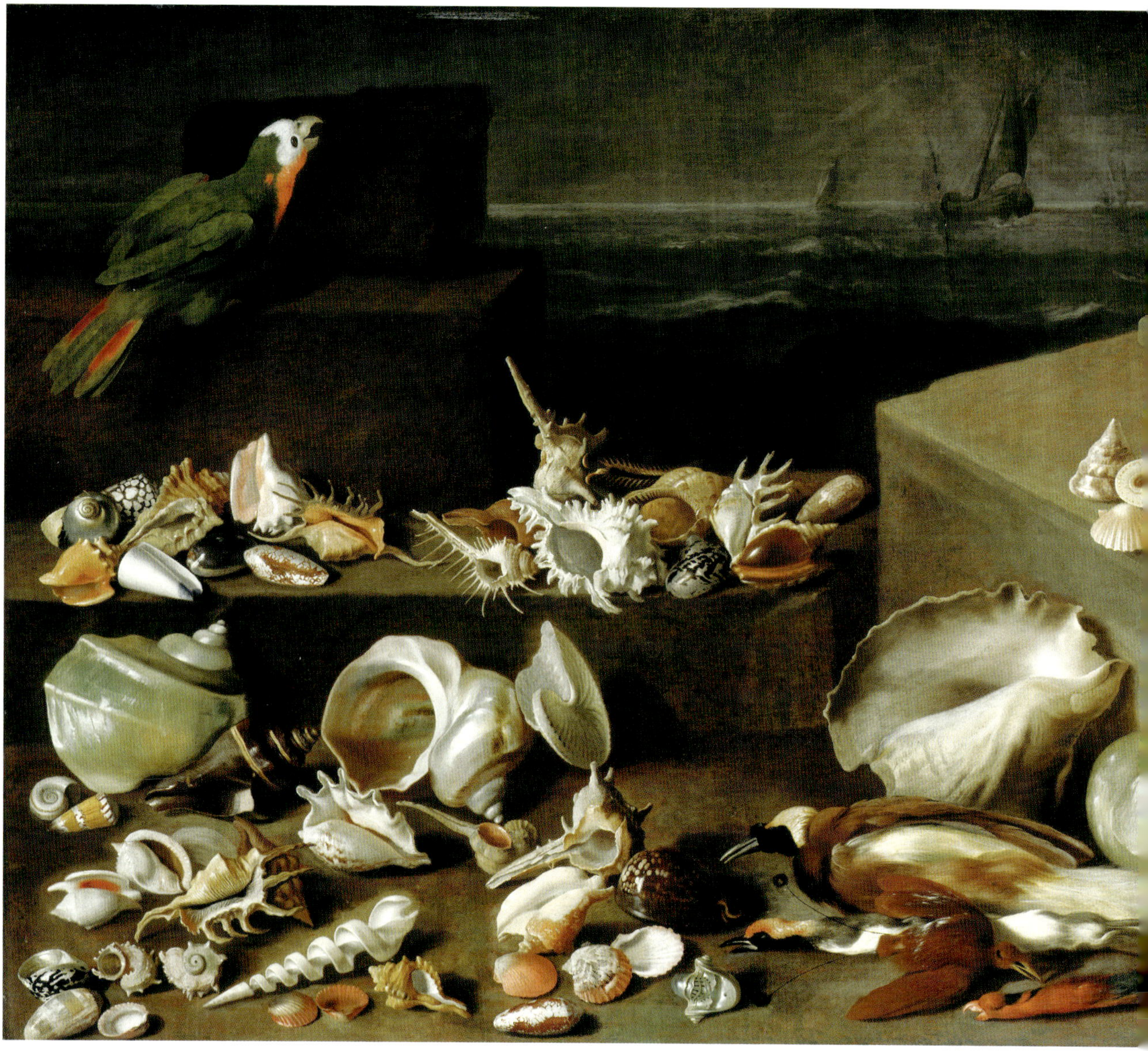

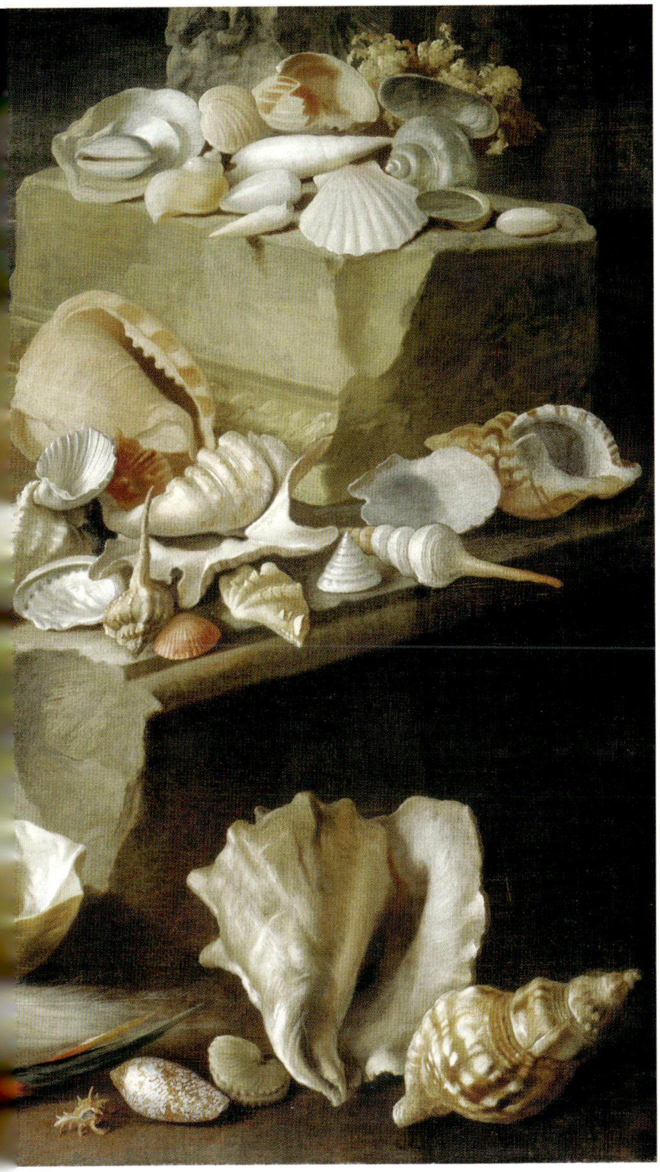

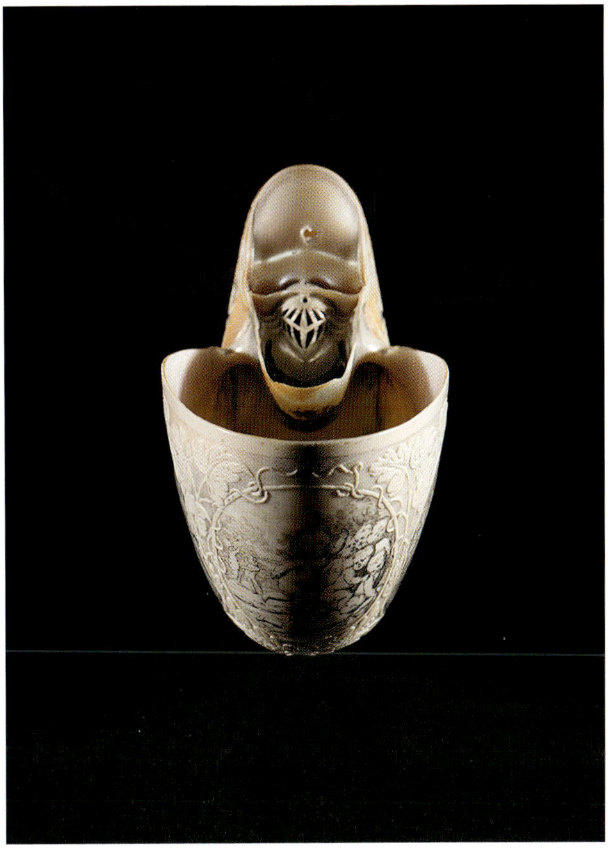

FIG. 47. Anonymous (Flemish?), *Scene of Shells and Birds*, c. 1640–50. Oil on canvas, 88.5 × 156 cm. Kunsthistorisches Museum Wien, Gemäldegalerie [9675].

FIG. 48. Jan Bellekin, *Carved Nautilus (The Sloane Nautilus)*, late seventeenth century. Natural History Museum, London.

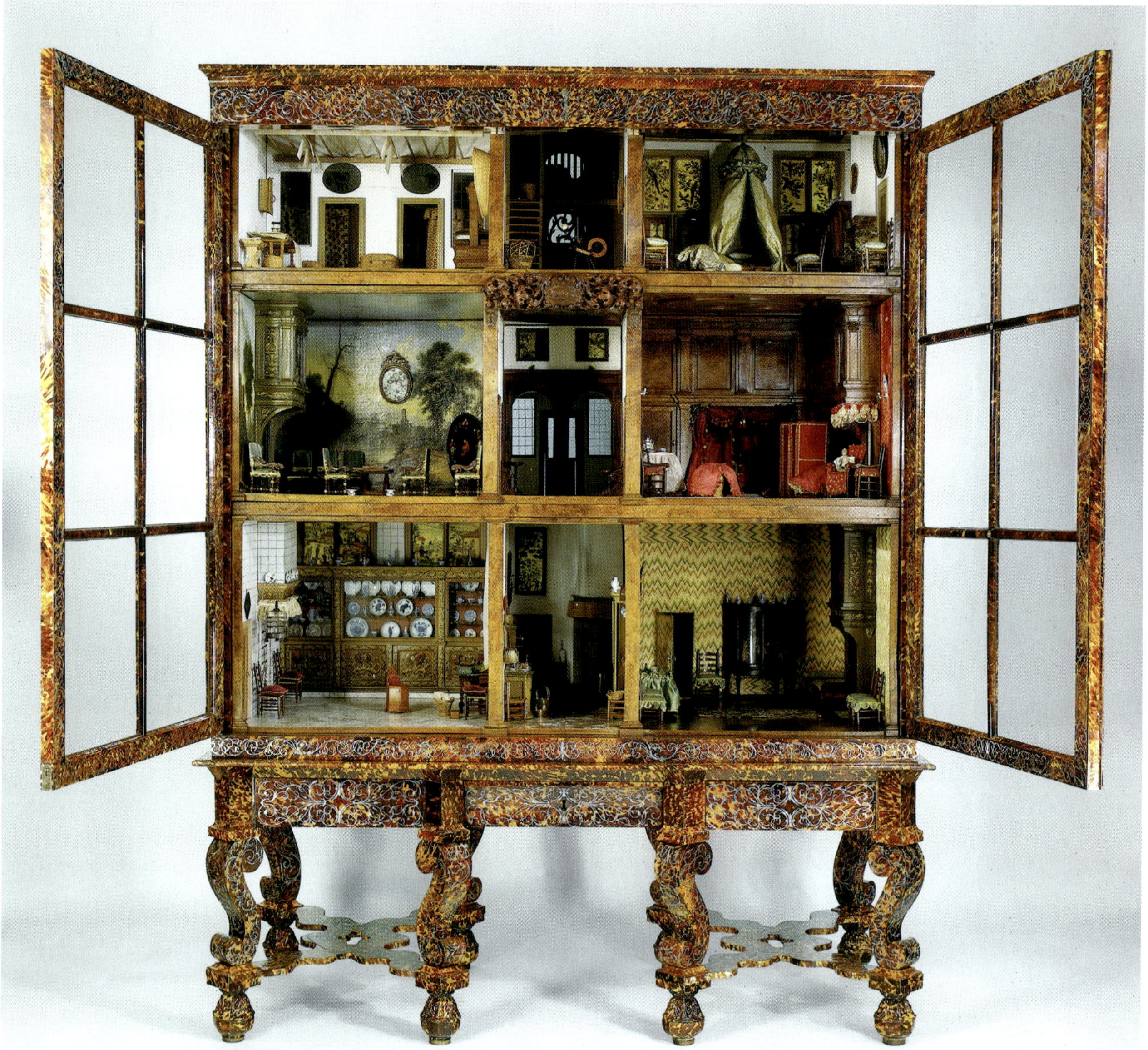

Hanneke Grootenboer

# Chapter 4
# Thinking with Shells in Petronella Oortman's Dollhouse

## "AN EXCEPTIONAL ART CABINET"

Among the largest, most ornate extant Dutch dollhouses is that of Petronella Oortman (1656–1716), currently on display in the Rijksmuseum in Amsterdam (fig. 49). Seventeenth-century dollhouses, commissioned by wealthy women to represent their homes, may be understood as the female counterparts of cabinets of curiosities, which were largely kept by men.[1] Miniaturization allowed women to assert ownership over household effects as collectors rather than users. Adorned with two glass doors and firmly standing on eight legs, Oortman's dollhouse has the appearance of an art cabinet, its rooms resembling compartments and drawers, making it—quite literally—a set of *Kunstkammern* or art rooms.

**FIG. 49.** *The Dollhouse of Petronella Oortman*, 1686–1710. Wooden case, various materials, 255 × 190 × 78 cm. Rijksmuseum, Amsterdam [BK-NM-1010].

Like seventeenth-century upper-class residences, Oortman's dollhouse was not without its cabinet of curiosities, which can be found in the so-called tapestry room at the bottom right, standing on legs, much like the cabinet-house that contains it (fig. 50). In an apt application of the Russian-doll effect, this miniature house resembling a cabinet is home to a miniature cabinet: when its doors, beautifully decorated in Japanese style, are opened, a collection of extremely small shells, neatly laid out on a few shelves, is revealed. Traditionally,

FIG. 50. *The East-Indian Art Cabinet with Miniature Shell Collection*, from *The Dollhouse of Petronella Oortman*, 1686–1710. Wood and faux Japanese lacquer and shells, 28.2 × 23.2 × 9 cm. Rijksmuseum, Amsterdam [BK-NM-1010-2].

the contents of cabinets of curiosity were divided between natural objects (*naturalia*) and human-made artifacts (*artificialia*). Oortman's dollhouse-cum-cabinet maintains this typical *Kunstkammer* division. All seven hundred items in her collection have been produced by human hands—apart from the shells. While all other items are small versions of larger originals, the shells are not, strictly speaking, miniatures. Nature knows nothing of miniaturization.[2] The tiny shells are remnants of deceased baby sea creatures whose deaths were premature. They offer evidence of lives cut short, future potential denied. They are, so to speak, the ossification of growing things whose lives were nipped in the bud. In a way, Oortman's dollhouse reinforces the impression that miniaturization is a process both of craftsmanship and of cultural appropriation. Whereas the miniaturization of the exquisitely handmade things affirms how extremely makeable they are, the coming-into-being of shells cannot be learned. Shells can be known through our senses: fingers may caress their smooth surfaces, eyes may peer inside their dark holes, and ears may listen to the sound of the waves that they have seemingly captured within them (see fig. 5). They can even be tasted: large shells were often transformed into drinking vessels. But they cannot be made.

This chapter explores how, within the context of the dollhouse as an art cabinet, these baby shells are particularly suitable things "to think with." The modernist poet Paul Valéry (1871–1945) wrote that the perplexing shape of shells gives rise to the question of how they were made. To explain them was in fact to *think* them, or rather to "remake their form in thought."[3] Gaston Bachelard (1884–1962) deemed shells equally thought-provoking, arguing that an empty shell, like an empty nest, invites us to fill it with our thoughts and dreams. In his phenomenology of intimate spaces, he examines shells alongside cabinets, attics, and dollhouses, proposing that these container-spaces have provided humanity with models of intimacy.[4]

The dollhouse was largely modeled after Oortman's own residence at the Warmoesstraat in Amsterdam, while the building initiative spanned a lifetime. Oortman was its mastermind, serving simultaneously as patron, architect, overseer, and curator.[5] This chapter will elaborate on the notion that Oortman played with various modalities of intimacy and interiority when she created a sequence of home, house, cabinet, and shell, each containing (a version of) the other. The dollhouse was a repository of all kinds of things—ordinary household effects as well as exotica—including numerous cases such as baskets, chests, cupboards, and sewing boxes that contained even more things such as linen, porcelain, books, and embroidery tools. In contrast, the shells are empty. Here in the miniature cabinet, they are lifeless display items, their previous lives now lost. In seventeenth-century moralizing literature, shells often served as emblems of decadent consumption; but I will demonstrate that they also served as tools for self-examination and introspection, their dark, spiraling

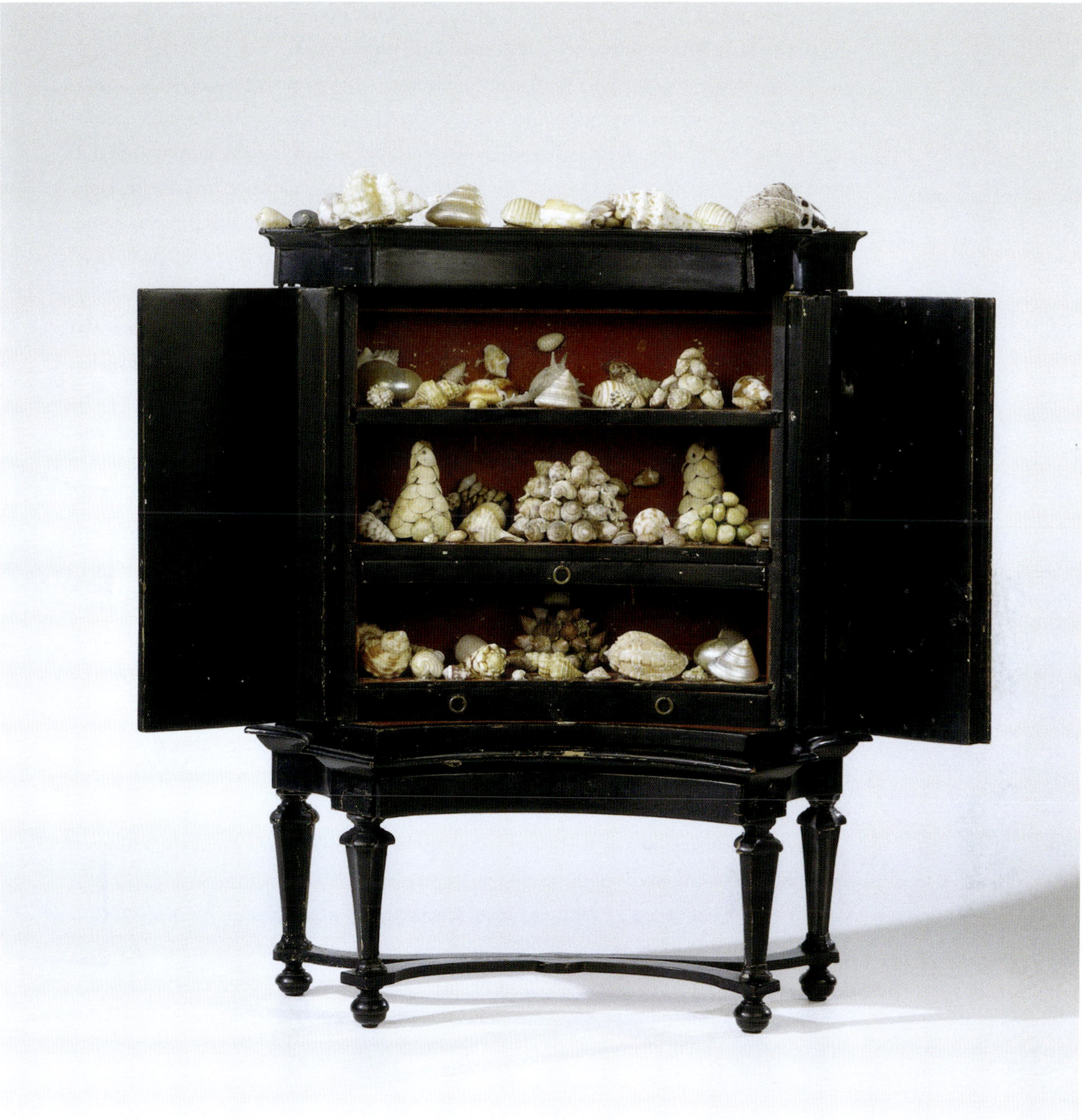

interiors evoking a quest into the reaches of one's own interiority. In this sense, they play a crucial role in the function and meaning of the dollhouse as a whole.

Along these lines, I consider Oortman's dollhouse as an ego-document of sorts, a testament to one woman's self-image, not in writing or artistic practice but through the assemblage of tiny material things that not only resembled her living quarters in miniature but also helped make up her inner world.[6] However impressive the patience, investment, time, and attention given to the dollhouse's design and management may be, it represents something greater than the noteworthy efforts expended in its assemblage and care. Both early modern and scholarly accounts of dollhouses explain their reduced size as a means whereby a devoted housewife could exercise control of an environment. While such moralistic interpretations are largely valid, they do not account for dollhouses' philosophical potential. I argue that Oortman's dollhouse enabled her to reflect on her domestic life as well as on the expanding world around it, by thinking deeply not only *about* the household effects and how they mark life's events, but also *with* them.

The idea that early modern people thought *with* things has been gaining ground in recent years in (art) historical studies.[7] In this chapter I will elaborate on this theme by relating craft to thought. Martin Heidegger famously compared a thinker to a cabinet builder, and the longevity of the idea in philosophy and history of thinking as a form of craftsmanship has been traced by Mary Carruthers and others.[8] I will consider Oortman as a thinker and a cabinet builder, and her dollhouse as a mental retreat, comparable to a study. Instead of expressing her thoughts via reading or writing, she would have thought *materially* and through the body, with the senses. She would have studied things by peering inside and imagining herself within the miniaturized environment. Taking out diminutive baskets, books, sewing boxes, or cabinets and opening them, she would have marveled at their insides. My starting point, inspired by Bachelard, is to consider this dollhouse as a model of intimacy that offers a figure of interiority. The dollhouse, with its intricate structure in which the container is contained, serves as a kind of "mental staircase"—a metaphor or mechanism that can assist a thinker to move ever deeper into the reaches of the mind. The presence of the shell collection reinforces these aspects of the dollhouse. The empty baby shells provide the key that unlocks this miniature world as a site of contemplation. Therefore, the shells are the things with which I begin thinking.

## "A CURIOUS PIECE"

Shell cabinets formed the core of most seventeenth-century collections of natural objects in Amsterdam. [9] It comes as no surprise, then, that Oortman's dollhouse cabinet followed this trend.[10] Her East Indian cabinet reveals a series of mollusks arranged on the shelves in elegant patterns (see fig. 50). Despite the tiny size of these specimens, some have been stripped of their skin to reveal a silky mother-of-pearl surface, the fruit of a labor-intensive process performed by skilled artisans.[11] In addition to a baby nautilus and a juvenile version of a marble cone or "Rembrandt's cone" (*Conus marmoreus*) discussed in chapter 6 of this book, the cabinet contains a miniature version of the very rare "spiral staircase" (*Epitonium scalare*) that, as we will see, serves a crucial role in the overall philosophy of the dollhouse. Even though most of these objects have come a long way from the seas in which they grew—the Indian and Pacific Oceans, for the most part—Oortman's shells seem very much at home in their snug arrangement, some heaped up and fixed with glue and wax. According to an early eighteenth-century eyewitness, Oortman acquired her shell collection from someone who had spent forty years in the East Indies, though recent research indicates that a small number of the shells originated from the coasts of West Africa and of South America.[12] Although only a few people would have been able to provide Oortman with such an extraordinary assortment, the identity of this particular individual remains unclear.[13]

Originally there were more shells and small pearls in the Indian cabinet's tiny drawers, "all very small and real," an eighteenth-century inventory assures us.[14] Whereas the contents of these drawers have disappeared, glue traces reveal that Oortman laid out her treasures in elegant patterns according to the trend set by Johanna van Breda, the wife of the famous collector Levinus Vincent, who received high praise for her sophisticated designs featured in the Vincent collection.[15] Pictures of Breda's work published in the illustrated *Nature's Theatre of Wonder* (*Wondertoneel der Nature*) in Amsterdam in 1706 (fig. 51) offer an indication of what Oortman's cabinet must have looked like. Apparently, it contained a curious piece in the shape of "a face of a bearded man," which, though no longer recognizable, must have resembled the composite heads in Filippo Buonanni's shell book, originally published in 1681 (fig. 52).[16]

Although exceptional in its setting, Oortman's collection of small-scale shells was not unique in Amsterdam. German scholar Zacharias von Uffenbach (1683–1734), who visited Amsterdam in 1710, viewed the shell collection of Jan de Jong, an accountant for the East India Company, whose wife possessed a small black cabinet standing on a table that might have been the same size as Oortman's.[17] It contained a wide variety of beautiful juvenile shells, arranged in eloquent patterns meant to serve as counterparts to the mature specimens assembled by her husband.[18] It is fascinating that this assemblage, presented as a female mini-version of the husband's collection, expresses gendering through differences in scale. Dutch dollhouses followed a similar pattern, whereby miniaturization gave women access to ownership over artifacts. In addition, Dutch dollhouses in general show other forms of typically female appropriation, for instance through the inclusion of a lying-in room. In Oortman's dollhouse, the presence of the lying-in room is connected to the shell collection, and it is this link—essential to her larger philosophy—I turn to now.

Pinac. 6. TAB. VI.

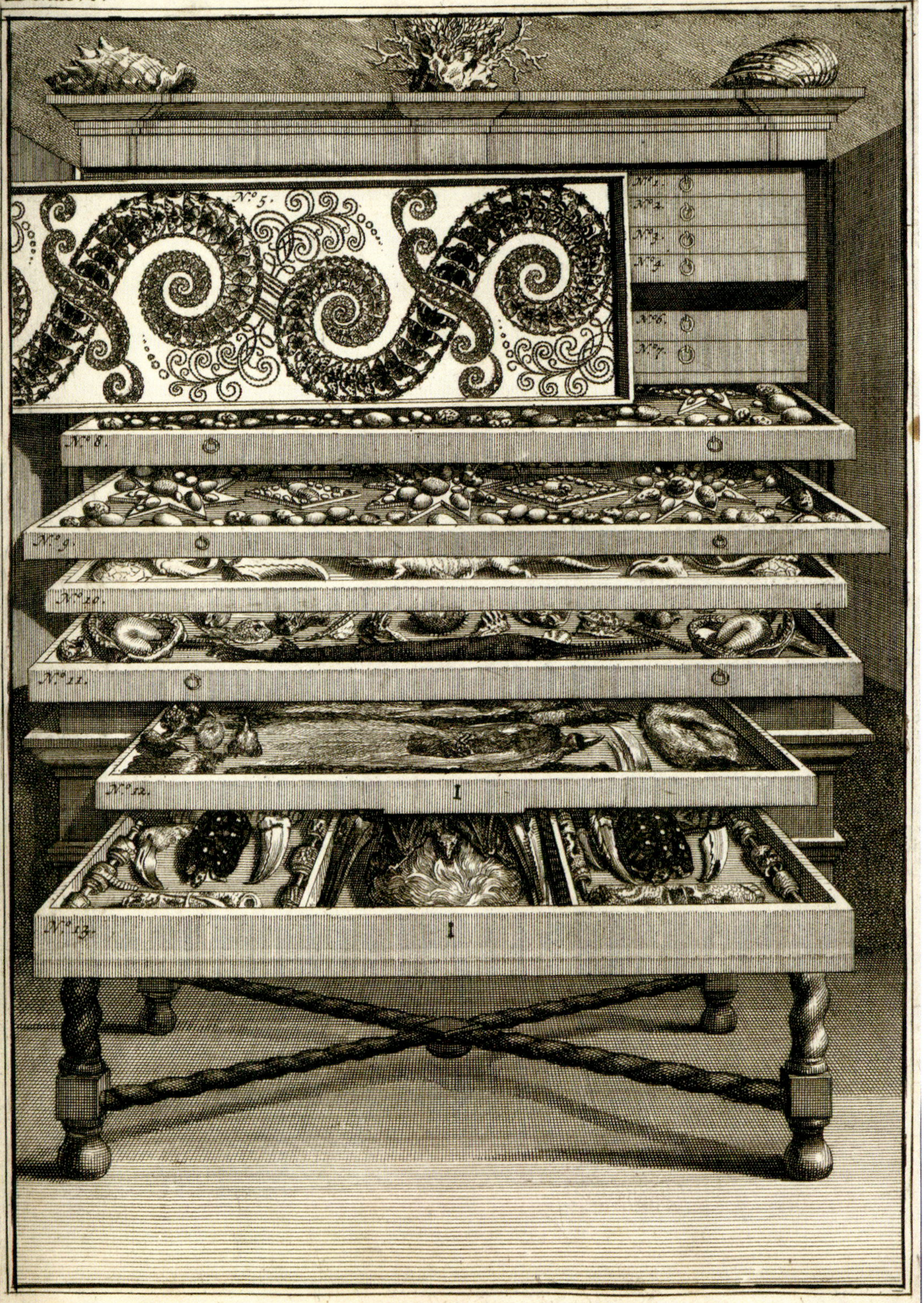

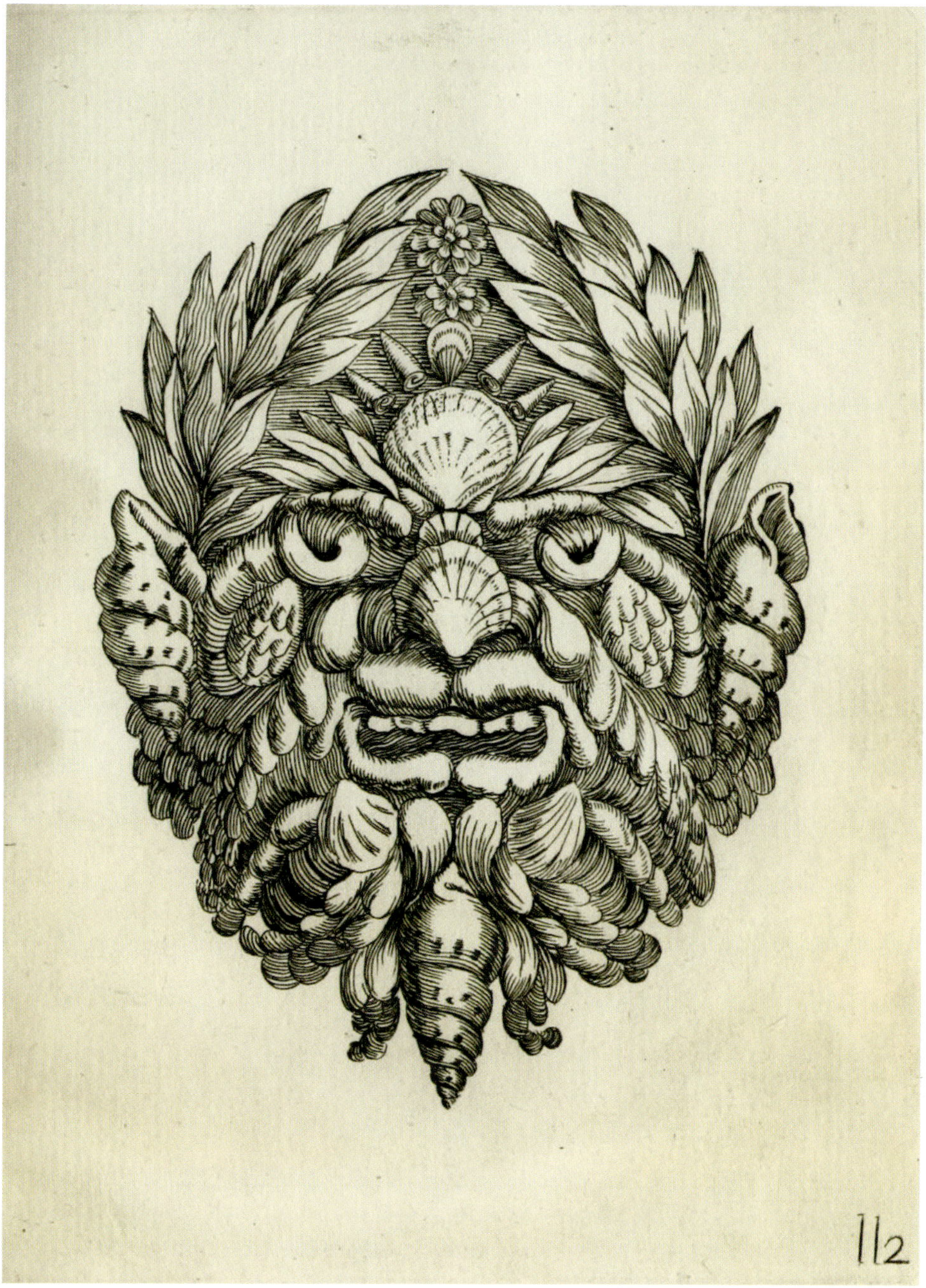


**FIG. 51.** Levinus Vincent, *Shell Cabinet (Tab. II)* from *Wondertooneel der Nature* (Amsterdam, 1706–15).

**FIG. 52.** Filippo Buonanni, *Heads*. Engraving, 22 × 15.5 cm. From *Recreatio mentis et oculi in observatione animalium testaceorum curiosus naturae inspectoribus*, 1684.

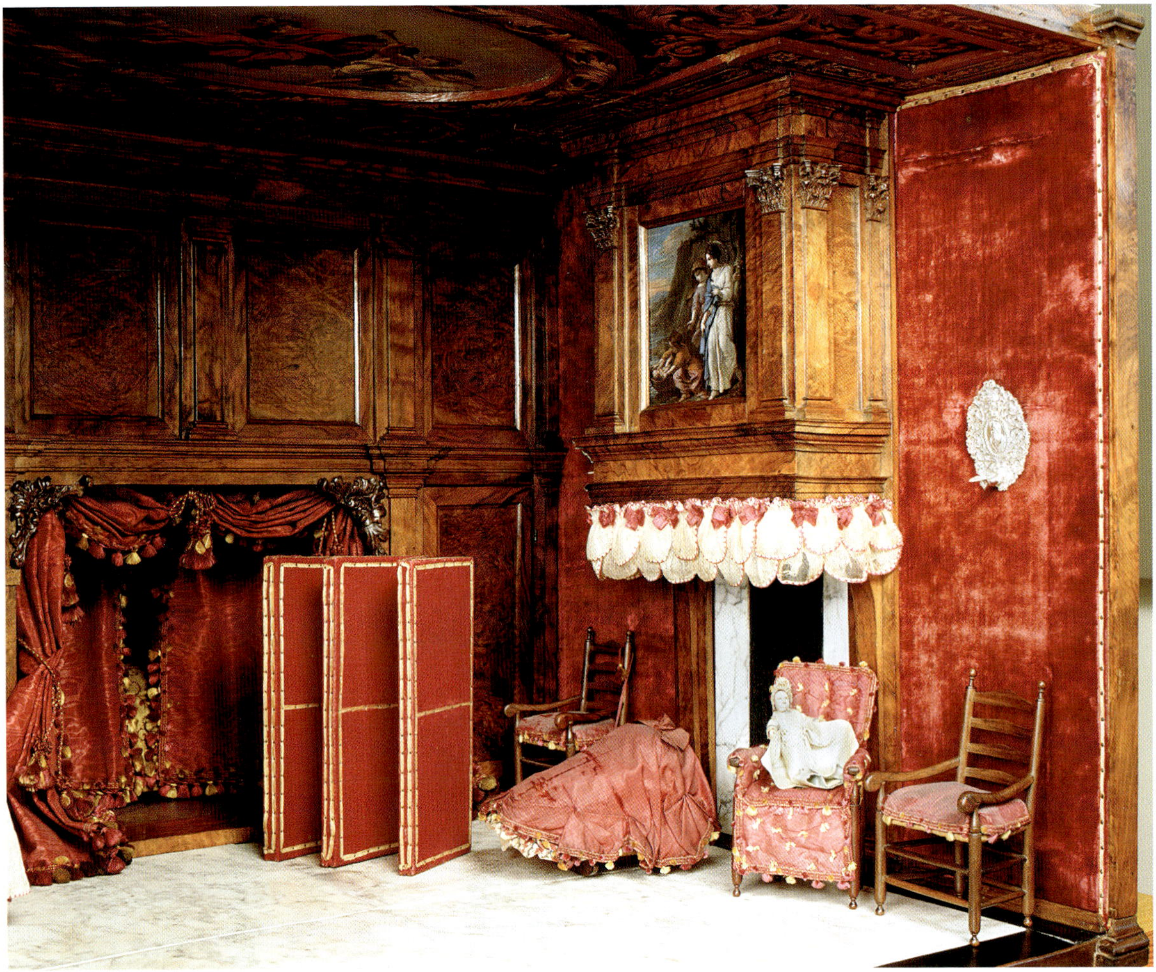

FIG. 53. *The Lying-in Room*, detail from *The Dollhouse of Petronella Oortman*, 1686–1710. Various materials. 55 × 68 × 70 cm. Rijksmuseum, Amsterdam.

## PROCREATION, FORMATION, FABRICATION: THE LYING-IN ROOM

Oortman began what became a lifelong project of re-creating her domestic environment in miniature in 1685, the year she married her second husband, the silk merchant Johannes Brandt (1654–1731). She was ambitious. Her dollhouse is large compared to others like it—as currently exhibited at the Rijksmuseum in Amsterdam, it has a bespoke staircase placed in front of it so as to allow visitors to peer inside. It is not only the largest but also the most luxurious of extant dollhouses. Its outer case has been lavishly crafted of tortoise and tin, reflecting a choice of rare materials that likely accounts for the exorbitant cost of the entire project: at somewhere between twenty thousand and thirty thousand guilders, it was comparable to the price of the four-story Warmoesstraat townhouse that was its home.[19] The cabinet was outfitted with two glass doors and comprises a total of nine rooms, including an attic for laundry and provisions, a nursery, a decorative kitchen with an exhibition of Chinese porcelain and silk screens painted with Chinese scenes, a cooking kitchen, a sitting room decorated with a panoramic Italianate landscape, and a lying-in room.

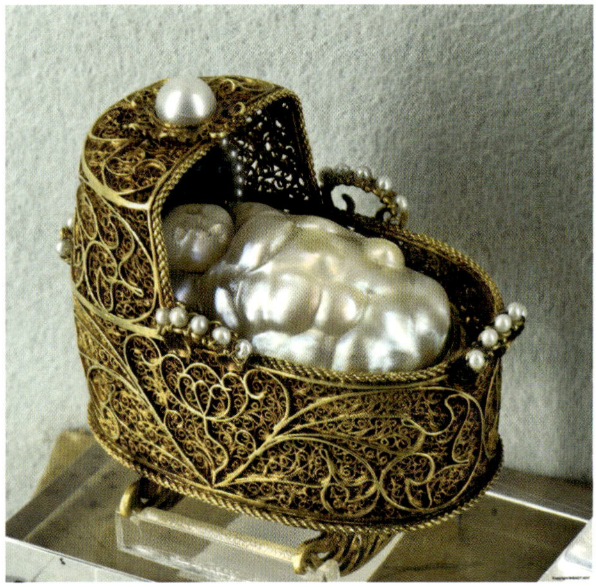

FIG. 54. Anonymous (Dutch goldsmith), *Golden Cradle with Child*, c. 1695, Gold, enamel, pearl, diamonds, blue silk quilted, 4.9 × 5.5 cm. Pitti Palace, Florence.

Whereas sixteenth-century German dollhouses, some owned by men, usually contain a nursery, Dutch dollhouses, which were exclusively owned by women, also always include a lying-in room.[20] Even in comparison to the other richly decorated spaces in the dollhouse, the furnishing of Oortman's lying-in room is lavish in the extreme, and occupies a central position in the house (fig. 53). A marble floor supports a four-poster bed, a braided cradle, a screen, and a rest chair, all rendered in coral-red textiles. The walls are covered with silk velvet of the same intense color. The painting on the mantelpiece, selected with iconographic precision, shows three female figures finding the infant Moses in a basket on the Nile.[21] This room's importance is reinforced through embroidery on the bed's coverlet, adorned with a woven monogram of Oortman and Brandt's intertwined initials. This same monogram can be found in the middle of the architrave of the dollhouse's outer frame as well as on both sides of its tortoise case.

This "branding" of the case and the textiles in the lying-in room suggests what many early modern sources support: familial identity and even dynastic ambitions were embedded in domestic trappings, and it is conceivable that the couple believed that paying such devotion to the lying-in room might be a symbolic means to help foster family life. However, there are indications of a wider connection linking childbirth, miniaturization, and marine material. In one of her poems describing her own art cabinet, Margaretha van Godewijck (1627–77) mentioned a miniature lying-in set including a cradle, a basket, and a footwarmer, all made of coral. Based on this poem, it is clear that traditional art cabinets would sometimes contain "dollhouse items." Incidentally, the poet, who was through her accomplishments known as the "pearl" of Dordrecht, also possessed a collection of shells.[22] In addition, an almost identical miniature version of Oortman's wicker cradle, rendered in gold filigree and containing a mother-of-pearl baby figure under a duvet, was produced by an Amsterdam jeweler around 1695 (fig. 54). The trinket was a precious gift from Palatine Elector Johann Wilhelm von der Pfaltz-Neuberg (1658–1716) to his wife Anna Maria Louise de' Medici (1667–1743), who carried the heavy burden of having to produce an heir, though ultimately the marriage was childless.[23]

The prominence of Oortman's lying-in room adds another layer to the theme of *technē*, or makeability, presented by the dollhouse through various levels of technology. First, Oortman's cabinet was seen as a technical marvel. It had a working fountain as well as an attached garden (now lost) containing an automaton mechanism whereby small dolls could "walk" around it. Second, what most surprised contemporaries was the amazing craftsmanship of the miniature items: not copies *resembling* their originals, they were genuine replicas in miniature, constructed exactly as their life-size models were. The eighteenth-century inventory formulates the matter precisely, emphasizing that the items' natural characteristics and materials had been honored and replicated: "what needs to be nailed has been nailed . . . glued, glued . . . welded, welded . . . just as would be in large."[24] Such astonishing exactitude contributed to the dollhouse's success as a tourist attraction. Serving as a kind of magnifying lens, miniaturization intensifies the awareness of the ingenious engineering of the items on display. Third, the artifice of makeable things contrasted sharply with two types of natural formation—of the shells, dead biological matter passing as artifacts, and of human beings, as evinced by the lying-in room. How do these different levels of fabrication relate to one another? How did the juxtaposition of various miniaturized items provoke thoughts about accommodating life within a house, things within a cabinet, and human beings within their home? We will find answers to these questions when we look at the dollhouse's conception as it is articulated in its narrative. First and foremost, this is a playhouse.

## THE DOLLHOUSE AS *DENKBEELD*

One of the most informative sources about Oortman's dollhouse is the aforementioned eighteenth-century inventory, which gives detailed information about its contents while praising its owner for her enormous effort and ingenuity. The anonymous inventory writer enthusiastically notes how visitors flocked to Oortman's treasured wonder to admire the artifacts' artistry and their philosophical and symbolic meanings expressed in so-called *denkbeelden* or thought-images. "[It] is the most perfect way that ever has been invented, to show foreigners a proper image (*denkbeeld*) of the renowned Dutch cleanliness and proper domestic economy."[25] The term *denkbeeld* here refers to an idea or example, in this case of typically Dutch cleanliness. Indeed, contemporary accounts confirm that Dutch burghers from all classes, male and female, were particularly house-proud. Polished interiors, with doorknobs and windows shining like precious stones in a jewel box, often amazed foreigner travelers. Good housekeeping was widely considered to be essential to good government. As the country lacked a ruling aristocracy, the family home was the place where the future governors of state were raised, and was thus understood as the nursery that had helped bring the young republic to maturity.

A *denkbeeld* may also refer to entries in emblem books such as those produced by Jacob Cats or Roemer Visscher, in which the combination of picture and explanatory text was intended to serve as a kind of moralizing puzzle, and a means of triggering thought. Dutch material culture, in particular with regard to the domestic sphere, was saturated with meaning, as is evident from the popularity of emblem books often featuring objects, from household effects to exotic commodities to tools and instruments. Jan Luyken's *Instructive Household Effects* (1711) is an example par excellence of the Dutch obsession with the allegorical reading of ordinary things. In this volume, Luyken has transformed fifty trivial objects, among them buckets, brooms, and cradles, into powerful spiritual metaphors, which were meant to entice people to reflect on the good, virtuous life through their daily labor. For instance, a tub used for washing out stains is presented as a vehicle for considering the difficulties of cleaning the soul; the couch providing rest between jobs elicits contemplation on how one should prepare the soul for its eternal rest; and the mirror, obviously, shows the importance of knowing one's inner self as opposed to one's outward appearance. Visitors to Oortman's dollhouse, well versed in such allegorical language, would almost automatically be moved to endow mundane things with deeper meanings. A *denkbeeld* could also suggest a complicated image (such as a *vanitas* still life) that would be thought-provoking but would not necessarily correspond to a fixed meaning or message. Such deliberately multifaceted images were intended to raise philosophical issues rather than deliver moral messages. The eighteenth-century inventory indicates that it was not the dollhouse as such but the arrangements in each of the rooms that offered a series of *denkbeelden*. Viewers were invited to engage not with single things (as in Luyken's emblems) but with elaborate tableaus that included figures interacting with things.

For Oortman's dollhouse was originally inhabited. Twenty-five elaborately dressed wax dolls, among them thirteen children, once occupied the small-scale residence. Each room was like a miniature stage presenting a carefully composed scene—reminiscent of Dutch genre paintings. The exact nature of such a composition has been preserved in a most unusual piece of evidence, a "portrait" of Oortman's cabinet painted by Jacob Appel in 1710 (fig. 55). Appel rendered each room in meticulous detail and, by using a clever perspective device, included even the elaborate ceiling decorations. The painting reveals that the cabinet used to be protected from the light by yellow curtains, and we can only imagine the anticipation of the visitors prompted by the theatrical gesture of drawing the curtains, which would precede the opening of the glass doors. Obviously, the pieces of furniture and the wax dolls were not meant to be toys, but they *were* played with. In 1718, Von Uffenbach and his family traveled through the Low Countries.[26] A remarkably detailed account of their sightseeing, preserved in manuscript form, includes an enthusiastic description of their visit to see the dollhouse. By 1718, Oortman had passed away, but the Von Uffenbachs were warmly received by the widower Brandt and their daughter Hendrina, who showed them around, telling them jokes and witty stories. Hendrina encouraged Von Uffenbach to examine the figures closely to see how they were dressed. She even lifted the skirts of a female doll to let

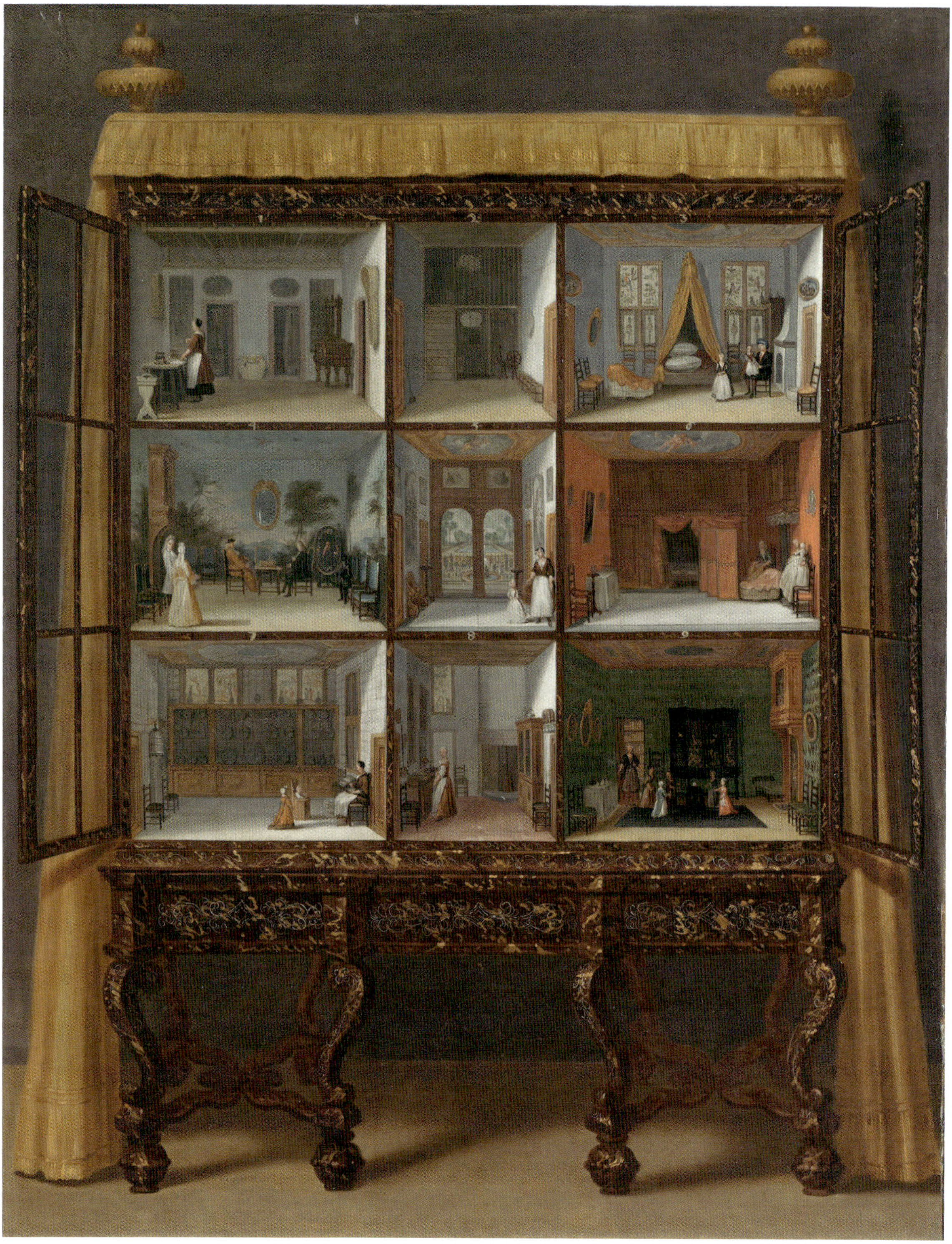

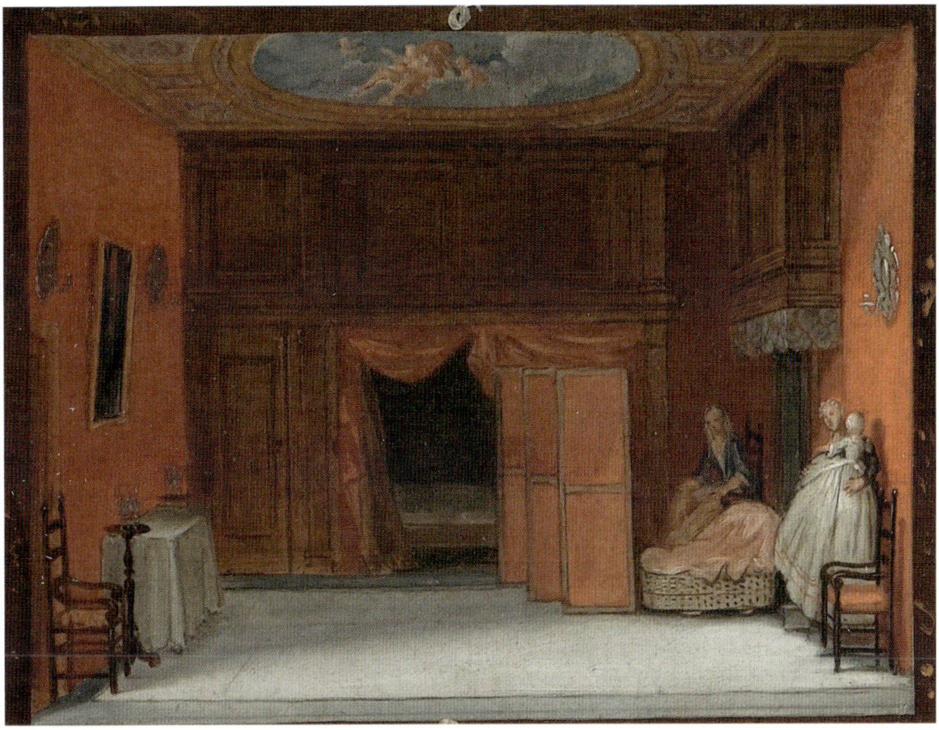

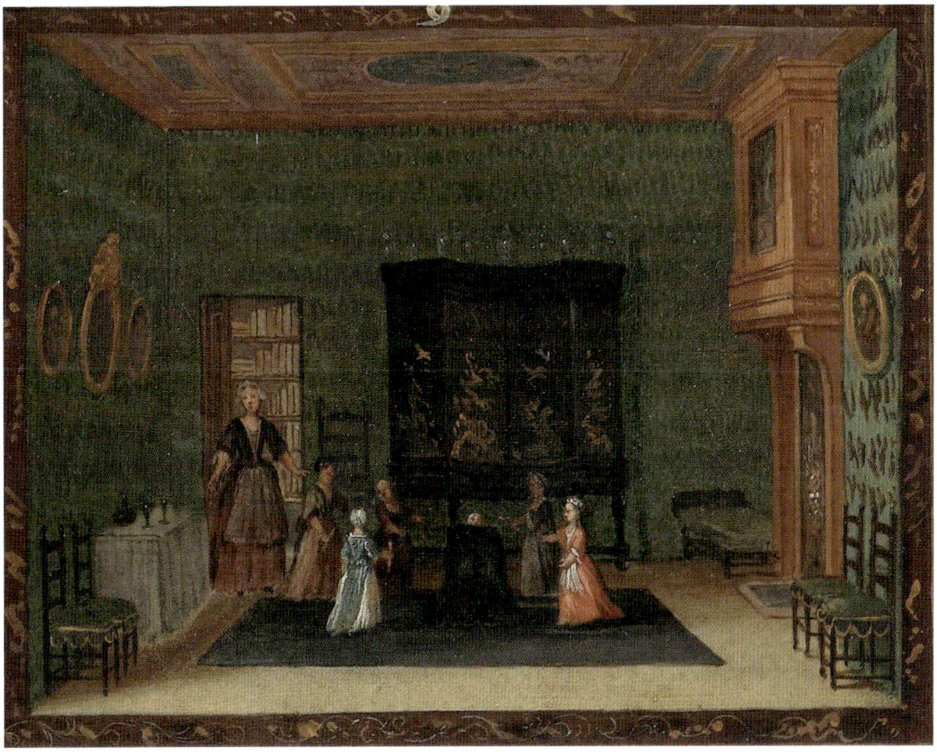

FIG. 55. Jacob Appel, *Painting of the Dollhouse of Petronella Oortman*, 1710. Oil on parchment and canvas, 87 × 69 cm. Rijksmuseum, Amsterdam [SK-A-4245].

FIG. 56. Detail of fig. 55 (The Lying-in Room).

FIG. 57. Detail of fig. 55 (The Tapestry Room).

him inspect her underwear. Von Uffenbach's eagle eyes missed nothing, not even the tiniest of details: he records that one child wore a tiny ring with a ruby, while another's neck and throat were graced with a little cross.[27]

Hendrina acted as a stage manager, guiding her visitors with a firm hand through the house, room by room. Von Uffenbach dutifully follows her directions in his report, describing each of the nine rooms from top left to bottom right and numbering them from one to nine. Even though there is no reference to Appel's painting, Von Uffenbach's descriptions of the various scenes enacted by the wax figures perfectly match the painted tableaus, almost uncannily so. Most remarkable is that Appel, too, applied the same numbering at the top center of each of the room "frames." There can be no doubt that Van Uffenbach has recorded an encounter with the actual dollhouse and not with Appel's painting in his recounting of how various objects, among them tiny books, were handed over to be scrutinized. He also jokes that his family members often had to sneeze when things were taken in and out of the dusty cabinet. Apparently, the dollhouse's nature was somewhere between that of a curiosity cabinet and what we would now call a participatory artwork.

Appel's and Von Uffenbach's matching numbered sequences point to an intentional, unfolding narrative. The house is, in fact, a three-story panorama chronicling seventeenth-century family life. The lying-in room stands at the heart of this story about the birth—and, as we will see, the death—of a child. The sequence of tableaus starts at the top floor with servants carrying out mundane household tasks, yielding to the central level and its presentation of the main event of the birth. In the drawing room on the left, decorated with a panoramic, Italianate landscape covering the walls and ceiling, we meet the lord of the house who, overwhelmed with emotion (Uffenbach informs us), has just stepped out of the lying-in room for a game of trictrac with the doctor. A Black servant boy, richly dressed and still very young, waits on them on the right. A couple to the left, mirroring our own position as we look in, have just entered from the left, probably with the intention of moving across the hallway, past the maid holding a child on a leash, straight through to the lying-in room, where they could congratulate the mother. The newborn, wrapped in linen, is being held up by a nurse, while the new mother rests behind the screen—as with most of the scenes, this one in particular seems to have been copied straight from a Pieter de Hooch painting (fig. 56). This joyful sequence meets a tragic end in the tapestry room on the bottom end, where we find five children who, under the guidance of their governess, are grouped around a small bier draped with black cloth, on which a dead child has been laid out (fig. 57). This gloomy room, filled with sadness, is the site of the shell collection.[28]

If we read Oortman's dollhouse as a *denkbeeld* or thought-image, we see how the beginning of life is intricately bound up with the presence of death, in the same paradoxical way as in vanitas still lifes. Often referred to as a thought-image, a vanitas painting combines the pleasures of life with the inevitability of death, the feast of the senses with the empty stare of the death's-head. In the tapestry room the contrast between life and death, joy and sadness, possession and loss receives its articulation but is not quite resolved. As in the lying-in room, a painting—here, *Christ Approached by the Children* hanging above the fireplace—is significant within the larger theme of the chamber. The reference to childhood innocence points to the undeserved and premature passing of someone so young, an event that nonetheless needs to be accepted as the will of God. The contrast between these two opposite emotional states, innocent mirth and sentiments related to the death of a child, is what Oortman may have experienced in playing with her shells. The shell collection takes on a larger meaning in the context of the tragic scene: both the deceased child and the sea creature living in a baby shell passed away before their time, their growth broken off. Although she had four children fathered by Johannes Brandt, Oortman lost her firstborn, a daughter, as well as her first husband and, eventually, one of her sons. For her, musing about the shells in the context of the scene with the bier may have been a way to reflect on what she possessed as much as on what she had lost. Handling and playing with shells was traditionally considered as an activity supposed to bring comfort and joy in stressful times.

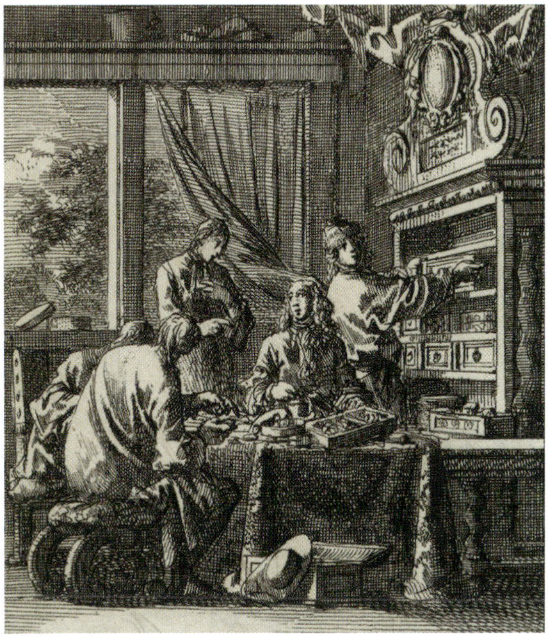

**FIG. 58.** Jan Luyken, "Het Kabinet" (XLVII). In *Het Leerzaam Huisraad*, 1711. Engraving, 9.5 × 8.6 cm. Rijksmuseum, Amsterdam.

## PLAYING WITH SHELLS

Conceived as a sort of serious arena for play for adults, the dollhouse resonated deeply with currents in seventeenth-century literary culture. One of the entries in Luyken's instructions for musing about household effects includes a cabinet of curiosities. The picture shows a large dresser surrounded by five figures inspecting its contents (fig. 58). Luyken is keen to wag a finger at those who preoccupy themselves with the idle contemplation of luxuries. In the poem complementing the image, the *liefhebbers*, or lovers of curiosities, are compared to children playing with "empty shells" (*leege schellen*). The poem concludes by stating that virtues are far greater possessions than exotica, and that real riches may be safely stored only in the cabinet of the heart. For Luyken, childlike behavior is undesirable in adults. However, other moralists such as Jacob Cats provide a more nuanced understanding of what it means for children to play. In Cats's long poem, "Children's Play" ("Kinder spel") written as a prologue to *Houwelick* or *Marriage* (1625), his popular, didactic work instructing women on how to carry out the various roles of girl, bride, mother, and widow, Cats recommends that his readers not reject children's games as senseless pastimes. Instead, he invites them to take a good long look at a playground so as to see that an entire world is reflected within it.[29] Playing, he insists, is for children a form of learning about themselves that allows them to develop the talents they need for their roles later in life. Cats's idea that the world is like a playground seems to build on the proverb (visualized by artists like Jan Steen) that the world is a stage on which all people play their roles.

By contrast, in another popular emblem book, Johan de Brune (1588–1658) moralizes children's games as caprice:

> All that one sees on earth
> Is doll's stuff and nothing else.
> Whatever a person thinks about
> They enjoy like a child
> What they love for a short time,
> They easily throw away later,
> So a person is, as one can see
> Not just once, but always a child.[30]

De Brune comments on the capriciousness of children, who lovingly adore a toy one day only to disregard it the next. In an appropriate gesture, Sara Rothé (1699–1751) selected this poem to be copied on a miniature scroll that was originally kept in the lying-in room of one of her two dollhouses.[31] Within the context of the dollhouse, however, the poem's meaning is turned on its head. Like Cats's playground, the dollhouse mirrors an entire world, and Rothé seems to suggest that by its very construction she has refuted De Brune's idea of children's inconstancy. Furnishing her treasured dollhouse, she had held on to her "doll's stuff"; by remaining a child, she has shown herself to be strong and unswerving, precisely the opposite of what De Brune suggests about human nature. Expanding Cats's view of the playground as a mirror of the world, Rothé asserts that we should, in the grown-up world, seek moments that mirror our childhood. To remain childlike, she suggests, is an essential aspect of being human.

A similar interest is at play in Oortman's dollhouse. Her miniature Indian cabinet contains items that, in Luyken's words, would literally be "empty shells," yet they are filled with meaning when people start to play with them—and think with them. In addition to the common conception of the shell as home, early modern Dutch literature demonstrates that shells were versatile metaphors used in relation to self-consciousness, tied up as they often were with notions of self-knowledge and self-restraint. A (still) popular figure of speech, "to crawl back into one's shell," evokes a sense of having overstepped one's boundaries and having been compelled to make a shameful retreat. In a section of *Houwelick* containing expressions and proverbs relating to the shared duties of couples, Cats warns, "Stay in your shells, as that will help you," which, addressed to women, translates as "don't be ambitious outside of the domestic sphere."[32] Like other women of her status and intellect, Oortman was forced to pursue her ambitions exclusively within the domestic sphere. She clearly made the most of the little space she had: within the confinement of her own house, she erected a miniature house that served as a monument to Dutch domesticity as well as a unique and highly valuable art collection that was widely respected as a genuine work of art. *Did* she stay in her shell? My view is that in addition to a monument and a curiosity cabinet, the dollhouse was also her private space, a philosophical retreat through which Oortman allowed herself to reflect. As a shell of her own making, it became not a space of confinement but instead one of solace and self-exploration.

Indeed, the image of a shell as a form of shelter was also understood by Oortman's contemporaries as a space for reflection on the self. The Dutch Protestant poet Dionysius Sprankhuysen, active in the first half of the seventeenth century, wrote that "it would be a good thing if man would turn into his own shell and think to himself: What was I before I was born?"[33] Both Cats and Sprankhuysen compare the shell to the human body as a kind of container housing the mind, and also as a space where one could withdraw so as to reflect on one's role and position in life. In the same period that their writings became much beloved, the ideas of René Descartes about the body as a container for a thinking "I" also gained ground.

Ontological issues evoked by shells would eventually resonate beyond the early modern period, in the work of Valéry and Bachelard and also of Heidegger. The latter philosopher gives the example of the snail forever bearing his shell as a vehicle to explain that we, as human beings, are on a continuing quest for our philosophical home.[34] The snail is for Heidegger the quintessential image of a living creature that is neither fully at home nor fully in the world but is always in a state of partial retreat. Even within the comfort of our own residences, we are also partly "nowhere," a sentiment shared by Cats and Sprankhuysen, as well as by Oortman.

Heidegger may have been inspired in his contemplation of the snail by Desiderius Erasmus's *Colloquy* entitled "The New Mother" (1526), which features the shell-bearing creature in connection with the spaces of dwelling and motherhood.[35] "The New Mother" is a conversation between a teacher and a new mother, Fabulla (whose name means "fable" or "story"). The dialogue starts with a discussion on the superiority of men, a claim that Fabulla, though feigning ignorance, masterfully refutes. She counters the observation that men go to war and are prepared to die for their countries with a statement about women who, through the experience of childbirth, always "engage with death at close quarters."[36] The key question for Fabulla is whether the female body is an instrument of the mind or the mind an extension of the body. Erasmus explains that the soul is bound to the body as a snail to its shell. He adds that people have the opportunity to make their "shells" more habitable for their souls, "just as in houses walls are plastered or hung with tapestries."[37] Was Oortman trying to make the dollhouse a more inhabitable dwelling of her mind? Was arranging her baby shell collection next to the bier with a dead child, and within the context of a joyful story of childbirth, a way of doing so? Like Fabulla, Oortman had engaged with death at close quarters.

Whether related to the story of childbirth or the housing of the soul, or employed as a metaphor for the premature death of a child, the baby shell collection in Oortman's cabinet is at the center of these narratives of self-examination. Whereas at the start of the seventeenth century the wonder evoked by curiosities was largely interpreted through the lens of Christian spirituality, parallel explanatory discourses emerged in the second half of the century that were based on natural philosophy, Cartesian thought, and humanist notions of the self.[38] Early modern collectors began increasingly to use their curiosity cabinets in general, and shells in particular, as sites of contemplation so as to see themselves reflected in the objects they collected, even though the devotional element never faded completely.[39]

Humanist writers such as Erasmus, one of the earliest possessors of a shell collection in Europe, and Michel de Montaigne had already referred to the benefits of shell collecting in relation to self-examination with regard to Cicero. In *The Orator* (46 BCE), Cicero recommends that city officials who were worn out by worldly duties could refresh their minds and free them from sorrow by roaming the beach to collect pebbles and shells.[40] "To collect shells" (*conchas legere*) during idle wanderings was an adage that became associated with the idea of *otium*, the established mode of remaining mentally occupied even while at leisure.[41] The fourteenth-century poet Petrarch quoted Cicero almost verbatim in *The Life of Solitude* (1346), in a passage emphasizing that shell collecting on the sands would displace all depressing thoughts.[42]

What is of specific interest here is Cicero's observation that shell gathering would allow grown-ups to "become boys again" (*repuerascere*).[43] For Cicero, and for the humanists after him, playing with shells resulted in feeling refreshed, or rather rejuvenated. As Cicero implies, such childlike play allows people to step out of their fixed behavioral patterns and modes of thought to revitalize their minds. That Oortman's collection comprised baby shells makes their link to childhood even more poignant. Finding the child within by playing with shells, or by building a dollhouse, as Rothé insists, is also a way of rejuvenation and of thinking new thoughts. Whereas activities like reading emblem books like Luyken's made people think about what they ought to know—or what others know—shell collecting evoked wonder and amazement that would trigger people to think about what they didn't know, and to think for themselves. Among Oortman's shells is one in particular that was especially useful to help plumb the depths of one's inner self: an extraordinarily rare, snail-like species, the precious wentletrap (*Epitonium scalare*).

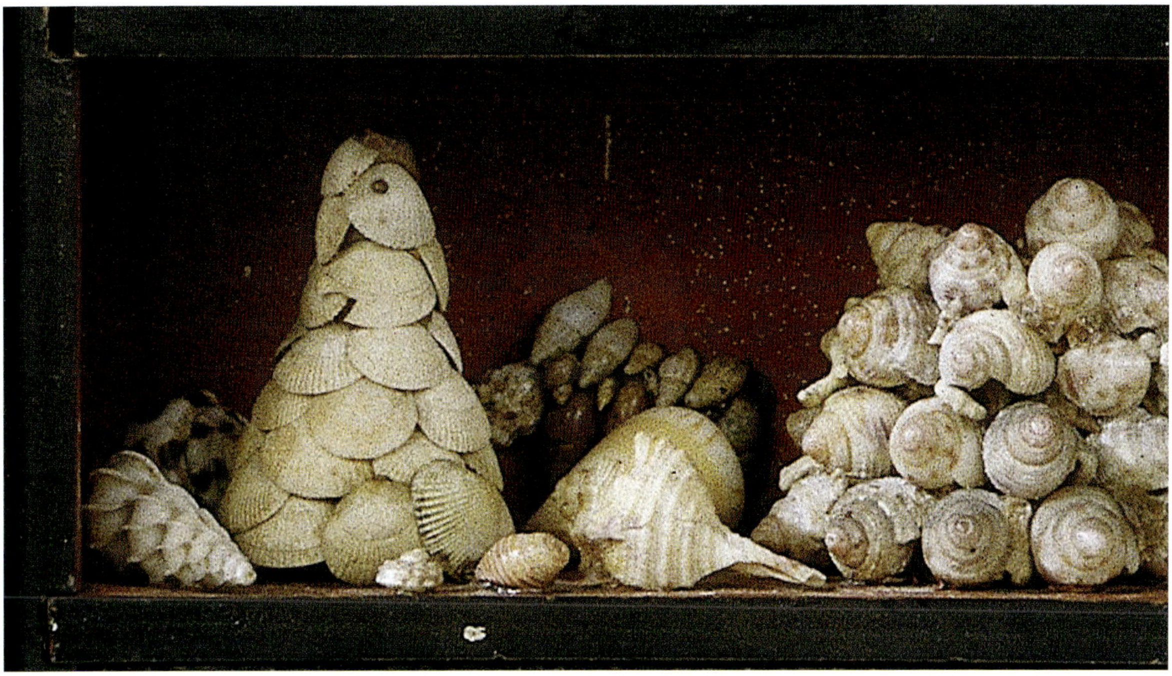

## THE PRECIOUS WENTLETRAP AS THE STAIRCASE OF THE SELF

**FIG. 59.** Detail of fig. 50 (*Epitonium scalare* or precious wentletrap).

As we have seen, Oortman worked out a series of elaborate metaphorical games in her dollhouse that provided steps for a visitor to follow. In addition to the sequence of narrative tableaus, she designed the Russian-doll effect that invited visitors to descend ever more deeply into the artwork's heart. Let us descend once more on this metaphorical staircase, spiraling downward from the Warmoesstraat residence to the dollhouse to the East Indian lacquer cabinet, where we will find, at last, its final rung in the shape of a specific shell. On the middle shelf, on the far left, we find a baby version of an extremely rare specimen, the so-called precious wentletrap that literally translates as "spiral staircase" (fig. 59). This exquisite item was widely celebrated as being one of the most perfect shell shapes ever encountered. One of the most magnificent curiosity collections in France contained one such shell, described as follows:

> the famous shell, called "La Scalata" or "L'Escalier (Epitonium Scalare L.)." This shell is the rarest of all those known; it is unique in Paris, and perhaps one cannot count half a dozen in all Holland, which is the general shopping place for this kind of curiosity; there it is usually pushed up to enormous prices, especially when it is as big as this one.[44]

At another auction in Paris, one of these shells was sold for the exorbitant amount of 1,611 livres, and similar specimens continued to fetch high prices throughout the eighteenth century.[45] The item is also sometimes listed in inventories as an Amboina wentletrap, and although its image was included in Rumphius's *Ambonese Curiosity Cabinet*, a description is lacking.[46] Its editor, Simon Schijnvoet, explains that the shell is so rare that it was possible that not even Rumphius had encountered it, and he admits his own failure to do it justice: the *Wenteltrap* [*sic*] "is bright white and of a miraculous making, and [hence] cannot be depicted as it actually is." Among the happy few who possessed such treasures, he writes, are the Grand Duke Cosimo (to whom Rumphius sold an entire collection of shells) and the chief constable of the city of Delft, Johan de la Faille, who was offered 500 guilders but refused to part with it. A third shell was apparently to be found in England, in the collection of the painter Jürgen Ovens (1623–78). If the Amboina wentletrap was too rare to have been seen even by Rumphius, we can only speculate on the singularity of its baby version in Oortman's collection.

Perhaps the earliest recording of this rare specimen occurs in the diary of Balthasar de Montconys (1608–65), who in 1663 paid a visit to the cabinet of Ernst Roeters in Amsterdam. Montconys described it as "a white snail, like a twisted, spiraling baton."[47] The almost mathematical precision of the shell's structure was aptly captured by Montconys through his use of the term *limaçon*, meaning "snail" but also denoting an algebraic curve, which Étienne Pascal (1588–1651) had described only a few years earlier in what became known as the *limaçon de Pascal*. The wentletrap's ridges are so perfectly aligned that their positioning appears to be the result of painstaking mathematical calculations.[48]

The "white snail" even featured in a contribution to a philosophical debate made by the poet Christoffel Beudeker in his long poem on the benefits of keeping a cabinet of curiosities, published in 1748. Beudeker argues against the claim of the controversial philosopher Baruch Spinoza (1632–77) that nature's beautiful patterns (of flowers or trees, or shells for that matter) have not been created by a God but are generated by their own "conception." Nature designs itself, Spinoza claims, and the cause for its existence, rather than being explained away through the existence of a transcendental God, must be found within nature as such. Firmly rejecting this idea, Beudeker uses the shell in general, and the wentletrap in particular, as evidence that God is the divine mechanical engineer who has designed and created the world:

> Behold the Wenteltrap [*sic*], to which no other kind compares, and tell me whose intellect is able to make a judgment, as to how she is stitched together, her circles chained together, forged by nature's links. Who has invented this artwork? you disbeliever! He alone who created all.[49]

It is significant that Beudeker uses the term "invented," indicating the shell's status as a technological marvel, which is, paradoxically, also "stitched" together as if by human hands. We cannot know what Oortman thought about how shells came into being. Aristotle's view that they were formed by mud dried by the sun had largely been rejected, but it was still unclear that a tiny mollusk would be responsible, building its own shelter around itself, step by step, according to its own design. The juxtaposition of human artifice, human craft, natural artifice, and human reproduction in her cabinet shows that she must have held some views on the matter. If we understand the elaborate furnishing of the dollhouse as an attempt by Oortman to make her soul more habitable, her miniature shell collection in general, and the precious wentletrap in particular, serve as strong reminders of how house, soul, and shell are entwined—in philosophical and spiritual texts as well as in the collective imagination. Perhaps Philibert van Borsselen, in his ode to shell collecting, evokes the strongest connection between human and shell when he compares a human being to the precious wentletrap, calling out to the Almighty: "Let your Creature be a spiral staircase."[50]

Montaigne famously embarked on his quest for self-knowledge by retiring to the cabinet in his tower. His writings attest that his interior world gradually took shape according to the model of the architectural structure of the tower—its spiral staircase slowly becoming the staircase of the self.[51] The staircase, a common medieval meditative device of ascension in the mind, enables him to travel up and down among various depths of thought. I suggest that through the building and the arrangement of her dollhouse, Oortman has designed a similar figure of interiority. In fact, she was a near virtuoso in expressing various levels of interiority through the *mise-en-abîme* of her project as a whole. What Montaigne did in his writing, Oortman achieved through her work on her dollhouse: both carved out a space of retreat that allowed for the experience of the core structure of the self. Whereas Jan Govertsz van der Aar was able to portray (as a means to promote) himself through his shell collection by associating with the largest item in his collection, Oortman decided on a distinctly more subtle route (see fig. 12). Operating in different registers of representation and signification, her shells are tools that enable and reinforce a contemplative life as a collector and curator as much as a wife and mother, celebrant and mourner.

Seventeenth-century grammar manuals defined "metaphor" as a "borrowed residence," a kind of house or empty shell that, when vacant, opens up to different meanings.[52] Significations, like creaturely inhabitants, move in and make themselves at home, until they die and leave the images vacant for other meanings to move in. Residences as much as metaphors, Oortman's baby shells were once cradles of marine life that came prematurely to an end of their biological purpose. Washed up on distant beaches, they were sent on a long journey to Amsterdam, finding a new home away from home, henceforth never to be filled with anything other than human thought.

# Part III

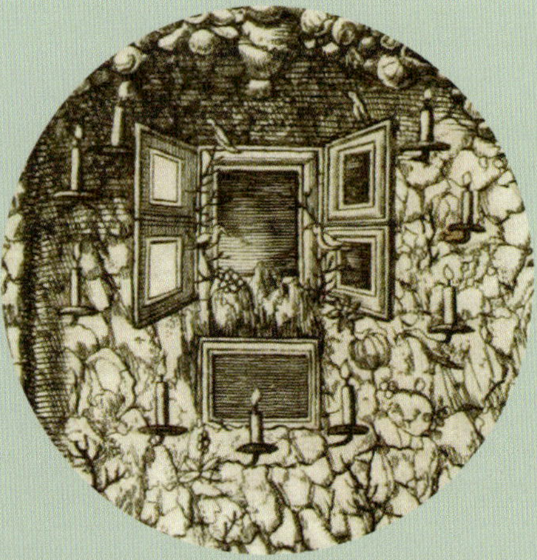

# THE MULTIPLE EXPERIENCED

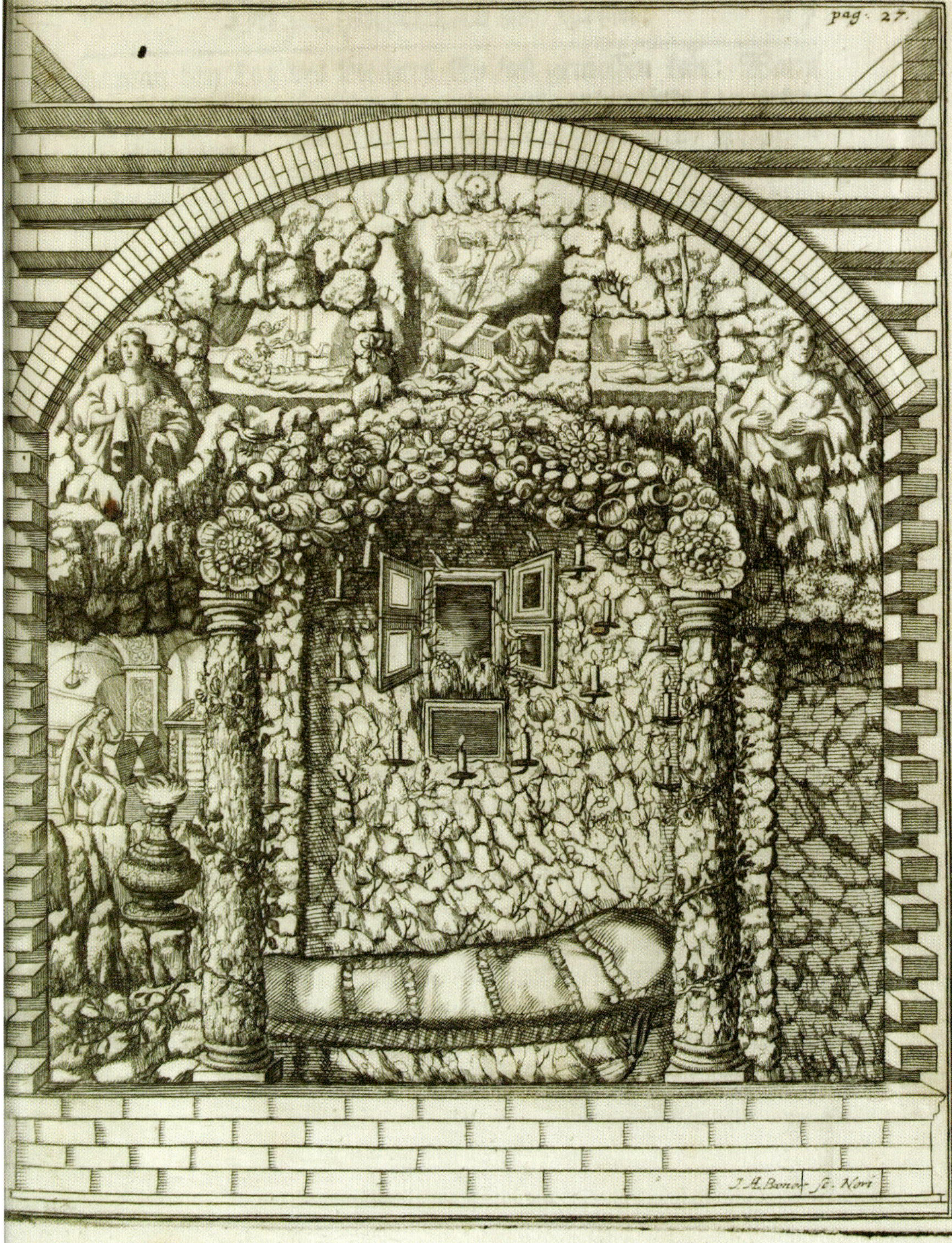
J. A. Boner sc. Nori

Róisín Watson

# Chapter 5
# Shells and Grottoes in Early Modern Germany

Between 1677 and 1679, Duchess Magdalena Sibylla von Württemberg commissioned a grotto for her rooms in Stuttgart designed to emulate the appearance of a subaquatic cave (fig. 60). Encrusted with rough stones, corals, and shells, the chamber was intended as a place in which Magdalena Sibylla could mourn her late husband, Duke Wilhelm Ludwig von Württemberg—following his untimely death—and prepare for her own eventual demise.[1] Stepping into the grotto, Magdalena Sibylla entered a space that blurred the boundaries between the natural and the artificial, as well as the material and the heavenly realms. Although the room no longer survives, we have a sense of its original appearance thanks to a set of engravings executed by the Nuremberg artist Johann Alexander Böner and published by the duchess' spiritual adviser Ehrenreich Weismann.[2] The focal point of the grotto was a wax model of Magdalena Sibylla's husband in his coffin that was surmounted by three obelisks; the central needle, directly above the coffin, was decorated with a series of emblems and images of Christ that provided visual commentary on the transience of life and acted as prompts to contemplate the nature of a good Lutheran death (fig. 61).[3]

**FIG. 60.** Johann Alexander Böner, Interior of Magdalena Sibylla von Württemberg's grotto. In Ehrenreich Weismann, *Christliche Betrachtung der Betrübten Zeit und freudenvollen Ewigkeit*, 1680. Engraving. Sächsische Landesbibliothek–Staats- und Universitätsbibliothek, Dresden [Theol.ev.asc.609].

Many histories of grottoes have focused on the automata and water features found within them as their most notable curiosities and as evidence for humankind's ingenuity. Early modern automata, motorized statues propelled by cutting-edge technology, might range from tweeting mechanical birds to representations of gods like Poseidon, riding around a fountain in his chariot as if across the ocean waves.[4] But Magdalena Sibylla's grotto contained no such contraptions. Nor did it contain elaborate fountains. It was characterized by stillness and contemplation, rather than the rhythmic push and pull of motorized statues and the frenzied delight these machines elicited. Such elements have informed our understanding of these spaces, often to the neglect of the shells that were also key and complex components of grotto design. By emphasizing the curiosity roused by shells specifically, this essay focuses instead on the ways that conchological decoration informed the function of grottoes, and how the spaces themselves guided the ways that shells were understood.[5] In looking at this reciprocal relationship between function and signification, we might understand more fully the multiplicity of associations prompted by the material object of the shell.

Böner's illustrations of Magdalena Sibylla's grotto depict shells framing images of Christ's Passion and representations of the duchess's relatives on their deathbeds (fig. 62). They also form an opulent canopy over a bed, where she may have rested following long solitary hours copying out prayers (see fig. 60). On the grotto's pillars, shells have been arranged to depict miscellaneous figures, contrasting with the more formal portraiture that also appeared within the space (fig. 62). The conchological decoration here was also found elsewhere in Stuttgart's ducal residence. The gardens included a series of grottoes, begun in 1610 by Duke Johann Friedrich von Württemberg (1582–1628) and continued by his successors, that contained statues bedecked and walls encrusted with shells.[6] Statues of Neptune, Andromeda, and a variety of nymphs also decorated the chambers, as well as a statue of Midas, the musician Orpheus, and a naked woman adorned with seashells.[7] All these statues were typical elements of grotto decoration, emphasizing the space's connection to the sea, to classical mythology, and to the notion of transformation. But, unlike Magdalena Sibylla's chamber, Johann Friedrich's commission was not a place for mourning. His grotto was meant as a site of courtly pomp and celebration, and the shells employed in its design supported that intended function.

These ducal commissions were not unusual. As we will see, the grotto was an almost indispensable part of any sophisticated garden or courtly residence in early modern Europe: an opportunity for the elite to display their refinement, erudition, and awareness of contemporary trends in architecture and design. The fashion for grottoes flourished first in Renaissance Italy, but by the late sixteenth century it had spread to the courts of northern Europe as noble families there sought to emulate—if not to surpass—the models of their southern counterparts.[8] This growth in grotto construction was matched by an increasing interest from architects and engineers in the questions of how to re-create such structures. The German architect Joseph Furttenbach the Elder (1591–1667), who worked on the borders of the Duchy of Württemberg in Ulm, was one artisan who provided answers to these questions. His guide on how to construct grottoes and the suitable use of shells within them affords us invaluable insight into the shell's function and signification from the perspective of an artisan.

Although the grottoes initially conceived of by Johann Friedrich and the mourning grotto of Magdalena Sibylla were constructed almost seventy years apart, in very different social and political contexts, their shared use of shells is striking. Comparison between them offers a salient opportunity for a broader exploration of the polyvalence of shells as design material, splendid decoration, and prompts for reflection within early modern grottoes. How did these two patrons understand the role of shells in their grotto projects, how did visitors engage with them, and how did these responses correlate with the intentions of architects? This chapter argues that shells invited plural responses and associations within both of the Württemberg grottoes in question. That shells were deemed as suitable to a space designed for pleasure as to one designed for mourning already signals that they were a versatile material in terms of their function and aesthetic appreciation. I am particularly interested in how encounters with conchological specimens within early modern grottoes ranged from curious to contemplative to celebratory.

Although contemporary commentary on visiting early modern grottoes is scarce, a note recorded by Leonardo da Vinci offers a suggestive reflection on the experience:

> After having remained at the entry [of the grotto] some time, two contrary emotions arose in me: fear and desire—fear of the threatening dark grotto, desire to see whether there were any marvelous thing within it.[9]

The curiosity, or the desire to learn about the unknown, demonstrated by Leonardo could have been stimulated and sustained by a number of activities, but in the early modern period, it was particularly associated with exploration of the natural world. It was in nature that one might encounter the unfamiliar and be compelled to uncover its secrets through sustained observation and experimentation. It was also commonly held that reflection upon the natural world might reveal divine truths inaccessible elsewhere.[10] For those constructing artificial grottoes, like Furttenbach, curiosity crystallized around the exploration of how natural materials might be used in human edifices. For those visiting these spaces, like Leonardo, their curiosity would have been piqued by the privileged access required to enter them and speculation about what they might contain. Once visitors were inside, their curiosity would have been amplified as they encountered the extraordinary natural objects sourced from places still unfamiliar to many Europeans: marvels that might incite both fear and desire. In some grottoes, sustained contemplation of the shell might make individual visitors reflect upon God's power of creation and thus turn to deeper contemplation of their own eschatological journey. In other cases, shells brought pleasure and embodied the status and

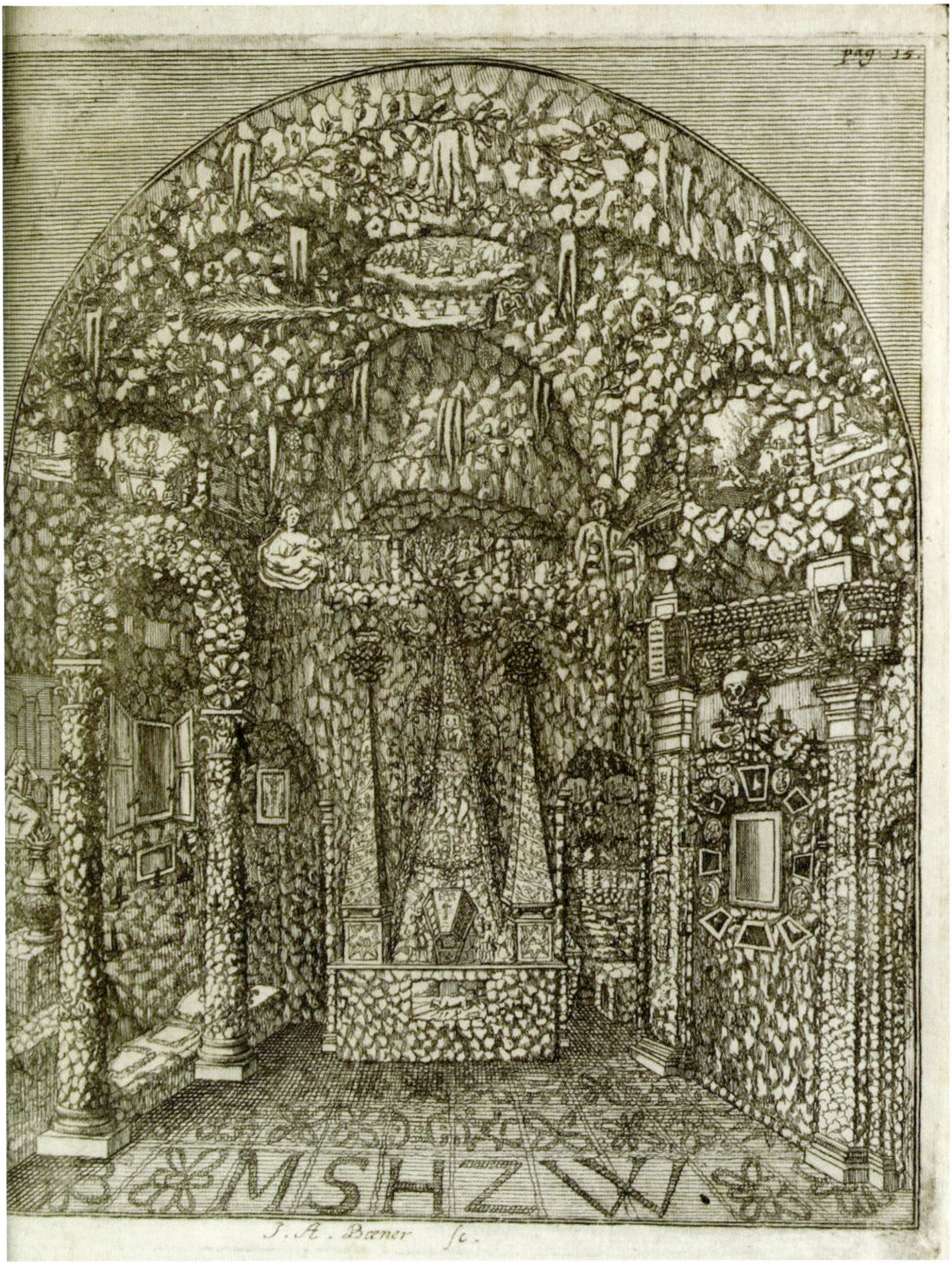


**FIG. 61.** Johann Alexander Böner, Obelisks in Magdalena Sibylla von Württemberg's grotto. In Ehrenreich Weismann, *Christliche Betrachtung der Betrübten Zeit und freudenvollen Ewigkeit*, 1680. Engraving. Sächsische Landesbibliothek–Staats- und Universitätsbibliothek, Dresden [Theol.ev.asc.609].

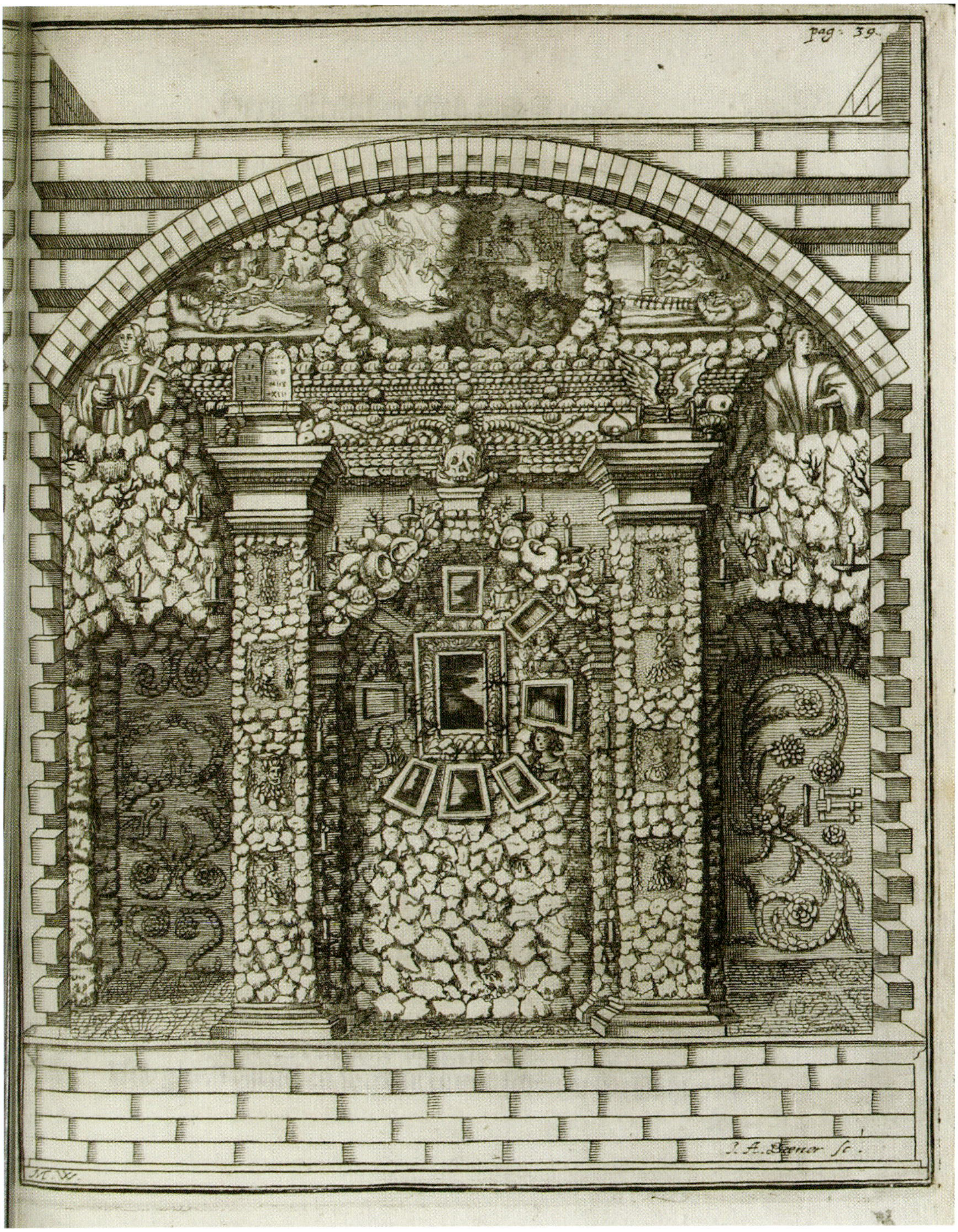


**FIG. 62.** Johann Alexander Böner, Interior of Magdalena Sibylla von Württemberg's grotto. In Ehrenreich Weismann, *Christliche Betrachtung der Betrübten Zeit und freudenvollen Ewigkeit,* 1680. Engraving. Sächsische Landesbibliothek–Staats- und Universitätsbibliothek, Dresden [Theol.ev.asc.609].

wealth of the grotto's commissioner. For Magdalena Sibylla, Johann Friedrich, and Joseph Furttenbach, shells were objects that triggered the imagination, but each individual's imagination might be transported in a different direction. Grottoes, as sites to build and as sites to inhabit, underline that shells were objects to think about, to think with, and to think within, not only as isolated specimens but especially when they appeared in multiples.[11]

## BUILDING GROTTOES IN EARLY MODERN GERMANY

In the ancient world natural grottoes at the sites of springs became associated with specific gods and evolved into places of religious devotion. The construction of artificial grottoes grew to accommodate the expanding religious cults of antiquity and their performances of ritual.[12] The revival of the grotto in sixteenth-century garden architecture was spurred in part by engagement with ancient texts that often discussed these sites, both real and imaginary. Ovid described the mythical grotto of the nymph Thetis as a cave of ambiguous making: "whether fashioned by art or nature is uncertain, but probably by art."[13] Ovid's lines capture the complexity and play of the Renaissance courtly grotto as an artificial space constructed using natural materials that have been manipulated in order to imitate a naturally occurring form. Certain early modern grottoes also retained the religious associations of ancient exemplars, acting as sites of spiritual withdrawal and hermitage. But the artificial grotto at the sixteenth- and seventeenth-century court was more frequently a site of socialization rather than seclusion.

The earliest examples in early modern Europe were found in Italy.[14] Travelers frequently commented on the beauty of the Grotta Grande in the Boboli gardens in Florence, designed by Bernardo Buontalenti between 1583 and 1593 for Francesco I de' Medici. Francesco's patronage was also responsible for the series of subterraneous grottoes at the Villa di Pratolino to the north of Florence, which drew international attention. Already in the sixteenth century these architectural forms had appeared north of the Alps as the courts of the Holy Roman Empire and elsewhere aspired to the artistic and architectural sophistication of their Italian counterparts. Duke Albrecht VII and his wife, Isabella, governors of the Spanish Netherlands, embarked upon a series of grottoes within their new garden complex at the Coudenberg Palace in Brussels at the end of the sixteenth century. The first Württemberg grotto had been inspired by Johann Friedrich's visit to the Coudenberg grottoes in 1606, as attested by a description of the Brussels grotto that remains among the plans for Johann Friedrich's grottoes in the Stuttgart archive.[15] Grottoes were common at German courts before the initiation of the Stuttgart project. Between 1581 and 1586, a grotto courtyard at the Wittelsbach residence in Munich was built by Friedrich Sustris at the instruction of Duke Wilhelm V of Bavaria and with the aid of Italian craftsmen.[16] The Hortus Palatinus in Heidelberg contained a series of grottoes carefully planned and built by French architect and engineer Salomon de Caus between 1614 and 1619 (fig. 63).[17]

Grotto construction continued at German courts throughout the seventeenth and eighteenth centuries. Two styles of grotto were popular—mannerist grottoes, which emulated the coarse and jagged qualities of a subaquatic cave, and those that adopted the oceanic materials of the grotto but presented them in a more elegant arrangement. For instance, at the Neuburg Palace in Bavaria, Philipp Wilhelm von der Pfalz included both a cave-like grotto (fig. 64) and a more refined reception room decorated with festoons of white shells (fig. 65) as part of his renovations in the 1660s. The grotto hall in the gardens of the Zwinger in Dresden was built between 1710 and 1714 as part of a larger complex to display mathematical and scientific objects, highlighting the link between the production of knowledge and the grotto. Frederick the Great's Neues Palais designed by Carl von Gontard in the second half of the seventeenth century echoed the Dresden example with a grotto hall that contained thousands of shells, semiprecious stones, and fossils (fig. 66). Noblewomen often had walls or rooms in their apartments decorated to mimic garden grottoes. The Electress of Bavaria Henriette Adelaide (1636–76) had one wall of her rooms decorated with tufa and shells in the Munich residence of the Bavarian electors. Like Magdalena Sibylla, the wife of Johann Georg II von Anhalt-Dessau, Henriëtte Catherine of Nassau (1637–1708), used a grotto in Oranienbaum to commemorate her husband.[18]

As in Italy, these garden constructions remained the preserve of the elite since the materials and expertise required to erect a grotto surpassed the resources of most. The ability to appoint a sought-after artisan like de Caus was a symbol of status for those who employed him. Elites must have been delighted to make the French engineer responsible for the grottoes and gardens at the courts in Brussels, Heidelberg, and Stuttgart.[19] De Caus published extensively on grotto interiors, but his writings focus primarily on automata and make little reference to shells and coral. One automaton, for instance, represented Galatea, a sea-nymph from Greek mythology sought by Polyphemus, riding a chariot constructed from shells in a pool of water, but the shells were ancillary rather than central to the overall design (fig. 67). De Caus briefly noted the possibility of using shells for decoration, stating that "when it comes to the ornament of said grotto, it can be made from various rocks, and rustic shells, or with frames of figures or grotesque decorations."[20] But his fleeting references to shells indicate that his scholarly interests lay firmly in the field of hydraulics rather than conchology. De Caus left little guidance for how grotto architects might use shells in their designs.

By contrast, De Caus's contemporary the German architect Joseph Furttenbach the Elder wrote in great detail about shells and grotto construction and provides an excellent guide for understanding the labor and expertise underlying the creation of chambers like those of Magdalena Sibylla and Johann Friedrich. Furttenbach was born in the free imperial city of Leutkirch in 1591 to an influential Protestant family in a confessionally mixed town. Although the architect wrote a lengthy autobiography, only the second volume detailing the final decades of his life has survived. We know that he traveled to Italy at the age of

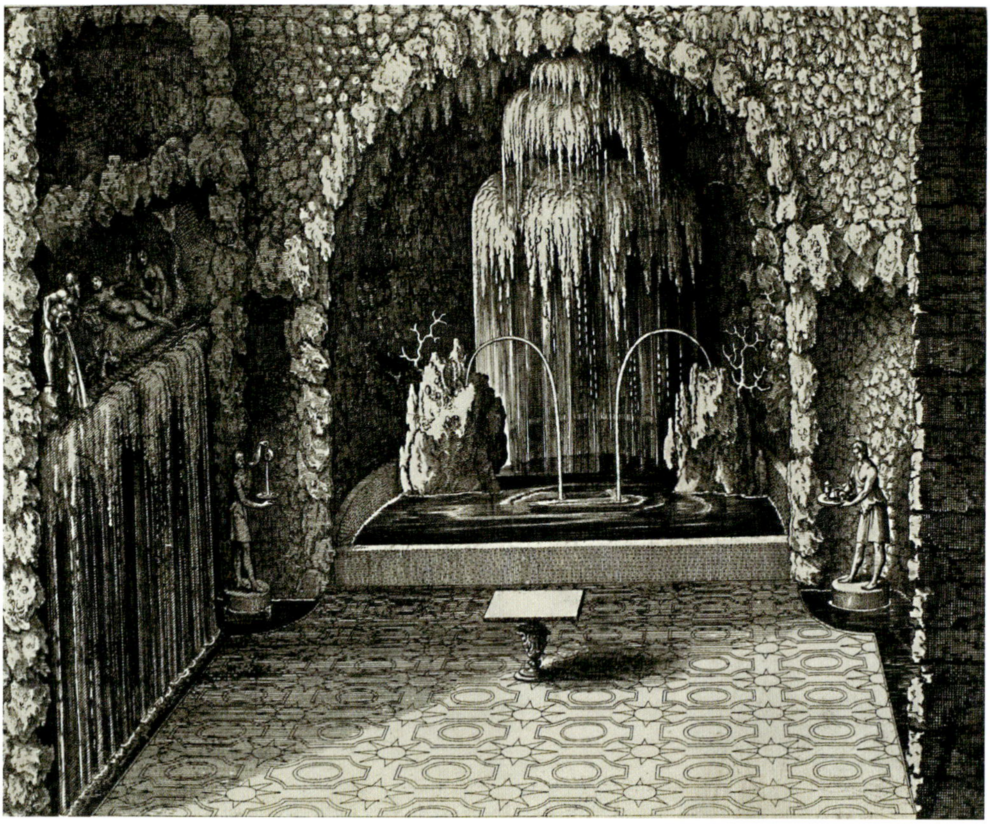

**FIG. 63.** Matthaus Merian, *Grotto in the Hortus Palatinus*. In Salomon de Caus, *Hortus Palatinus*, 1620. Engraving.

**FIG. 64.** The *Grottenhalle* in Schloß Neuburg an der Donau, 1667. Bayerische Verwaltung der staatlichen Schlösser, Gärten und Seen.

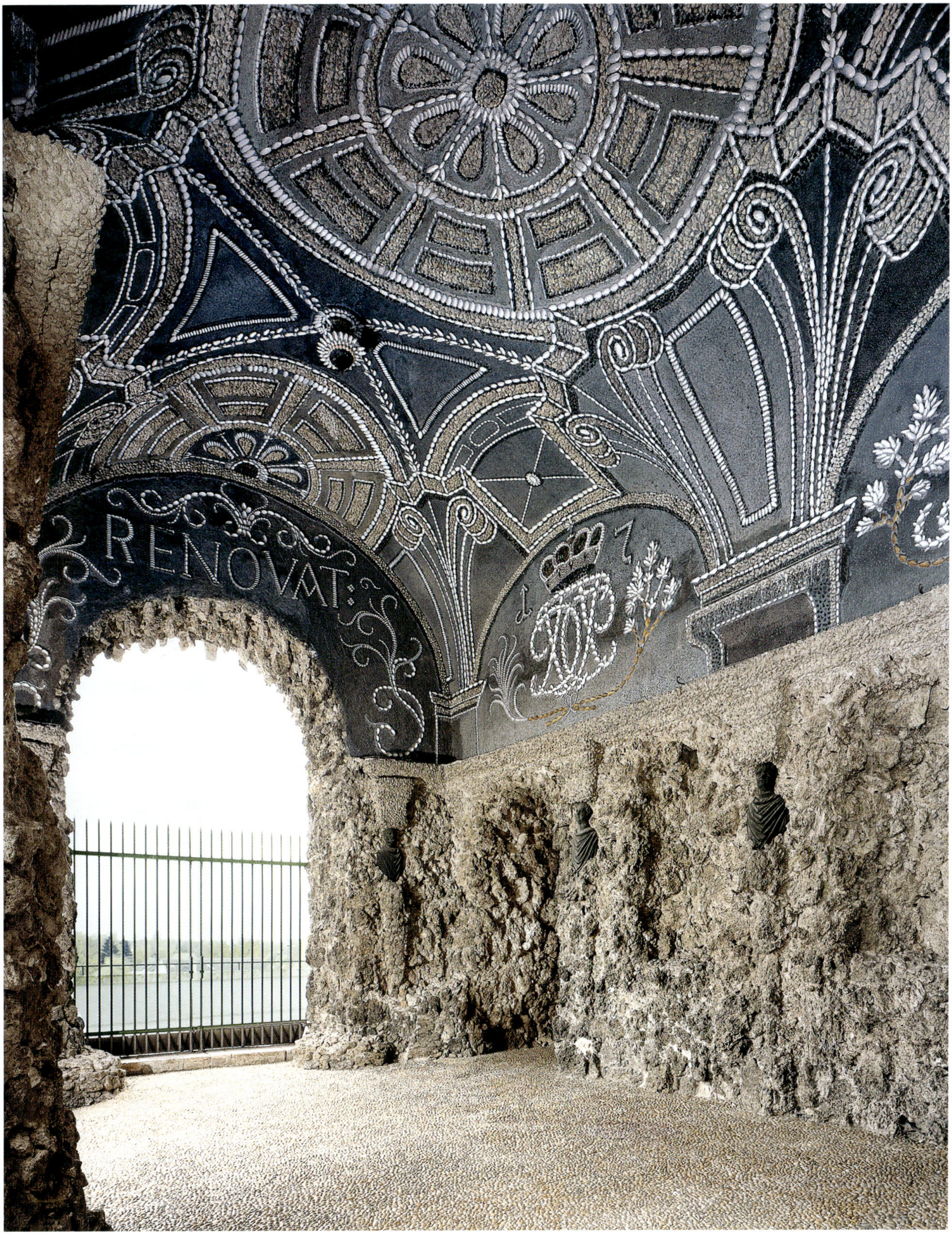
RENOVAT:

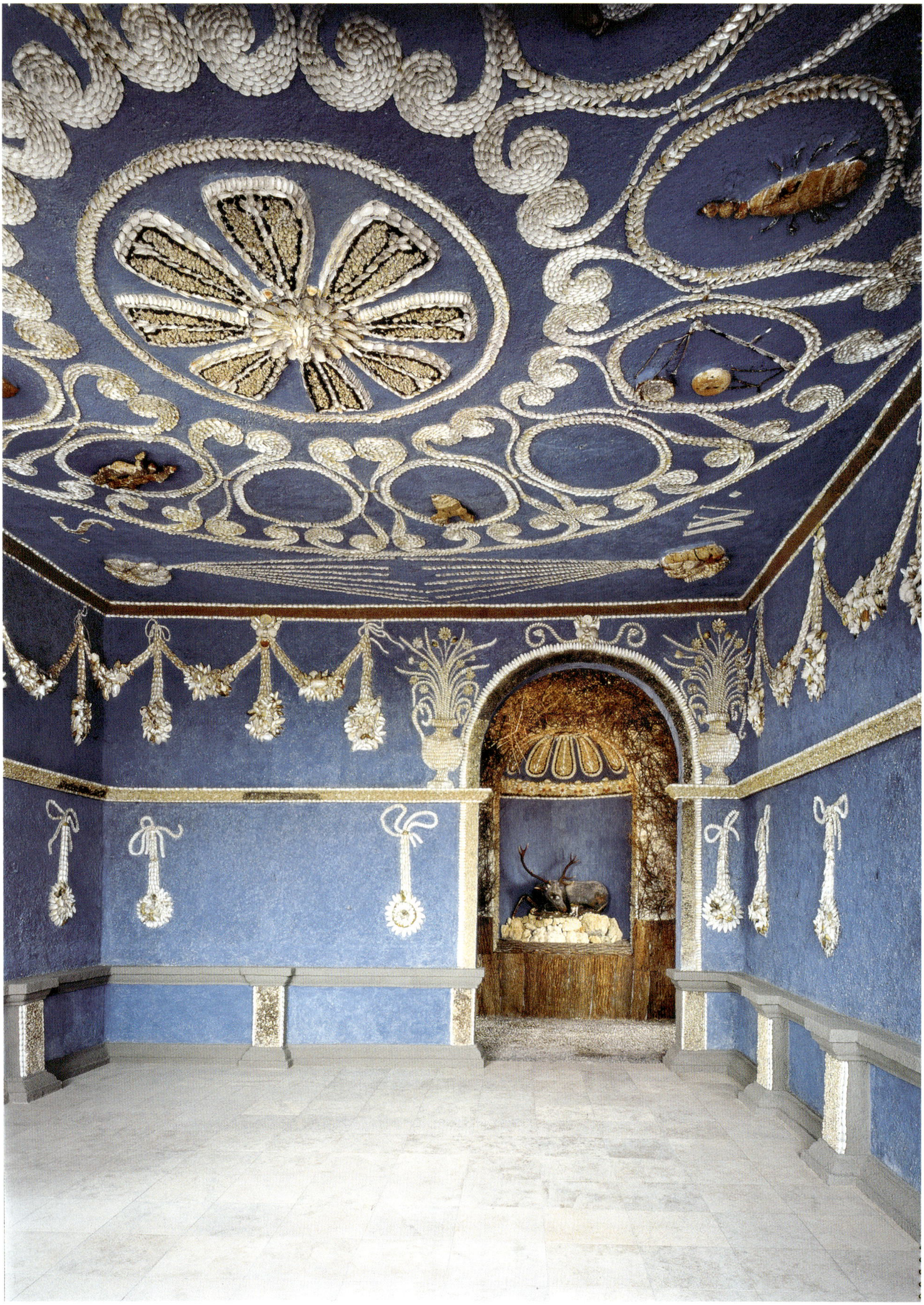

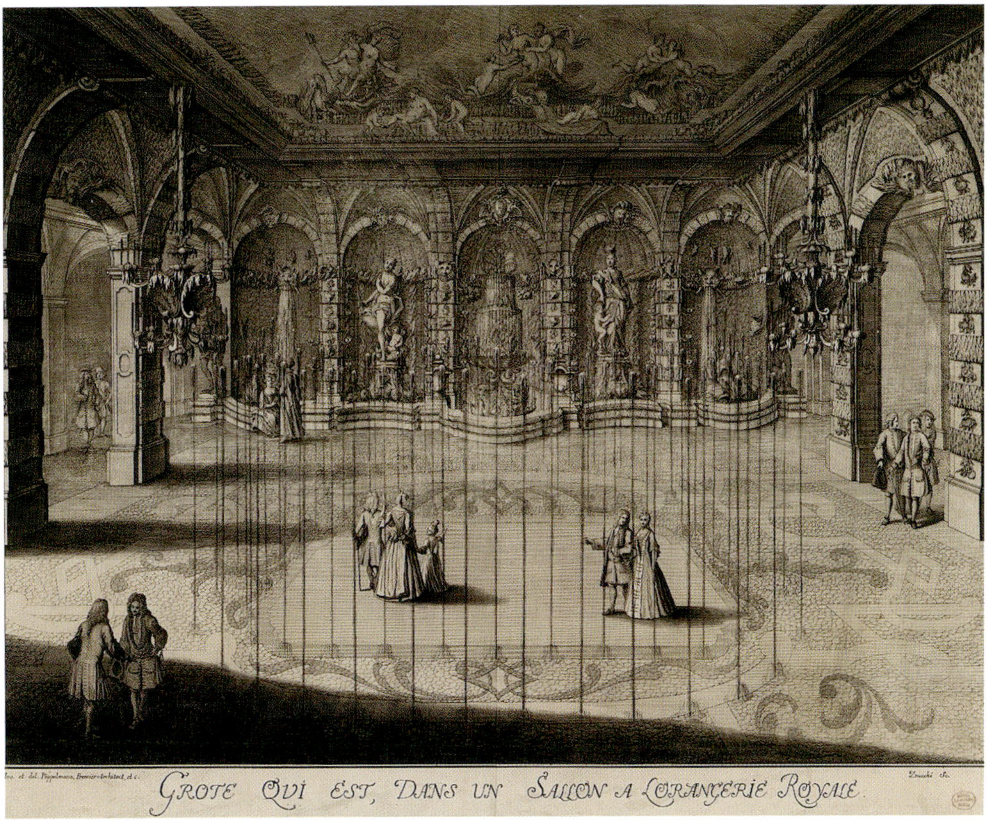


**FIG. 65.** The *Höhle des Cosmos* in Schloß Neuburg an der Donau, 1667. Bayerische Verwaltung der staatlichen Schlösser, Gärten und Seen.

**FIG. 66.** Lorenzo Zucchi after Matthäus Daniel Pöppelmann, *Grote Qui Est, dans un Sallon a L'Orangerie Royale*, 1721. Engraving, 58 × 44 cm. Sächsische Landesbibliothek - Staats- und Universitätsbibliothek Dresden [SLUB/KS B1986].

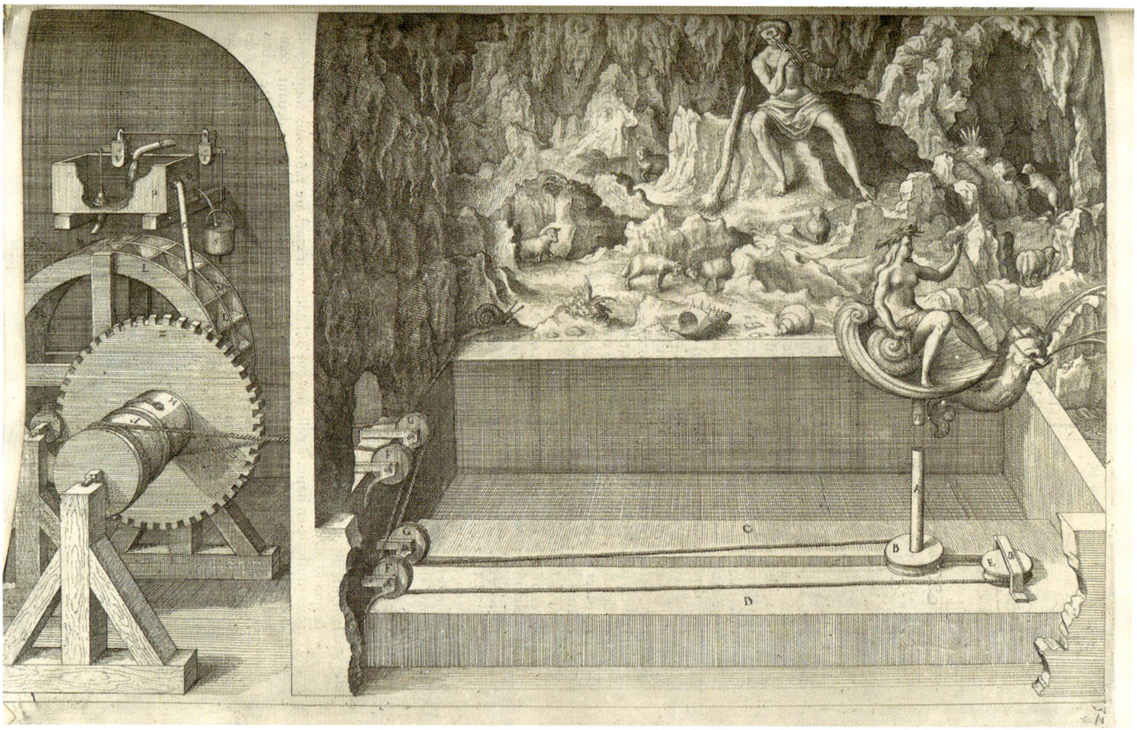

**FIG. 67.** Problem XXIIII: *A Machine for a Moveable Galatea in a Grotto (Plate 32).* In Salomon de Caus, *Les Raisons des Forces Mouvantes*, 1615. Engraving. Metropolitan Museum of Art, New York. The Elisha Whittelsey Collection, The Elisha Whittelsey Fund, 1949 [49.122(3.32)].

sixteen, where he trained himself so thoroughly in the theoretical and practical dimensions of architecture that when he returned to Germany in 1629, it was only two years before he was appointed chief municipal architect in Ulm, a free imperial city on the borders of the Duchy of Württemberg. In this role he was responsible for the upkeep of the city's municipal buildings and public gardens. In 1636 his status in Ulm was confirmed by his election to the city council. Alongside his civic service, Furttenbach sought to find broader audiences for his extensive architectural knowledge. Between 1627 and 1663, he published a number of architectural tracts whose topics ranged from shipbuilding to theater design.

Furttenbach demonstrates the attention that an architect might apply to decoration of a grotto. As with many of his peers, it was his tour of Italy that had fueled his interest in the grotto as a site for innovative architectural design. There he met the most skilled engineers and viewed the most famous grottoes of the day, describing them in a travel diary that he published in 1627. Other southern German artisans made similar trips, among them the Württemberg engineer Heinrich Schickhardt, who paid particular attention to the hydraulics in the gardens he visited. But it was the shells that often caught Furttenbach's particular attention. While many were amazed by the façade of the grotto in the Pratolino, Furttenbach snubbed the exterior, stating that it was not "particularly outstanding." Instead it was the splendor of the shells and corals within it that made the grotto one of the finest in Italy (fig. 68).[21] He noted that the Italian origins of the Renaissance grotto arose in part

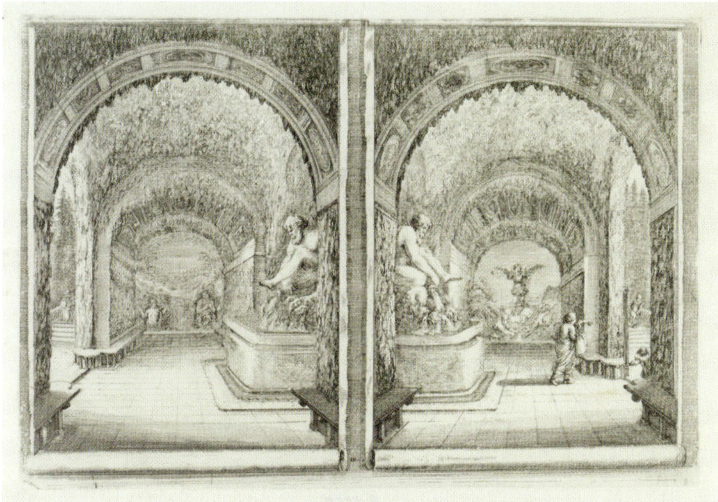

**FIG. 68.** Stefano della Bella, *View of a grotto in the grounds of the Villa Pratolini*, c.1653. Etching, 24.6 × 36.7 cm. Victoria & Albert Museum, London [E.1511–1898].

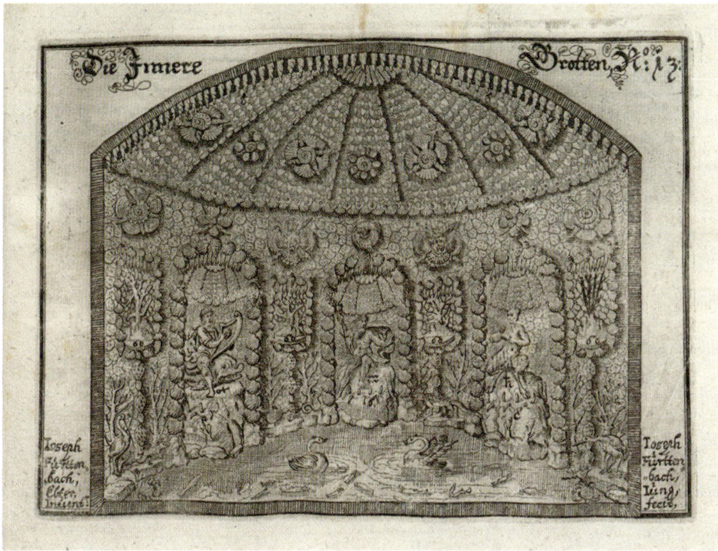

**FIG. 69.** Interior of Joseph Furttenbach's grotto in Leutkirch. In Joseph Furttenbach, *Newes Itinerarium Italiae*, 1627. Engraving. Bayerische Staatsbibliothek, Munich [12495381].

from the fact that the region possessed the "correct material" for constructing them, "not only the rough rocks, but also the diverse varieties of marvelous shells, mollusks, [and] corals."[22] Genoa had especially plentiful grottoes, since the coastal town had singular access to coral.[23] His knowledge here was a product of practical experience. In the Ligurian town he had learned the art of grotto construction alongside the engineer Paolo Rizio for seven years.[24] In Germany Furttenbach built at least two grottoes, one at his family home in Leutkirch (fig. 69) and another at his residence in Ulm. The former demonstrated all the conchological skills that he outlined in his manuals: shells tightly packed together to form the grotto's canopy; shells arranged to create flowers and grotesque faces; and shells used to frame statues of mythological characters. Furttenbach's reputation as a grotto architect may not have been as widespread as that of Salomon de Caus, but his skill caught the attention of the Margrave of Baden-Durlach Friedrich V (1594–1659), who invited him to build a grotto in Durlach. Furttenbach had a steady stream of guests to his home in Ulm from as far afield as China and America.[25] They visited his *Kunstkammer*, a cabinet of curiosities common in early modern Europe, designed to display objects, natural and human-made, but also to showcase the range of a natural material from its rawest to its most artificed form. After viewing the *Kunstkammer*, guests would visit Furttenbach's grotto. While the former displayed thousands of shells in drawers, the latter presented shells at work, showcasing their effects when exposed to the natural environment of the architect's garden.[26]

Furttenbach's writing on grottoes is dispersed across a number of his works. His travel diary of his years in Italy mentions the famous examples he visited. His treatises *Civil Architecture* (*Architectura civilis*) and *Domestic Architecture* (*Architectura privata*), published in 1628 and 1641, respectively, also contained instructions for building grottoes that are especially relevant to this essay.[27] While his discussion of these spaces makes up only a small fraction of his published output, it is noteworthy in its approach. For Furttenbach, shells were not a secondary consideration; they were the very sites where the meaning of a grotto was generated. He often began the building process by thinking about shells. Rather than taking the edifice's foundations, walls, and roof as the point of departure, in *Civil Architecture* and *Domestic Architecture* Furttenbach leaped to advise the reader on the preparation of shells for the space.[28] The architect wrote that it was not enough simply to create the structure of a grotto, but that the person charged with the project must have the knowledge and experience to adorn it effectively with objects from the sea.[29] Shells, given their material and sensory qualities, required significant thought and attention; as a consequence, they had to be considered in grotto design from the outset. Furttenbach's architectural manuals help us understand the decisions that artisans had to make when decorating their grottoes with shells, including those involved in the Württemberg projects.

## FROM "ROUGH AND VERY UNSIGHTLY WARES"

In Stuttgart, the engineer Gerhard Philippi was responsible for the day-to-day construction of Duke Johann Friedrich's grotto. In 1615/16 Philippi ordered 1,150 kg of shells from Venice for the enterprise.[30] Shells were one of the few materials that had to be imported, since Philippi had received privileged access to Württemberg mines for the sourcing of other decorative material such as marble, jasper, and black amber.[31] Obtaining shells required connections with conchological experts in maritime cities and with the merchants who could best negotiate with the sailors who brought shells to port.[32] The sourcing of Württemberg's shells from Venice demonstrates Philippi's preoccupation with quality, since Furttenbach noted that the best shells were to be found there.[33] In 1617 Philippi himself made trips to Italy and France in order to procure shells, blue marble, and white coral for the Stuttgart grottoes.[34] This process of procurement was beneficial for him in other ways, allowing him to visit Italian grottoes and make contact with craftsmen, such as the artisan responsible for the Pratolino grottoes Giovanni da Bologna. The engraver and goldsmith Esaias van Hulsen also frequently provided shells for the Stuttgart project. In 1614/15 the ducal coffers paid him forty-two gulden for shells and other material. He was subsequently paid thirty-seven gulden in 1615/16 for "two jars full of shells and plants" that he had sourced from Brussels and Amsterdam and a further eighty-seven gulden for "shells and other things" bought for the grottoes.[35] This method of collecting shells was a common practice. Duke Wilhelm V of Bavaria also sourced shells from Venice, Naples, and Amsterdam.[36] It appears that by the mid-eighteenth century, shells were also sourced locally by the Württemberg dukes. During repairs to the ducal grottoes in 1743, officials observed that shells might be found in lakes and rivers near Ludwigsburg, Lauffen am Neckar, and Cannstadt.[37] Perhaps this was an indication of the declining prestige of Italian shells or of the need to minimize expenditure.

While the types of shells and the decoration used in Johann Friedrich's seventeenth-century grotto remain unknown, an inventory survives from the mid-eighteenth century, listing those conchological items bought for repairs to the existing grottoes in the ducal gardens and for the "creation of shell-figures and all kinds of adornments with shells and snails" within the space.[38] The four-page catalogue from 1743 lists at least nineteen different types of shells, frequently identified by both their technical and their colloquial names. These included the *Strombus*, which is also described here as the "screw shell," and the *Pinnidae* or "pen shell." No further subspecies were given. The catalogue also listed the size and color of the shells, since these were the qualities that most concerned those artisans charged with using them in the grotto's decoration. The use of both Latin and colloquial names for each shell suggests the differing levels of expertise among those employed on the project—on the one hand, individuals familiar with developing scientific taxonomies and, on the other, those without knowledge of Latin nomenclature. Writing around the same time as this catalogue was composed, the German theologian and conchologist Friedrich Christian Lesser maintained that colloquial labels were used only by the "inexperienced."[39] Yet according to the Württemberg catalogue, those involved in grotto construction used both terms interchangeably. Any grotto architect sourcing shells, Furttenbach noted, needed to be able to judge whether the "rough [and] very unsightly wares (which many common people would scarcely want to pick up from the ground)" might be transformed into an object of necessary beauty—a skill that derived not just from books, but from the experience of using oceanic material.[40]

Once the shells arrived in Stuttgart, and before they were used in the grotto, they had first to be returned to their "natural colors." Preprepared shells could be bought but cost three times the price of those direct from the ocean.[41] Furttenbach provided directions for their preparation. First those that were already open were to be covered with wax on

FIG. 70. Johann Walther, *Interior of Count Johannes of Nassau's grotto at Idstein*, c. 1650–1670. Watercolor on paper. Bibliothèque nationale de France, Paris.

the inside so that they were protected from the corrosive effects of the cleaning process. Furttenbach's recipe for the necessary corrosive fluid required human urine, quicklime, and two handfuls of salt. The shells were left for six to eight days in this solution before being cleaned in fresh water and returned to soak in nitric acid. Here each shell was to remain until its "unclean skin [had come] away."[42] The shells were now ready to be polished. This latter process required chalk and "Venetian soap" in equal measure, both of which were mixed with water and then used with a white cloth to burnish the shells until their true luster had been revealed.

Once Philippi and Van Hulsen's shells were prepared, they could find their place in the grotto. They might be placed individually within the space or broken up and embedded within the walls. The shells would have been brightly colored rather than the muted tones seen now in surviving early modern grottoes. Johann Walther of Strasburg's depiction of the grotto at Idstein in Germany, belonging to Count Johannes of Nassau, demonstrates just how colorful the experience of entering a grotto could be (fig. 70). So bright were the colors of shells in a grotto in Utrecht that Thomas Goldney II visited during a tour of Europe in 1725 that he first thought its ceiling designs were painted.[43] However, Furttenbach hinted at the limitations involved in sourcing certain types of shells; he advised that not everybody might be able to find mother-of-pearl shells, which were prized for their reflective surface, but architects could seek out other variations that might mimic the translucence of the favored material. While availability determined which shells were used, the physical properties of the objects also dictated how and where in a grotto they were employed. In Stuttgart, Philippi and de Caus disagreed over the appropriate use of shells. Philippi criticized de Caus's plan to decorate the walls of the entrance gallery with shells, since, he argued, they were not weather resistant. The Stuttgart engineer preferred to use tufa, a rock formed from volcanic ash, believing that it better withstood the depredations of time.[44] Perhaps this was a moment of professional competition rather than a true reflection of the material of the shell, given that elsewhere shells were regarded as being remarkably durable.

Once the mollusks had formed the backdrop, the artisan could move on to the finer decoration, which for Furttenbach consisted of creating flower patterns from a range of specimens. In both his manuals and his Italian diary, Furttenbach provided engravings of his planned shell-flowers (fig. 71). The letters marked on the engravings show that these were first and foremost practical guides, since they allowed readers to clearly identify the objects required for re-creating Furttenbach's designs. These shell constructions were elaborate, to say the least. One example required thirty shells to make a single flower (fig. 72). While the shapes of the sculptures sought to imitate the appearance of real flowers, their colors—here red, yellow, blue, brown, and white—presented an array of colors not found in botanical specimens. Duke Johann Friedrich's grotto may have included shells likewise sculpted into representations of other natural forms. His engineer's estate contained images of "a pair of wild life-size figures [made from] shells," and Esaias van Hulsen's papers contained

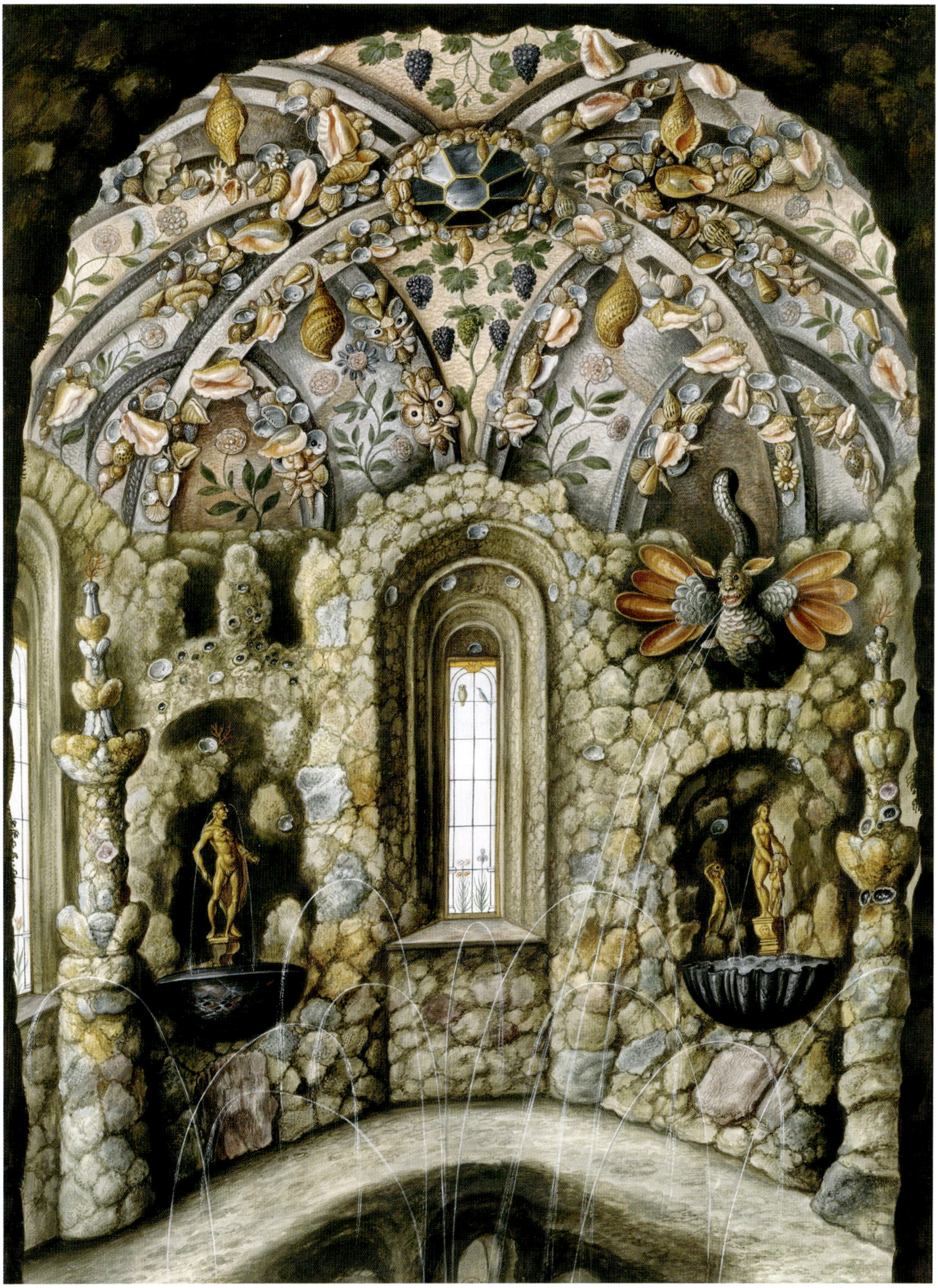

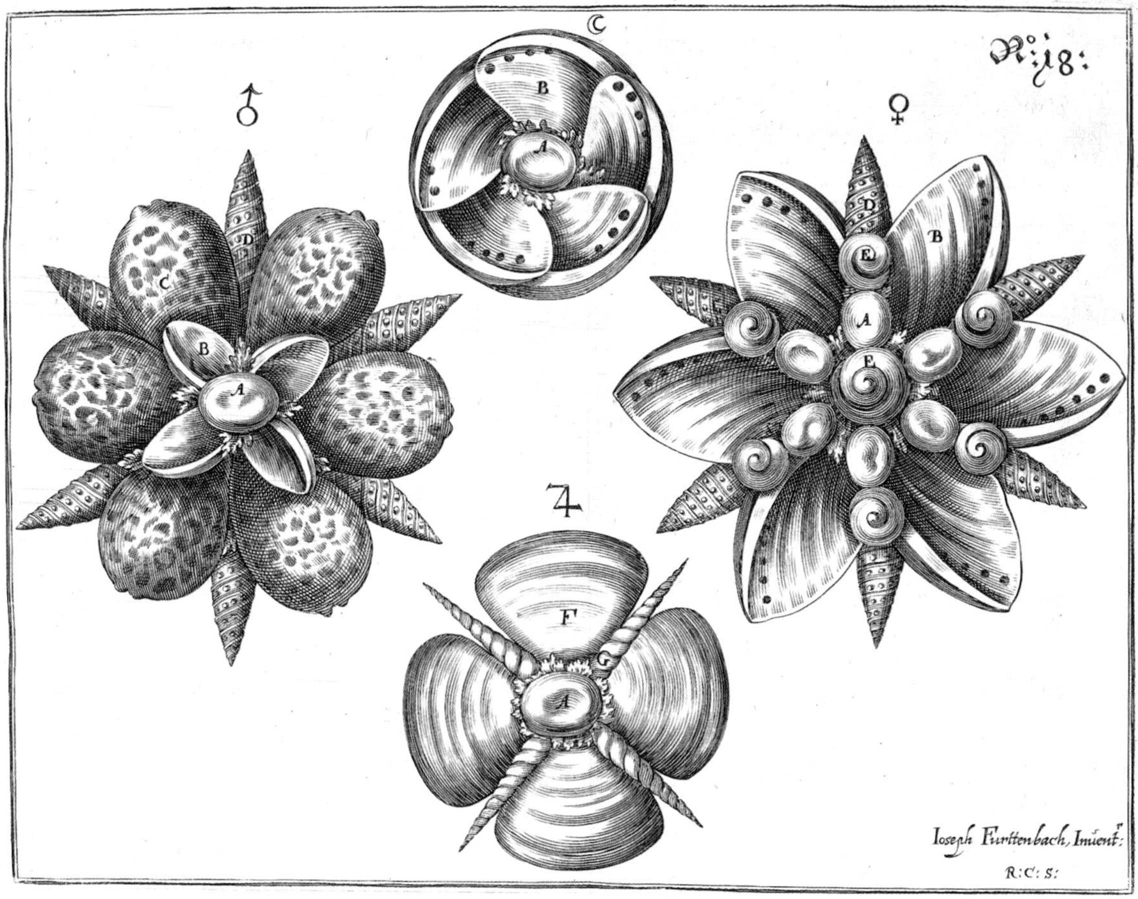


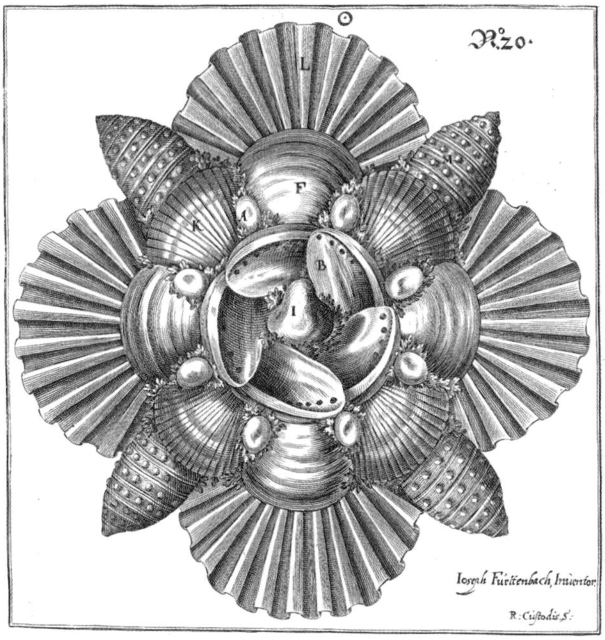


**FIG. 71.** Selection of flowers made from shells. In Joseph Furttenbach, *Architectura civilis,* 1628. Engraving. Universitätsbibliothek, Heidelberg [T 2269 RES].

**FIG. 72.** Flower made from shells. In Joseph Furttenbach, *Architectura civilis*, 1628. Engraving. Universitätsbibliothek, Heidelberg [T 2269 RES].

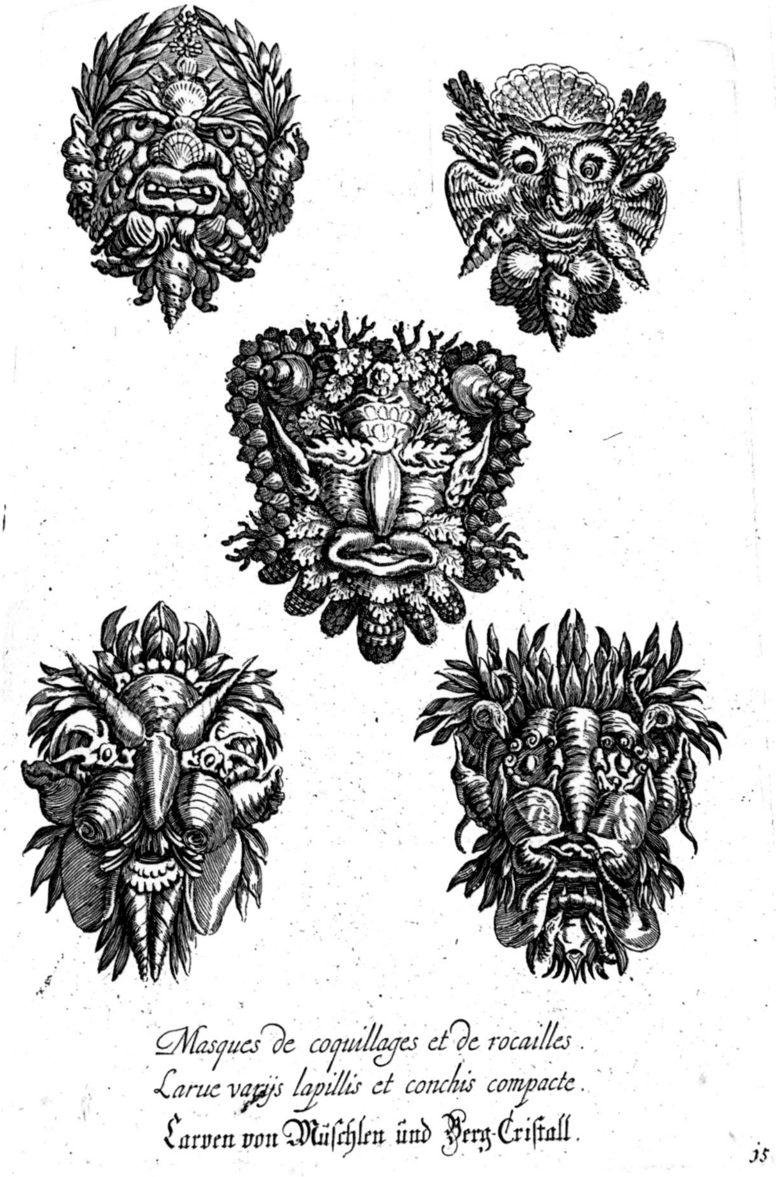


**FIG. 73.** Masques de coquillages et de rocailles. In Johann Ulrich Kraus, *Description de la Grotte de Versailles oder: Beschreibung der Grotten zu Versailles*, c. 1700. Engraving. Bayerisches Staatsbibliotheek, Munich.

plans for "21 grotesque faces [made] of shells," as well as plans for shell sculptures depicting birds, other animals, artichokes, and pinecones.[45] Perhaps these were designs for Johann Friedrich's grotto. Magdalena Sibylla's grotto from the 1670s included shell flowers atop the columns on either side of her bed as well as busts adorned with shells (see figs. 60, 62). Since neither of the grottoes in Stuttgart has survived, engravings from another contemporary grotto can provide an impression of the intended appearance. For example, in the grotto of the seventeenth-century palace of Versailles outside Paris, grotesque faces composed from shells appeared throughout the space (fig. 73).

How were these fantastical composite flowers and masks to be viewed? Visitors might zoom in to look at their detail, contemplating and studying the features of the individual shells that made up the arrangement, comparing them to specimens they had encountered elsewhere either in *Kunstkammern* or in conchological catalogues. Other guests might zoom out and consider the overall effect of the composition, struck by the theatricality of the object and the ability of natural material to create such bizarre and astonishing figures.[46] Visitors, led by their own interests and motivations, might alternate between these two modes of viewing, with each reinforcing the other. The theatricality of the composition overall might encourage viewers to contemplate the object's component parts. Similarly, close observation of individual shells might compel viewers to step back and compare the characteristics of individual shells to others within the composition.

## A "MIRROR AND SIGN OF GOD'S BOUNDLESS POWER"

Beyond their use as construction materials, shells also had sensory qualities and cultural associations that might take their viewers in multiple contemplative directions. Visitors to grottoes might reflect upon the shell as a curiosity in ways that moved beyond the intentions of the patron and architect. Responses to shells within courtly grottoes, in other words, were not always circumscribed by the immediate context in which they appeared. Early modern grottoes had a great variety of associations, some inherited from their classical ancestors and others particular to the ceremonial practices of individual courts. Shells might support a grotto's purpose as a space of courtly celebration, amusement, and noble representation, but they could equally support the grotto's use and appeal as a site of curiosity, observation, and spiritual contemplation.

Furttenbach wrote that the grotto was intended as a space where one "ate, played pleasant games, and enjoyed amusement."[47] In practice, these areas often were the sites of court celebration that emphasized the notion of abundance, both the richness of the natural world and the wealth of the courtly one. For instance, Duke Johann Friedrich's grotto had been a part of the celebrations for the duke's son's baptism in 1616, while another provided the backdrop for the wedding celebration of the Duke de Joyeuse and Marguerite de Vaudemont at the court of Henry III of France in 1581.[48] The Farnese Palace in Rome had a dining room with grotto decoration on one wall, while the Pratolino's grottoes contained a chamber especially for dining. In these contexts, shell-encrusted spaces became sites that represented the abundance of the natural world, a place where the elements of the earth combined to provide the necessities for human life.[49]

Although writing before the development of comprehensive conchological taxonomies, Furttenbach recognized that it would take multiple volumes to list all the shells known to humankind. Indeed, contemporaries believed that the never-ending variety of shells demonstrated the bounty of nature. An educated visitor to Johann Friedrich's grotto may have been prompted by the sculptures of shell flowers to consider the parallels between botanical and conchological specimens. Lesser referred to shells as "the flowers of the sea," following the classical natural historian Pliny the Elder, who had drawn parallels between the great variety of both flowers and shells.[50] The shell flowers found in grottoes could be seen as more than amusing constructions: as a conscious statement on the richness of the natural world.

The representation of abundance was an important political tool for rulers in establishing the nature of their authority. In the case of the grotto, the natural materials within it attested to the resources at a ruler's disposal. The grotto courtyard at the Wittelsbach court in Munich has been interpreted not only as an architectural triumph, but also as a space where Duke Wilhelm V of Bavaria might demonstrate his command of both the natural and the human realms. The natural materials in the space, including the shells, revealed the duke's wide-ranging interests, material collections, and knowledge.[51] A noble's authority over the natural world was made clear through a personal assemblage of shells, since they were often sourced from beyond the borders of the ruler's territory and cost significant sums. The acquisition of shells not only demonstrated rulers' wealth but also manifested their connections to exclusive networks of trade.

However, the noble status of rulers was not enhanced simply by their owning distinctive natural material sourced from elsewhere, but also by how they used that raw material once it was in their possession. Shells allowed their owners to showcase their own knowledge via their connections to innovative and ingenious craftsmen, who possessed the expertise to manipulate and shape natural materials into visual spectacles. A description of the grottoes in the gardens of Versailles by German engraver Johann Ulrich Kraus noted that what was impressive was not the cost of the material or the labor that had been required to construct the space, but the ingenuity demonstrated by the artisans when creating decorative images from shells.[52] The portraits or composite faces composed from shells, so Kraus argued, were more impressive than anything plated with gold, since they transformed these small objects into artistic inventions. While nature provided the material, it was their ingenious artifice that made them marvels.

Ingenuity and curiosity went hand in hand, especially when it came to the experience of something innovative or unfamiliar that confounded expectations.[53] As Leonardo's note on his experience of visiting a grotto already suggested, the inaccessibility of these spaces piqued the visitor's interest. Access to Magdalena Sibylla's grotto was tightly regulated, with three locks on the entrance to the space. Furttenbach controlled limited access to his grotto in Ulm so that strangers were barred from entering and possibly damaging the shell decoration within it.[54] Even once a visitor was inside, the interior decoration of the grotto was not always visible. The distribution of light within the space was key to triggering a desire to explore the space and its contents more fully. Within the imagined grotto depicted in Furttenbach's treatise *Civil Architecture*, light was intentionally limited. As Furttenbach wrote, "a grotto should be somewhat gloomy, so that it resembles a wilderness." Windows were closed off with grilles, with light entering only through a number of small holes. This restricted light also intensified the visual impact of the shells within the space. Their reflective surfaces became all the more impressive as rays intermittently danced across them. Their mystery, as objects only partially seen, stimulated curiosity.

Furttenbach saw the visual restrictions in place as encouragement to the visitor to inspect the shells "with greater fervor and deeper intent."[55] In Magdalena Sibylla's grotto, the windows became mirrors when closed, restricting light within the space. The lack of illumination compelled the viewer to study the shells and other decoration that surrounded them with greater effort and focus. But darkness was not always considered advisable in noble grottoes, as other sources also tell us. During the design of Johann Friedrich's grotto, a dispute arose as to whether Philippi's or de Caus's architectural designs were best for the project. Each engineer accused the other of designing spaces that were too dark.[56] In the architects' opinion, darkness detracted from the functionality of the space and meant that the impact of the interior decoration might be lost. In a space that was primarily for courtly celebration, rather than individual contemplation, illumination had its advantages.

Yet according to Furttenbach, shells were inherently bewildering, regardless of how they were displayed. The shell had a particular capacity to take viewers' minds away from the everyday and to confound them in innovative ways. The symmetry of shells, the repetition of their forms across multiple specimens within a grotto space, could cause a sense of discombobulation; according to Furttenbach, "the viewer is somewhat perplexed and consternated over [the shell patterns], therefore they barely know where they begin, much less are they able to figure out the end."[57] The shell, in other words, had the ability to make a familiar space appear unfamiliar. We might imagine this experience on Magdalena Sibylla's part. She would have passed through the more traditionally appointed rooms at the Württemberg court into an entirely different realm—a kind of wilderness—decorated almost exclusively with rough stone, coral, and shells. With only a few steps, she was transported from the ceremonial rooms of the Stuttgart court into another world beyond the regulations of etiquette and protocol—a realm where she might legitimately be alone. This sense of separation from the everyday was particularly important in grottoes designed for relaxation and seclusion, allowing one's mind to turn from mundane preoccupations. By "feel[ing] so great a pleasure" in their distance from their usual world and environment, visitors to the grotto, Furttenbach believed, would forget themselves and where they were. In particular, the reflective surfaces of the shells presented visitors with multiple reflections of themselves and their surroundings.

No less significant than their ability to stimulate curiosity about the natural world was the way that contemplation of shells prompted reflection on the heavenly realm. Lesser wrote that the more he observed shells, "the more [his] eyes of my soul were opened to the larger theater of new imprints of divine wisdom and omnipotence."[58] Although written in the following century, Lesser's statement helps us understand the grotto of Magdalena Sibylla. This was not the place for amusement that Johann Friedrich's had been, but a grotto to memorialize her husband's death and contemplate her own. The material emphasis of the space on its own interiority made it a place for solitude and withdrawal. Magdalena Sibylla followed a tradition of the grotto as a site to deal with collective loss. The first room of the Grotta Grande in Florence has been viewed as representing the destructive floods of the river Arno, turning the grotto into an environment in which the community might confront and memorialize environmental devastation.[59] But grottoes could speak to more personal loss as well. Magdalena Sibylla's grotto was particularly distinctive in the way it tied her bereavement to the contemplation of Christ's sacrifice on the cross, a recognition of which was required for salvation. The model of her husband's coffin as the heart of the space signaled that the shells here were to prompt a different form of reflection and centered the space on the duchess' individual grief. The juxtaposition of the shells with traditional memento mori, such as the skull atop a festoon of shells, reinforced the material similarities between the objects; both the shell and the skull were empty vessels, once animated, but now relics of previous lives (see fig. 62).[60] In this context the shell was transformed into a reminder of death.

As a seventeenth-century Lutheran, Magdalena Sibylla believed that the messages of scripture might be read in the landscape. The natural world was viewed as a microcosm of the heavenly realm and a space in which the mysteries of the divine might be revealed, "a mirror and sign of God's boundless power."[61] Grottoes in the Bible had sat in a liminal space between the earthly and divine realms, where contact might be made between the two. Jesus had been born in a grotto in Bethlehem, subsequently known as the Grotto of the Nativity, and had also ascended into heaven from a grotto.[62] The shell, too, could sit in this liminal space. It was a material object, altered by the cycles of the natural world, but it also might reveal divine truths and act as an interface between earth and heaven. Furttenbach noted that within the shell one might sense God's omnipotence.[63] The very existence of the shell appeared to some to be miraculous. Although scholarly opinion began to diverge from this view in the sixteenth century, the belief that these objects were produced ex nihilo persisted into the early modern period.[64] While the shells in Johann

FIG. 74. Shell goblet with dragon foot, mid-seventeenth century. Shells and gilded silver, 26 × 19 × 14 cm. Landesmuseum Württemberg, Stuttgart [KK hellblau 13].

Friedrich's grotto were celebrated for the ways in which they demonstrated human ingenuity, Magdalena Sibylla may well have seen in her shells the hand of God. Human ingenuity might be able to create new forms from shells, but it was unable to re-create the shell itself. The inability of the human hand to replicate its form transformed the shell into a symbol of God's supremacy. Friedrich Christian Lesser wrote that "neither the greatest monarch" nor "the most skillful artist" could replicate all the features of the shell.[65] Perhaps this is why there were no automata in Magdalena Sibylla's grotto: this was not a space to showcase the ways humans might play with nature, but instead one that explored the power of nature and the inevitability of decay. The notion of irreplicability, that the human hand could not produce a shell, was supported by the sheer diversity of mollusks. No mortal could devise such great variety. This could be supported only by divine inspiration.

## "PRODUCE OF THE SEA SO MASTERLY AND MAGNIFICENTLY MADE"

When Furttenbach wrote that shells were "produce of the sea so masterly and magnificently made," he spoke of the forces of nature and the powers of divine artifice.[66] But his own manuals for grotto construction spoke also to the ways in which a shell's ultimate magnificence could emerge only through its expert use by a skilled artisan. Within the context of the early modern grotto, both ways of understanding the shell's embodiment of mastery and magnificence were equally important. The shell was at once an object that attested to the power of divine creation and one that offered an opportunity to showcase the ingenuity of humankind.

This was as true in the case of grottoes as in other early modern works such as shell vessels. A seventeenth-century shell cup from the Württemberg *Kunstkammer*, for instance, juxtaposes the raw material of the spider conch—here the cup's base and the body of a dragon—with the turban shell at the vessel's summit, which has been polished to reveal its inner layer of mother-of-pearl (fig. 74).[67] Within grottoes, unpolished shells used to imitate the rough surfaces of a cave likewise appeared next to those that were zealously burnished and arranged into elaborate shell structures. The shell portraits and flowers could reveal the ways that natural products and human artisanal practice combined to create something beyond nature's imagining.[68] But the enduring form of the shell might also stand as a symbol of its ability to withstand the passing of time in a way that humankind could not.

In many ways, the grotto itself functioned as a sort of *Kunstkammer*. The grotto, like the *Kunstkammer*, revealed the interaction between the natural world and the technical proficiencies of humankind.[69] Both Furttenbach and the dukes of Württemberg kept shells in their cabinets that were eventually intended for their grottoes.[70] Within both

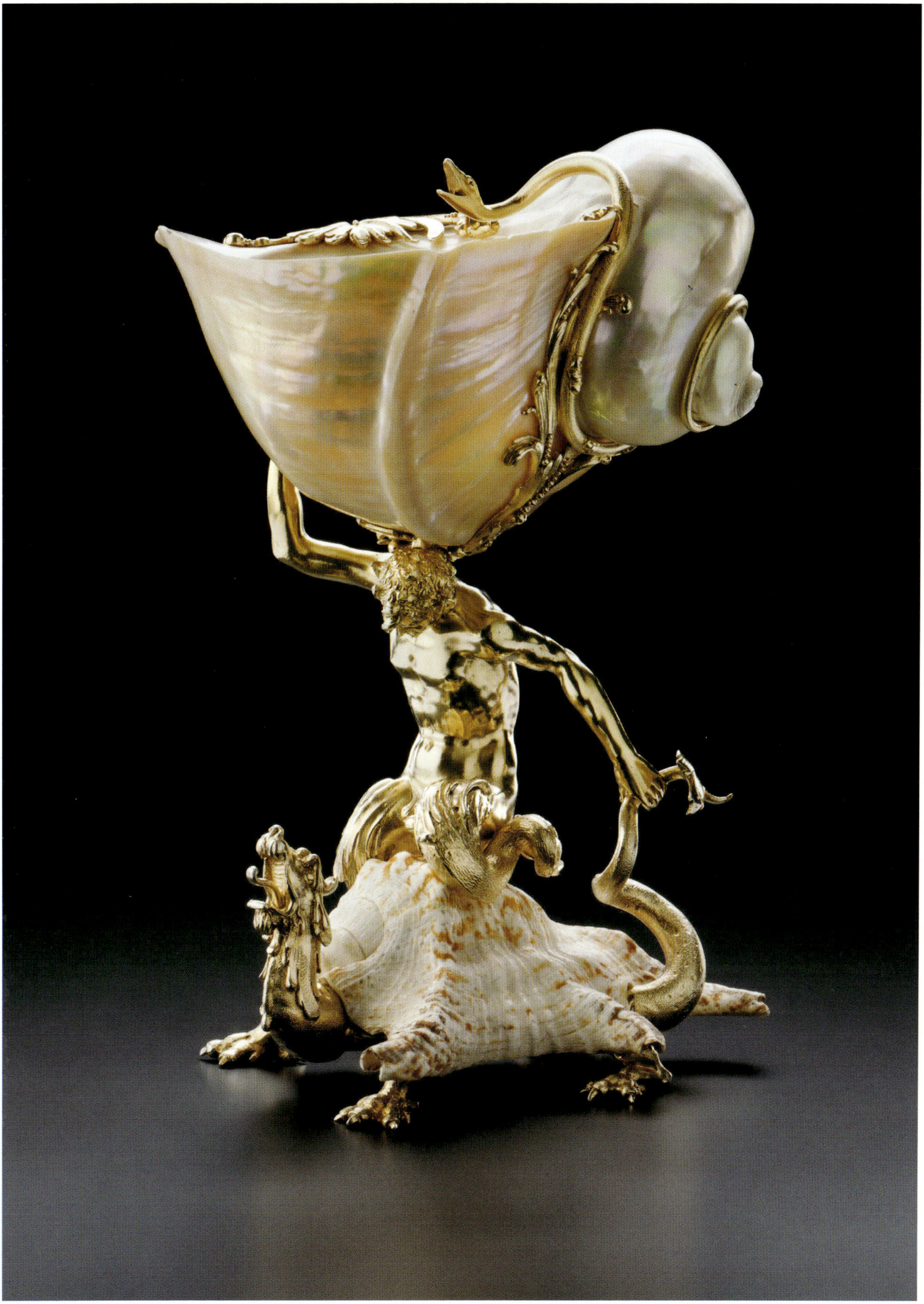

settings, shells were also presented in similar ways. The catalogue of Dutch pharmacist Albertus Seba's *Kunstkammer* depicted his shell collection in the same festoons and faces of shells that we have already seen in the grottoes discussed above. The link between the space of the grotto and that of the collector's cabinet was often achieved through proximity as well as content. Johann Friedrich, for instance, located his shell-adorned grotto in the same complex as his *Kunstkammer*.

However, this comparison can be taken only so far: stepping into a grotto was an immersive experience unlike that of viewing shells in the cabinets and drawers of a *Kunstkammer*. While viewing natural objects in a collection meant seeing them out of their original context, and in a space protected from the elements, visitors to a garden grotto entered an artificial realm designed both to imitate the patterns and processes of nature and to generate thought, whether fleeting or sustained. In many ways the shell can be viewed as a sort of emblem. The emblem had been a popular form of visual communication since Andrea Alciato's emblem book published in 1531. The tradition was diverse and strict definitions of its constituent parts have aroused criticism, but at its most fundamental the emblem was characterized as a literary or visual representation of a moral truth, whose meaning was intended on first sight to be opaque, requiring viewers to reflect upon it and unravel its meaning using their own knowledge of the symbolic associations embedded within the image.[71] These symbolic associations were as diverse as the emblematic forms that contained them, so elucidatory text frequently formed part of the apparatus to guide viewers to the moral truths concealed within. The image was a vessel for multiple meanings that were given greater clarity through accompanying text. The shell in the grotto worked in a similar fashion. As an object, it was not immediately readable, concealing, as it did, its interior magnificence and evading simple typologies, given its endless variety. Its meanings were multiple and dependent on the context in which shells were viewed and by the intentions of those who commissioned their use. While there was no elucidatory text to help fix the meaning of the shell in the grotto, the space's supplementary decoration and the activity that took place within it guided the viewer along certain interpretative avenues.

Like the emblem, the shell was valued not simply for what it looked like, but for what it could mean. It was understood by the early modern viewer as an object made, whether by human ingenuity or by God, but we can see that it was also an object that made meaning within the spaces it inhabited. Johann Friedrich's guests may have thought about courtly abundance when viewing conchological decoration. Joseph Furttenbach hoped visitors would be enraptured by the material qualities of the shell, that they would be transported to unfamiliar realms and consider nature. Magdalena Sibylla may have viewed shells as a commentary on the transience of human endeavor and on God's authority. The story of the grotto shell is one of multiples—shell decoration relied upon multiple varieties in order to draw out their contrasts and emphasize maritime abundance. The grotto's versatility as an architectural form, facilitating celebration, curiosity, and contemplation, stemmed in part from the shell's multiple meanings. Finally, the shell lived in multiple realms, as simultaneously an object of the divine, an object of nature, and an object of artifice. This is why we might understand shells as "produce of the sea so masterly and magnificently made."

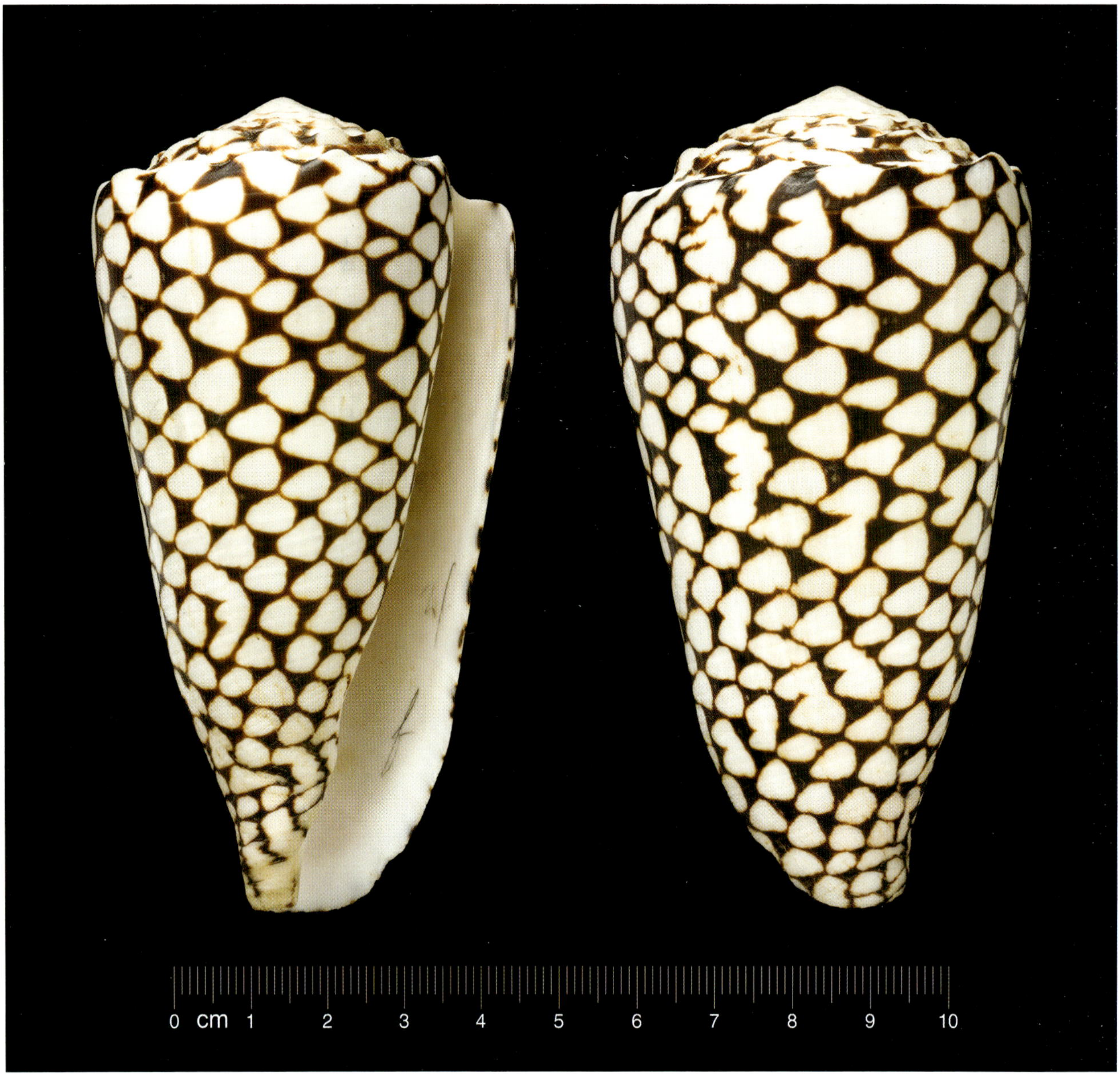
0 cm 1 2 3 4 5 6 7 8 9 10

Stephanie S. Dickey

# Chapter 6
# Shells, Prints, and the Discerning Eye

The Dutch master Rembrandt van Rijn (1606–69) was an exceptional painter, but his prints first secured his international fame. In a corpus of over three hundred prints treating diverse subjects, one small etching from 1650 is his only still life. It depicts a single specimen of the *Conus marmoreus*, or marbled cone, an Indo-Pacific snail whose shell is easily recognized by its reticulated surface pattern (fig. 75).

**FIG. 75.** *Conus marmoreus Linneaus 1758*, recto and verso. Natural History Museum, London.

Rembrandt was both an avid observer of the world around him and the owner of a renowned curiosity cabinet.[1] It is likely that he recorded a shell in his own collection, possibly sketching its form directly onto the plate. In the first stage of work (the first state), he combines etching with subtle touches of engraving and drypoint to describe the cone's geometric shape, bold coloring, and smooth sheen. A sharp cast shadow marks the ground on which it rests (fig. 76). In the second state, dense crosshatching surrounds the shell, suggesting a dimly lit space, perhaps the interior of a cupboard (fig. 77). A third state adds definition to the cone's serrated tip but renders its central apex less convincing. Perhaps Rembrandt recognized that this was not an improvement, since only a single impression of this final state is known (fig. 78).[2]

What was it about the marbled cone shell that caught Rembrandt's interest, and how can this simple yet compelling image shed light on the relationship of art and nature, collecting and curiosity in early modern Europe? Like many of his contemporaries, Rembrandt must have valued the shell's elegant form as a manifestation of God's ingenuity, a prompt for both aesthetic and devotional contemplation.[3] Here, however, I am interested in its status as a commodity.

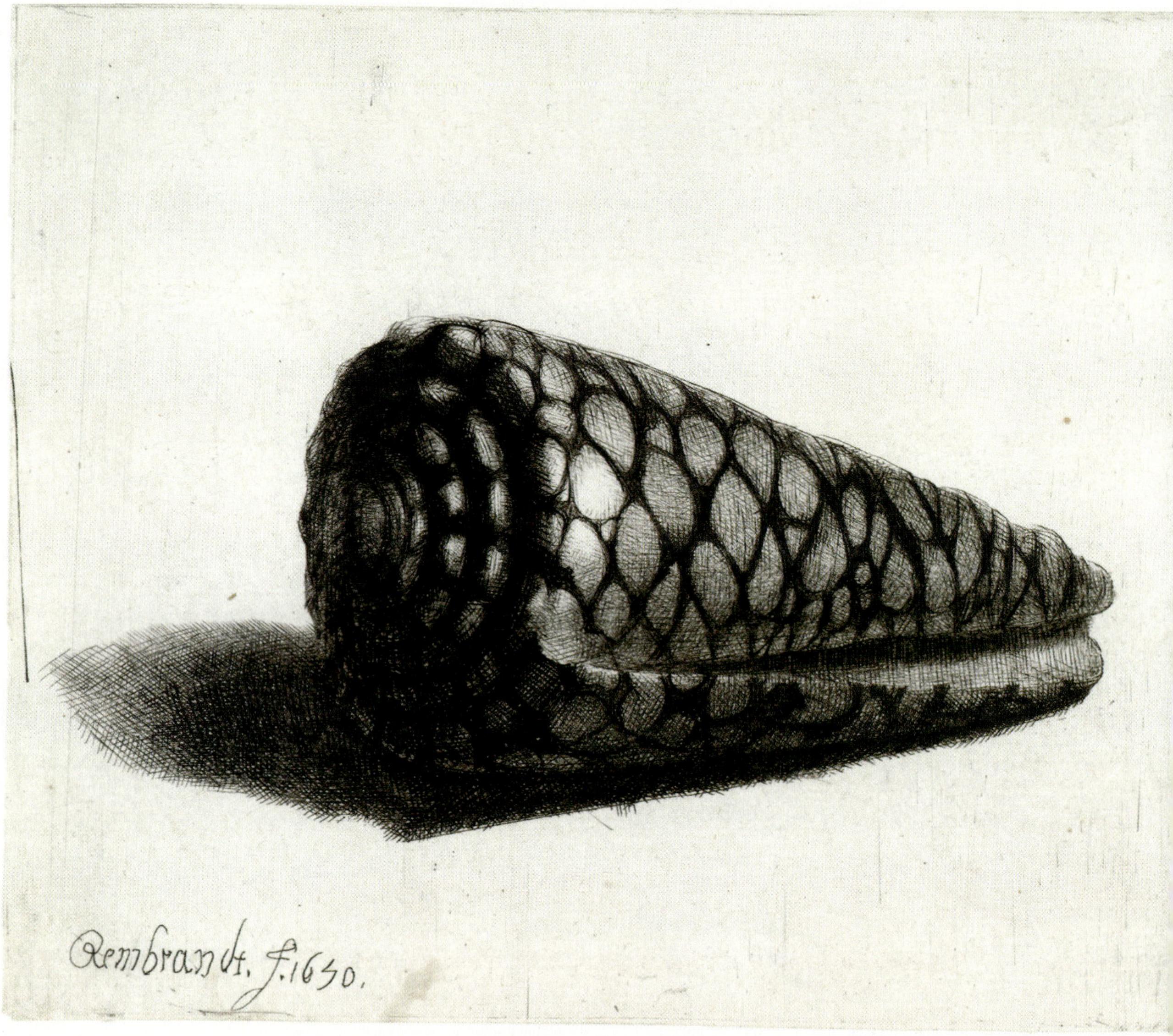


FIG. 76. Rembrandt, *The Shell (Conus marmoreus)*, 1650. Etching, engraving, and drypoint, first state. 9.9 × 13.2 cm. Rijksmuseum, Amsterdam [RP-P-O-241].

FIG. 77. Rembrandt, *The Shell (Conus marmoreus)*, 1650. Etching, engraving and drypoint, second state. 9.7 × 13.1 cm. Rijksmuseum, Amsterdam [RP-P-OB-242].

FIG. 78. Rembrandt, *The Shell (Conus marmoreus)*, 1650. Etching, engraving, and drypoint, third state, 9.6 × 13.2 cm. Rijksmuseum, Amsterdam, Mr. and Mrs. De Bruijn-van der Leeuw Bequest, Muri, Switzerland, 1962 [RP-P-1962-63].

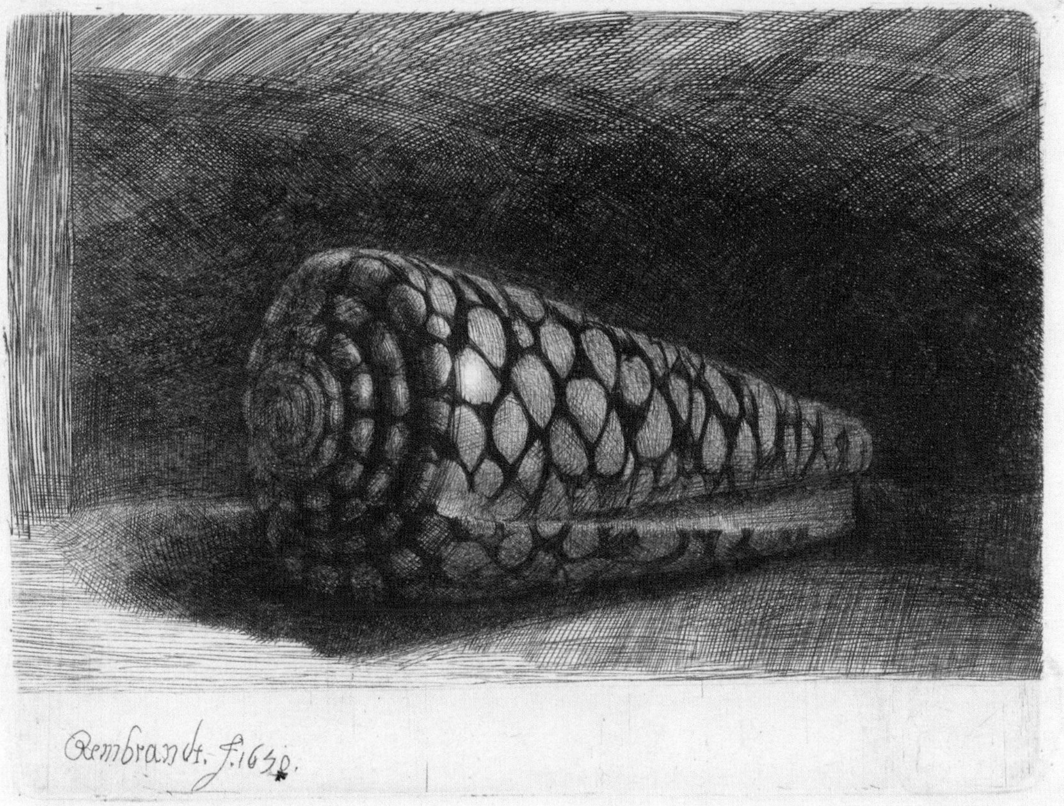
Rembrandt. f.1650.

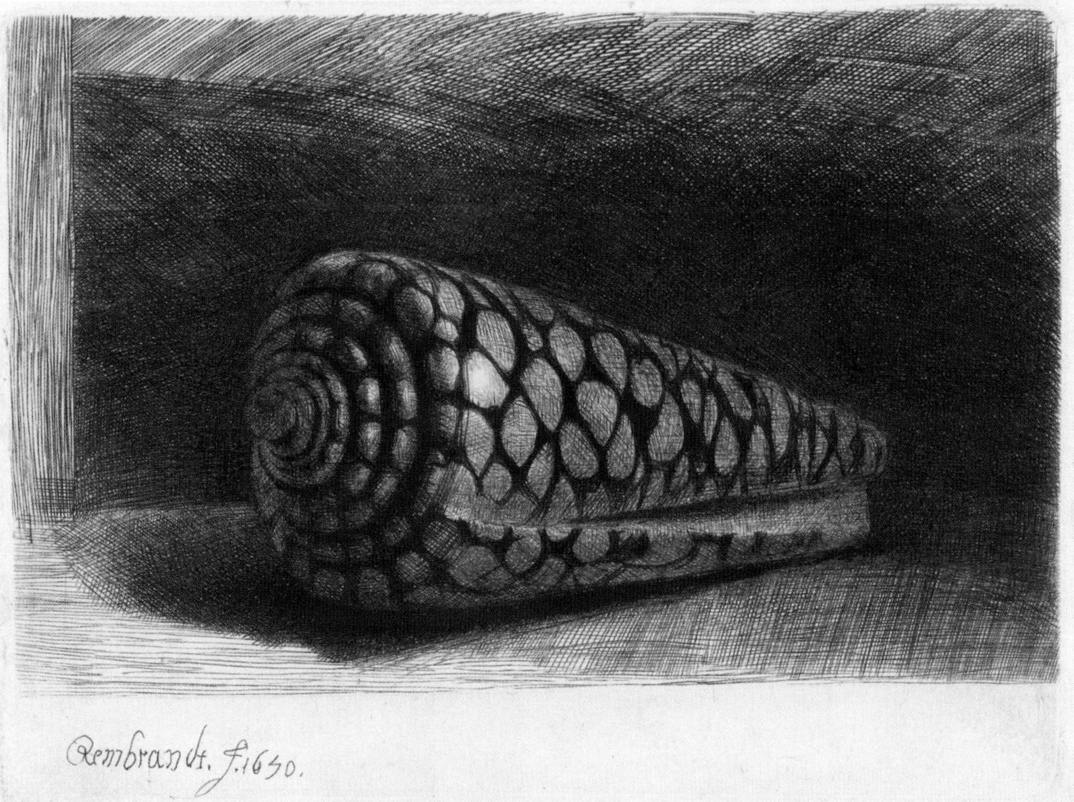
Rembrandt. f.1650.

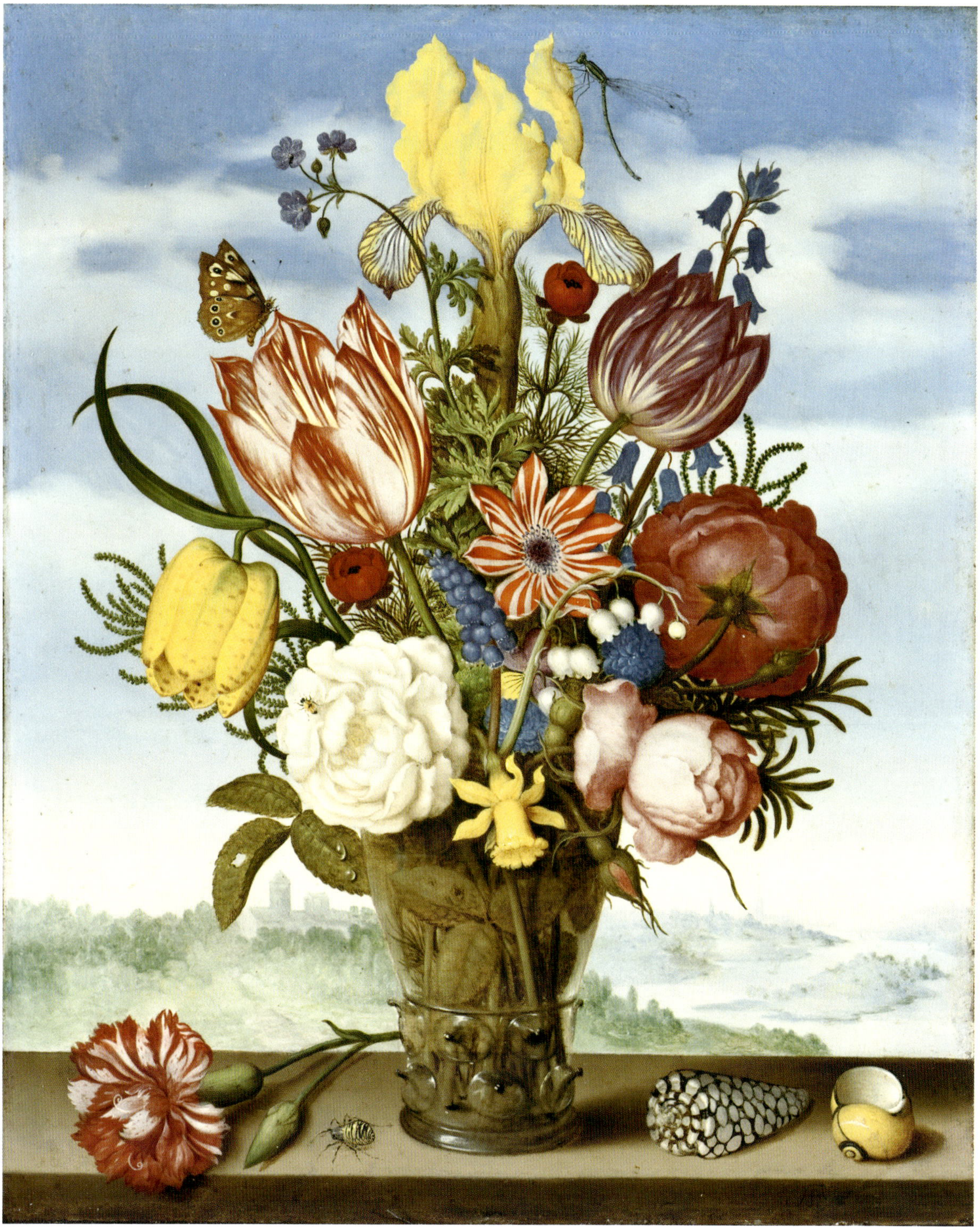

This chapter situates Rembrandt's etching at the intersection of two seemingly disparate types of material object, prints and shells, and explores the development of parallel criteria for appraising their desirability and market value. Both prints and shells were avidly collected in early modern Europe as objects of curiosity available to prosperous middle-class consumers. Thanks to the mechanical efficiency of the printing press on one hand and nature's abundant fertility on the other (abetted by global trade), both became accessible in greater numbers, and at lower prices, than art forms such as painting that could not be mass-produced. For the present analysis, the key point is that they were available as multiples: objects circulated in multiple specimens whose basic features are designed (by manufacture or genetics) to be essentially the same, but whose individual characteristics may vary slightly in ways that affect monetary or aesthetic value. This category encompasses other types of objects collected in early modern Europe, such as tabletop sculptures, coins and medals, and porcelain.[4] The accessibility of such objects contributed to the democratization of collecting as a recreational practice, a trend that continues to this day.[5] My goal here is to trace a distinctive network, stretching from the seventeenth-century Netherlands to England and France a century or so later, in which shells and prints (especially those by Rembrandt) were sought and valued, often by the same collectors and in curiously similar ways. Connoisseurs of both kinds of objects cultivated a discerning eye for qualitative distinctions between individual specimens and shared their discoveries through study, comparison, and exchange.[6]

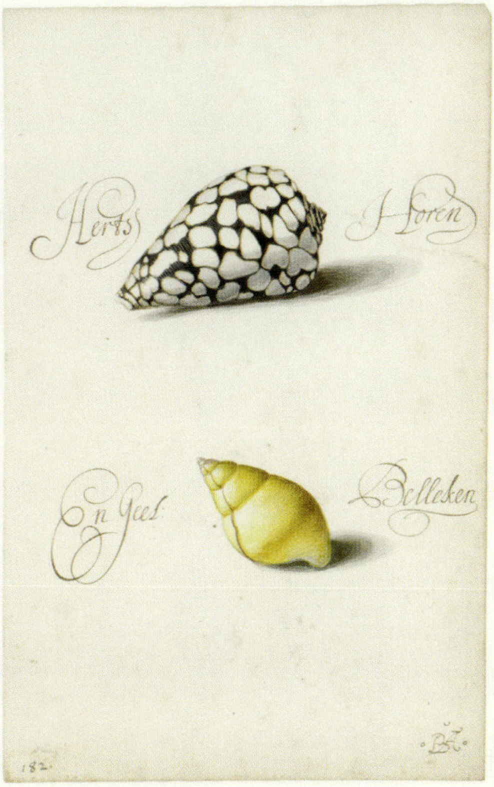


**FIG. 79.** Ambrosius Bosschaert, *Bouquet of Flowers on a Ledge*, 1619. Oil on copper, 27.94 × 22.86 cm. Los Angeles County Museum of Art, Gift of Mr. and Mrs. Edward W. Carter [M.2003.108.7].

**FIG. 80.** Balthasar van der Ast, *Study of Two Shells ("Herts Horen" and "Geel Belleken")*, mid-seventeenth century. Album leaf, pen and ink and brush in color on parchment, 31.3 × 20.2 cm, Frits Lugt Collection, Fondation Custodia, Paris [6534-47].

## SHELLS AND PRINTS IN THE CURIOSITY CABINET

The *Conus marmoreus* was one of the first shells to be widely appreciated by European collectors and recorded by artists. Inhabiting shallow waters from Madagascar to Fiji, it must have been a relatively easy target for explorers, at least for those who learned to avoid its poisonous sting. Its striking combination of strong color contrast and sculptural elegance may have been what endeared it to collectors and artists alike: in his ode to shells published in 1611, the Dutch poet Philibert van Borsselen compares its surface pattern to the beauty of flowers.[7] By 1620, the marbled cone was already a familiar feature in curiosity cabinets and in paintings. In a brilliant flower still life by Ambrosius Bosschaert (1573–1621), a single marbled cone sits beside a golden land snail from the Caribbean (fig. 79). A similar pairing occurs in a drawing by Balthasar van der Ast where the marbled cone is nicknamed "Stag Horn" (*Herts Horen*) and the snail "Little Golden Bell" (*Geel Belleken*) (fig. 80). These two

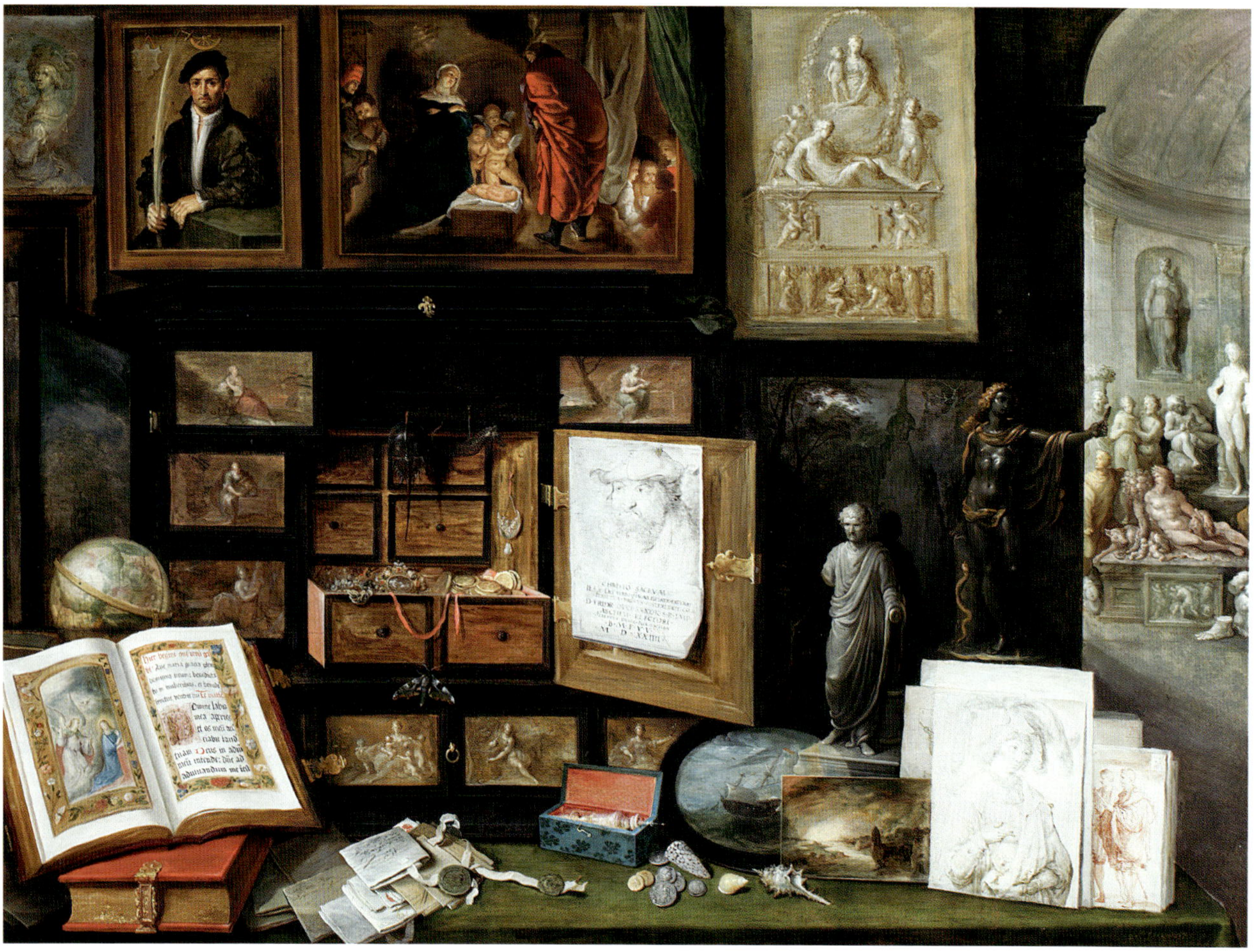

species are divertingly different in form and color, but informed viewers might also have recognized them as two mollusks, one marine and the other terrestrial, brought together from opposite sides of the world.[8]

In a *Kunstkammer* still life by Frans Francken, a marbled cone, a spiny murex, and a snail shell are surrounded by diverse examples of human artifice and two other aquatic specimens: a horseshoe crab and a seahorse (fig. 81). A moth, the single living creature, perches below an open drawer overflowing with coins, pearls, and a jeweled cameo, objects that transform natural materials through human ingenuity. The prominent placement of the three shells naturalizes their status as collectibles commensurate with the artworks around them, including several types of art on paper. A dog-eared impression of the 1524 engraved portrait of Frederick the Wise, Elector of Saxony, by Albrecht Dürer (1471–1528) is tacked to the cabinet door, while at lower right, an engraving by Lucas van Leyden (1494–1533), *Young Man with a Skull* (c. 1519), rendered to look more like a drawing, lies atop a red chalk study signed with the monogram of Heinrich Aldegrever.[9]

**FIG. 81.** Frans Francken, *A Curiosity Cabinet*, 1617. Oil on panel, 89 × 120.5 cm. Collection of the Duke of Northumberland, Alnwick Castle.

In an oval painting just behind, a seaborne galleon flying the Dutch flag hints at the journeys that carried exotic shells to the European market. Yet early collectors at home in their curiosity cabinets rarely concerned themselves with living mollusks or with the tedious process of preparing shells for sale, described by Georgius Rumphius (1627–1702) and others.[10] Cleaned and polished, with their slimy inhabitants left far behind, shells were easy to appreciate as inanimate material objects presented for aesthetic contemplation. Like prints by Dürer, Lucas van Leyden, and Rembrandt, they inspired practices of discernment and acquisition that linked collectors across Europe.

## REMBRANDT AS A COLLECTOR: PRINTS AND SHELLS IN THE NETHERLANDS

Rembrandt is the best known of numerous Dutch artists and amateurs who collected *naturalia* alongside works of art. In a survey of ninety collections documented in Amsterdam between about 1600 and 1750, at least sixty included both prints and shells.[11] The estate sale of the painter Jan Bassé (c. 1571–1637) in March 1637 featured over two thousand lots including prints, drawings, paintings, minerals, silver and gold medals, and *zeegewassen* (a term comprising seashells and other marine life). Rembrandt purchased thirty-nine lots of prints, two albums, a plaster sculpture, four lots of drawings, some paper, and four lots of two or three shells (*horens* or *horentjes*) for which he bid between 2.7 and 4.2 guilders. Similarly, most of the prints were combined lots that fetched between 1 and 5 guilders per lot. His most expensive purchases were a single print "by Raphael" (likely an engraving after the master), for which he paid 12 guilders, and a conch shell (*kokielje horen*) for 11 guilders.[12]

Rembrandt's bid for Bassé's conch was the highest recorded price for a single shell in over forty years of Amsterdam auctions.[13] Yet, as demand rose, the rarest shells could be valued much higher. The wealthy cloth merchant and collector Jan Volkertsz (1578–1651) reputedly turned down an offer of five hundred guilders (the equivalent of a master painting) for his precious wentletrap (*Epitonium scalare*), a species of which only three specimens were then known in the Netherlands. One belonged to the painter Jürgen Ovens (1623–78), a renowned shell collector who studied with Rembrandt's pupil Govert Flinck (1615–60).[14] Similarly, while most prints were inexpensive, the market featured a few special rarities. Another Rembrandt pupil, Samuel van Hoogstraten (1627–78), wrote that some print collectors (including Rembrandt himself) were ready to pay exorbitant prices for early prints by Lucas van Leyden.[15] At the Bassé sale, an album of prints by Lucas brought the astounding price of fl. 637.10.[16] One of Rembrandt's most accomplished etchings is still known to collectors as *The Hundred Guilder Print* because an impression sold in 1654 for that price, a hundred times what an average print would fetch.[17]

The sensational prices paid for the rarest specimens of shells and prints have long commanded attention. Yet, by emphasizing the exceptional over the typical, we overlook the significance of larger patterns of collecting. The Bassé auction offers evidence that already by the 1630s, some collectible shells were available on the secondary market at modest prices. Furthermore, many prints sold for similar prices; scarce impressions by Lucas van Leyden or Rembrandt occupied the pinnacle of a market dominated by commercial production. By the mid-seventeenth century, relatively plentiful species of shells, along with prints and other manufactured collectibles, were affordable to, and collected by, middle-class citizens such as artists, merchants, civil servants, and apothecaries.[18]

Rembrandt prospered in the 1630s, but he was a poor steward of his resources. By 1656, he declared insolvency and was forced to sell off his collections.[19] His bankruptcy inventory takes us room by room through the house on the Sint Anthonisbreestraat in Amsterdam he had purchased in 1639. Upstairs, across from his painting studio, a *kunstkamer* held paintings, drawings, dozens of albums of prints, and objects ranging from classical statuary, Venetian glass, Japanese weapons, and Indian basketry to minerals, stuffed carcasses, and "a great quantity" of *zeegewassen*.[20] This room must have resembled the cabinet of the Danish physician Ole Worm (1588–1654), who studied medicine in Rembrandt's hometown of Leiden, where a catalogue of his collection was published in 1655 (fig. 82).

Dutch connoisseurs such as Valerius Röver (1686–1739) referred to Rembrandt's etching as "The Little Shell" (*het hoorentje*).[21] Its singularity within his oeuvre was first noted in 1859 by the French art critic Charles Blanc (1813–82), who suggested that Rembrandt may have created this print to appeal to a friend who was also a shell collector.[22] While there is no evidence of a specific commission, the motif has obvious appeal for conchophiles. Another purchaser at the Bassé sale in 1637 was the tax receiver Jan Wtenbogaert (1608–80), who helped Rembrandt obtain payment for a commission from the Dutch stadtholder Frederick Hendrick (1584–1647) and was rewarded with an etched portrait.[23] Wtenbogaert inherited

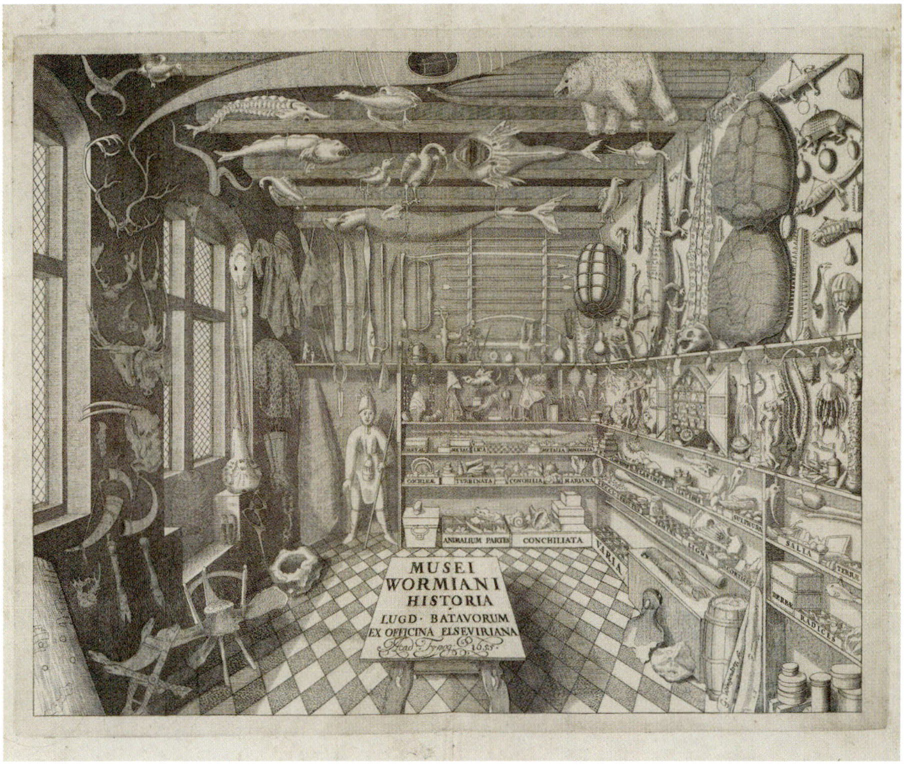

**FIG. 82.** Frontispiece, double-page engraved plate, artist unknown. In Ole Worm, *Musei Wormiani Historia*, 1655. Folio, 34.5 × 22 cm. Smithsonian Libraries, Washington, DC.

much of his own collection, which included prints, paintings, and shells, from a relative and fellow conchophile, the artist Jacques de Gheyn III (1596–1641).[24] By 1667, his holdings were impressive enough to merit a visit from Cosimo III de' Medici (1642–1723), Grand Duke of Tuscany, who also called on Rembrandt during his visit to Amsterdam.[25] In 1682, Cosimo purchased a collection of three hundred of the choicest shells gathered by Georgius Rumphius in Ambon.[26]

## FROM SLOANE TO GERSAINT: PRINTS AND SHELLS IN BRITAIN AND FRANCE

While Amsterdam offered collectors a hub of international trade, the founding of the Royal Society in London in 1660 established a nexus for intellectual exchange. Among the learned Dutchmen who became members were the conchophiles Levinus Vincent (1658–1727) and Albertus Seba (1665–1736). Both corresponded with the British physician and Royal Society president Hans Sloane (1660–1753), whose collections would form the foundation of the British Museum.[27] Sloane became interested in shells in the late 1680s when he spent time in Jamaica; he later published a book on the flora and fauna of the island. Simultaneously, he built a print collection within his extensive library. In 1702, he inherited the collections of his friend William Courten (1642–1702), whose curiosity cabinet in London had been open to visitors until 1698, when the press of visitors became overwhelming.[28] The Sloane papers in the British Museum contain numerous notes made by Courten. One is a list of some of his "best stamps," including Dürer's famous engraving, *Adam and Eve* (1504), valued at two pounds, and a less distinguished Rembrandt etching, *The Rat Catcher* (1632), valued at only nine shillings. A list of "Things bought" for May through December 1689 includes several conch shells with purchase prices between one and thirty shillings. Shells from Courten's *naturalia* collection served as models for the conchology guidebook compiled by Martin Lister (1639–1712), whose talented daughters etched the illustrations.[29]

The artist and art dealer Arthur Pond (1701–58) began collecting Rembrandt prints in the 1730s, amassing one of the most important collections in Europe. Some were acquired through the Dutch printmaker Jacob Houbraken (1660–1719), with whom he collaborated on a series of engraved portrait heads. After 1750, Pond turned his attention to shells, socializing with Sloane and other naturalists such as Emanuel Mendes da Costa (1717–91).[30] At the same time, another Londoner, Rev. Clayton Mordaunt Cracherode (1730–99), began forming a collection renowned for both Rembrandt etchings and shells.[31]

In April 1786, the estate sale of Margaret Cavendish Bentinck, Duchess of Portland (1715–85), drew crowds to her townhouse in Privy Gardens, Whitehall, where a lifetime's acquisitions were on display. In a sale lasting thirty-eight days, thirty were devoted to shells, but the duchess also owned art and antiquities, including the famous Portland Vase and prints by Dürer, Rembrandt, and other masters.[32] The listings in the sale catalogue drew upon findings by the duchess's curator, Rev. John Lightfoot (1735–88), and the botanist Daniel Solander (1733–82), whose research she had funded.[33] In its descriptive language we find equivalent qualitative judgments applied to shells and prints, as will be discussed below.

Today, the Portland Vase is in the British Museum, along with many shells and prints bequeathed by Sloane and Cracherode. On the verso of Cracherode's impression of Rembrandt's *Conus marmoreus* is the collector's mark of the French art dealer Pierre Remy (1715–97), with a date of 1749.[34] In Paris in 1757, Remy presided over one of the last auction sales to bring exceptional prices for rare shells, the collection of the Marquis de Bonnac, French ambassador to The Hague.[35] As an auctioneer trading in both art and *naturalia*, Remy followed in the footsteps of the *marchand-mercier* Edmé-François Gersaint (1694–1750), whose efforts to educate French collectors transformed the market for both shells and prints, especially the work of Rembrandt.

## GERSAINT IN PARIS: FROM SHELLS TO PRINTS

Renowned as a connoisseur and purveyor of luxuries, Gersaint is best known for his association with the rococo painter and printmaker Jean-Antoine Watteau (1684–1721), whose last painting, *L'Enseigne de Gersaint*, graced his shop on the Pont Notre-Dame.[36] Gersaint

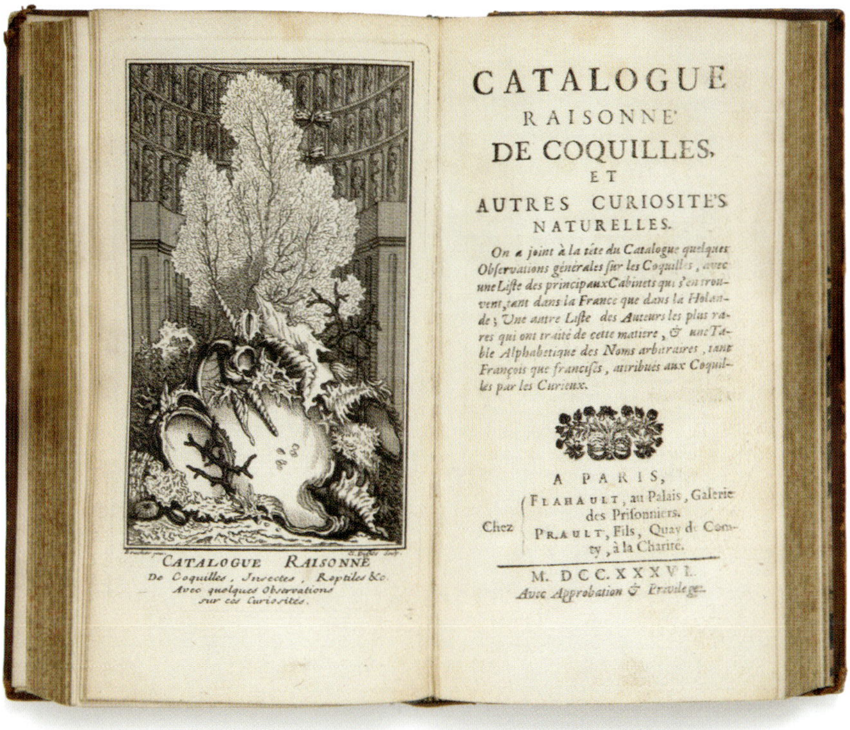

**FIG. 83.** Frontispiece engraved plate by Claude Duflos after François Boucher, and title page. In Edmé-François Gersaint, *Catalogue raisonné de Coquilles, Insects, Plantes marines, et autres Curiosités naturelles*, 1736. Duodecimo, 16.5 × 9.5 cm. Robin Halwas, London.

traveled frequently to the Netherlands; on a trip to Haarlem in 1730, Vincent's widow gave him a tour of their artfully composed curiosity cabinet, celebrated especially for its shells.[37] Gersaint introduced the French market to auction sales as a form of polite entertainment that was already popular in the Netherlands and Britain, and fostered the collecting of shells and prints.[38] In November 1735, the *Mercure de France* announced that Gersaint had brought back from Holland "a considerable collection of shells of every kind, in excellent condition," the fruits of an exacting search for the "rarest and most agreeable" specimens; the auction, scheduled for the following January, would be accompanied by a descriptive catalogue. The *Mercure* was also a popular vehicle for promoting newly published books and prints; this news item followed directly after a listing of the latest "New Prints" (*Estampes Nouvelles*).[39]

Gersaint's 1736 *Catalogue Raisonné de Coquilles, Insects, Plantes marines, et autres Curiosités naturelles* was the first of eleven descriptive sale catalogues published before his death in 1750. As the first of six presented as catalogues raisonnés, it launched a literary genre.[40] Enhanced with an etched frontispiece designed by François Boucher (1703–70), himself a conchophile, it was both a sale catalogue and an attractive handbook for the encouragement of beginning collectors (fig. 83).[41] While the text mentions scientific authorities such as Lister and Rumphius, Gersaint was clearly more interested in cultivating aesthetic appreciation for shells among collectors already attuned to works of art. Quoting the 1681 treatise

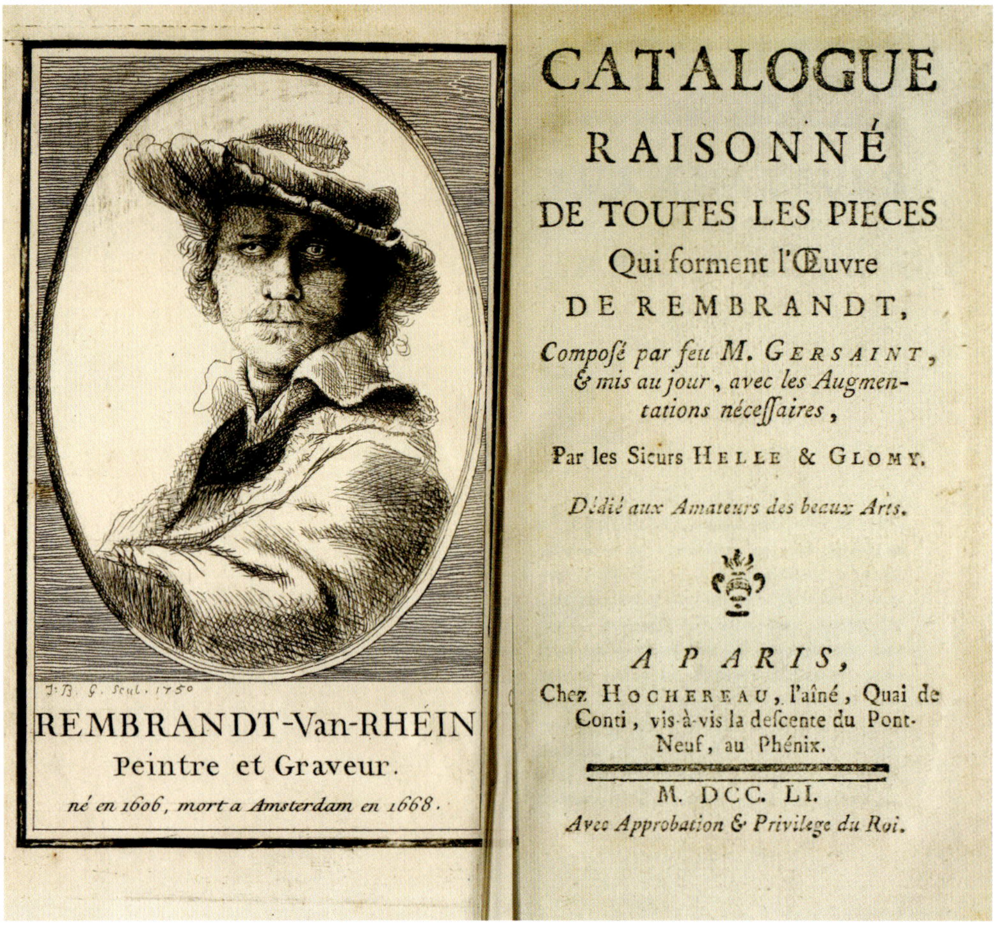

REMBRANDT-Van-RHEIN
Peintre et Graveur.
né en 1606, mort a Amsterdam en 1668.

CATALOGUE
RAISONNÉ
DE TOUTES LES PIECES
Qui forment l'Œuvre
DE REMBRANDT,
Composé par feu M. GERSAINT,
& mis au jour, avec les Augmentations néceſſaires,
Par les Sieurs HELLE & GLOMY.
Dédié aux Amateurs des beaux Arts.
A PARIS,
Chez HOCHEREAU, l'aîné, Quai de Conti, vis-à-vis la deſcente du Pont-Neuf, au Phénix.
M. DCC. LI.
Avec Approbation & Privilege du Roi.

**FIG. 84.** Frontispiece, engraved plate by Jean-Baptiste Glomy after *Self-Portrait* by Rembrandt, and title page. In Edmé-François Gersaint, *Catalogue Raisonné de toutes les Pièces qui forment L'Oeuvre de Rembrandt*, 1751. Octavo, 18 × 10.5 cm. Getty Research Institute Library, Los Angeles.

on shells by Filippo Buonanni (1638–1723), he promoted the value of shell collecting as "recreation for the spirit and the eyes."[42] Thus it was a natural progression from this aestheticized account of shells to a sequence of catalogues describing art objects, and specifically prints.

Because prints are usually signed and inscribed, they are easier to classify than paintings or drawings, but it was their status as multiples that motivated merchants like Gersaint to take the lead in documentary analysis and description.[43] Gersaint's final catalogue raisonné, left unfinished at his death and completed by his colleagues P.C.A. Helle and J. B. Glomy, was not a listing of objects for sale but rather a comprehensive catalogue of Rembrandt's prints (fig. 84). The French edition of 1751 was followed almost immediately by a translation for the English market, where Rembrandt prints were being avidly collected.[44] Gersaint offers the first published description of Rembrandt's *Shell*, here transcribed from the English edition:

> This is nothing more than the representation of a common shell known by the name of the Damier [checkerboard]. The back Ground is deeply shadowed, and at the bottom of the Left Hand is written, Rembrandt, f. 1650. / There is an Impression [state] of this Piece in which the Back-Ground is white; and trifling as this Subject is both these Impressions being very scarce, have been valued at an exorbitant Price, tho' it must be confessed that they are finely executed.[45]

Such a dismissive appraisal of Rembrandt's choice of motif from a merchant who promoted the collecting of both shells and prints is intriguing. Gersaint deems the marbled cone of "trifling" importance because it is common. Meanwhile, he finds Rembrandt's print overpriced because it is scarce. His description thus sets up the two objects—shell and print—as commodities whose value is assessed according to at least one shared criterion: rarity. Examining the descriptive language deployed in his sale catalogues reveals further parallels between shells and prints as multiples sought by early modern collectors.

## EVALUATING SHELLS AND PRINTS: RARITY, CONDITION, BEAUTY, AND PROVENANCE

Between the *Catalogue raisonné de Coquilles* and the *Catalogue raisonné* of Rembrandt's etchings, Gersaint published several descriptive catalogues designed not only to advertise auctions but also to advance the knowledge of prints, shells, and other collectibles, thus building a community of educated consumers. In their introductory *Avertissement* in the Rembrandt catalogue of 1751, Helle and Glomy observed that, just as Gersaint's catalogues for the collection sales of Louis Quentin de L'Orangère and others had sought to educate potential collectors and thus build the market for *naturalia*, especially shells, they "inspired new amateurs, and lifted the curiosity for prints out of the lethargy . . . into which it had fallen."[46]

These publications offer opportunities to compare the evaluative criteria by which prints and shells were judged by connoisseurs. Setting aside taxonomic descriptions, my focus is on recurring qualitative adjectives that distinguish individual specimens as worthy of note. As it turns out, they are virtually identical, centering on three characteristics: rarity, condition, and beauty, or some combination thereof. The diverse criteria for beauty cannot be addressed here, although common elements such as symmetry can be identified.[47] For the present analysis, the key point is that the perceived beauty of a specific specimen, whether shell or print, depended largely on how it measured up to established criteria for its type. In addition, a distinguished provenance could add prestige by affirming authenticity and value.

The Quentin de L'Orangère estate auction took place in March 1744. Like the earlier *Catalogue raisonné de Coquilles*, Gersaint's sale catalogue had a larger mission, this time to advocate for the collecting of prints, which he recommended as more affordable than paintings, more accessible than drawings, and useful for everything from instructing the young to entertaining the old.[48] Only three lots featured prints by Rembrandt;[49] rather, Quentin de L'Orangère favored French printmakers, especially Jacques Callot (1592–1635). His extensive holdings inspired Gersaint to compile a complete catalogue raisonné of Callot's prints, bolstered by research in other collections and prefaced by a biography of the artist.[50] Thus Rembrandt scholars have incorrectly claimed Gersaint's 1751 account of Rembrandt's etchings as the first catalogue raisonné devoted to the work of a single printmaker. More accurately, it was the first such book to be published independently, outside the context of a sale. Its aims were largely the same as Gersaint's more commercially driven publications: to aid potential purchasers in making judicious selections. The English edition provides "A List of the best Pieces of this Master [Rembrandt] for the Use of those who would make a Select Collection of his Works."[51]

The Quentin de L'Orangère sale also featured "un Coquillier fait avec choix," a cabinet containing nine drawers of shells, auctioned one drawer at a time.[52] Thus the catalogue juxtaposes Gersaint's descriptive language for shells and prints. Choice impressions of prints are signaled with epithets such as *rare, rarissime* (extremely rare), *presque unique* (almost unique), *difficile à trouver belle* (difficult to find in a beautiful example), *d'une épreuve parfaite* (an impression in perfect condition), *très-belle épreuve* (very beautiful impression), *premier épreuve* (first or proof state), or *bien choisi* (judiciously chosen). Shells are *beau* (beautiful), *très-beau* (very beautiful), *vive ou varié de couleur* (vivid or varied in color), *rare, singulier* (singular), *très-bien conservée* (well-preserved), *parfaite* (perfect, or in perfect condition), *jolie* (attractive), *bien choisi* (judiciously chosen).[53] These judgments promote the objects' value in the saleroom according to criteria of scarcity, condition, and aesthetic merit.

Gersaint's descriptors may have set a standard or simply codified existing discourse among dealers and collectors. Equivalents can be found in numerous other sources. To cite just one example, the catalogue of the duchess of Portland's sale in 1786 describes distinctive shells with adjectives such as "rare," "very scarce," "beautiful and rare," "curious and scarce," "fine," "rich in colour," and the like. The same designations apply to her prints, including a lot of "nine scarce etchings, by Rembrandt, very fine."[54] Paradoxically, for both prints and shells, "undescribed" is an especially enticing (and lucrative) designation in such catalogues, because it indicates a novel item not mentioned in any previous source.[55] While none of these criteria are especially surprising, their congruence is striking.

## RARITY IN A PLENTIFUL MARKET

Early curiosity cabinets were often described as collections of rarities. Here, I am concerned with rarity as a criterion of value. While rarity can apply to a class of objects, its market value adheres to individual specimens, which become more desirable the more difficult they are to obtain. In a print trade supplied largely by commercial publishers, Rembrandt was exceptional in marketing small numbers of proof impressions of his etchings, such as the first state of *The Shell* (see fig. 76). In 1718, the Dutch art historian Arnold Houbraken (1660–1719, Jacob's father) observed that serious collectors were not satisfied until they had acquired every known state; he even accused Rembrandt of printing proofs to increase sales.[56] Early states and specially printed impressions have never ceased to be prime desiderata. Aesthetically, they are admired because they reveal the artist's creative process, but commercially, their appeal lies in their rarity. Here we may recall Gersaint's observation that both states of *The Shell*, "being very scarce, have been valued at an exorbitant Price."

By the nineteenth century, printing limited editions to stabilize prices had become standard practice among fine printmakers, but no such limits could control the market for *naturalia*. In the early days of global trade, the first examples of exotic shells to reach the European market fetched sensational prices. By the later eighteenth century, as methods of harvesting, preparation, and transport became more regular and specimens more widely available, the bottom fell out of the market, just as it had for tulip bulbs in Holland in the 1630s.[57] The rarity of desirable shells was a factor more of market availability than of the ecological rarity of the relevant species.[58] The hobby of shell collecting still thrives today, but only a few species continue to be unattainable rarities.

## CONDITIONAL BEAUTY

When a type of object is not rare, the condition of an individual specimen can strongly impact its value. Connoisseurs of multiples develop criteria for what constitutes a prime example, and the avid collector must cultivate familiarity with compromising flaws. Judgment requires close inspection and discernment informed by knowledge of established standards—an archive of visual memory.

Multiples can manifest not only consistent baseline characteristics but also naturally occurring variations. For example, the reticulated pattern of the *Conus marmoreus* is a fundamental trait, but like the patches on a giraffe's coat it varies from one individual to another. The darker color can also vary from nearly black to orange. Judging from its representation in works of art (see fig. 79, fig. 80), early collectors preferred the striking contrast of black and white. Similarly, prints, and Rembrandt's in particular, can vary in character through richness of inking or choice of paper, as already noted by Houbraken and other early connoisseurs.[59] Fresh, early impressions are often stronger in contrast, since copperplates wear down with reprinting.

Specimens that show wear and tear—whether a chipped shell or a torn print—are less pleasing, and less desirable, than those closer to a pristine state of preservation. Gersaint advised that any shell picked up on the beach was likely to be damaged by waves, sun, and sand; perfect specimens required catching the animal alive. The ensuing process of cleaning and polishing could also reveal flaws.[60] Imperfect shells could be restored, overpainted, or otherwise manipulated to mask damage or simulate rare varieties.[61] Likewise, retouching with ink could improve imperfect prints or disguise common ones as rarities; holes and tears could be repaired, worn plates reworked. By the mid-eighteenth century, reprints and deceptive copies were also circulating.[62] Gersaint's catalogues were the first of many publications intended to help collectors navigate this complex terrain. Today, condition remains a fundamental criterion. Unsightly spots of foxing (paper damage) marred an impression of *The Shell* that sold in 2001 for well below what a fine impression would bring.[63]

A curious parallel is that both prints and shells can fade with exposure to strong light. To preserve them, early collectors stored their shells in drawers or specially built cabinets (*coquilliers*) like those used by Quentin de L'Orangère and Levinus Vincent, while prints were pasted or tucked into albums.[64] The Amsterdam collector Michiel Hinloopen (1619–1708) owned fifty-two such albums (but, surprisingly, only one print by Rembrandt). After his death, his collection was reorganized by Simon Schijnvoet (1652–1727), the editor of Rumphius's *Ambonese Curiosity Cabinet*, and displayed in the Town Hall of Amsterdam.[65] Album pages seldom reached the artfulness applied to the drawers of shell cabinets.[66] However, many extant examples resemble plates of shell handbooks by Rumphius, Lister, and others, manifesting a shared impulse to order and classify. These practices of selection, aggregation, and storage set shells and prints apart from paintings and sculpture, which typically functioned as visible decoration in the home. When visitors arrived, cabinet drawers and albums could be opened to provide an experience that was both privileged and intimate, and often tactile as well as visual.

## PROVENANCE AS SOCIAL CAPITAL

Gersaint recognized that global trade brought unique opportunities for Dutch collectors. However, he writes, "The ease with which they can acquire them does not prevent the Dutch from paying high prices for shells as long as they are in perfect condition and have a certain reputation." He cites as example the beautiful admiral cone (*Conus cedonulli*) in the cabinet of Johan de la Faille of The Hague, formerly owned by the king of Portugal.[67] "Reputation" here signals both the fame of this rare cone shell and the royal provenance of de la Faille's particular specimen. Possessing such an object brought the buyer into relationship with the good taste, acumen, and distinguished social status of the previous owner.[68] Through dealers, agents, or estate sales, the choicest specimens often passed from one major collection to another. Wealthy English collectors such as Reverend Cracherode and the Duchess of Portland acquired important shells from estates auctioned in Paris.[69]

The same was true for prints. The British Museum possesses several of Rembrandt's rarest proofs that were owned successively by connoisseurs such as John Barnard (1709–84), Sir Edward Astley (1729–1802), and Reverend Cracherode.[70] British collectors benefited from Arthur Pond's contacts with Jacob Houbraken, who acquired select impressions from the family of Rembrandt's patron Jan Six. On his visits to Holland, Gersaint tapped the same source; his catalogue of 1751 is based largely on impressions studied in Houbraken's collection.[71]

## CANONICAL EXAMPLES

Several weather-beaten specimens of *Conus marmoreus* are preserved today in the Rijksmuseum in Amsterdam. What makes them museum worthy despite their poor condition is their provenance: they were recovered in 1977 from the wreck of *De Witte Leeuw*, a VOC ship that sank in the South Atlantic in 1613. Not listed on the official cargo manifest, they must have been brought aboard by sailors hoping to sell them to curious landlubbers back home.[72] More often, legendary specimens of both prints and shells have lost their history when collections donated to museums are broken up and albums cut apart. Historians of conchology have made efforts to trace holotypes, the first specimens of newly discovered species and those that served as models for accounts published by authorities such as Buonanni, Rumphius, and Lister.[73] Historians of print collecting have likewise searched for the impressions of Rembrandt's prints owned or studied by early cataloguers such as Gersaint in Paris, Daniel Daulby (1745/46–98) in Liverpool, or Adam Bartsch (1757–1821) in Vienna. Like shell holotypes, impressions described in published catalogues established baseline characteristics against which all subsequent examples were measured.

While the rise of published descriptions enabled beginners to educate themselves, Gersaint freely admitted that no single author could solve every question. Each observer had the potential to make new discoveries.[74] Budding connoisseurs in France, the Netherlands, England, and elsewhere embraced this challenge. As collectors of multiples found themselves chasing similar objects at public sales and in dealers' shops, comparison and exchange offered opportunities for both learning and sociability.[75] Shell connoisseurs frequently traded specimens, and the first shell collectors' club was founded in Dordrecht in 1720.[76] Likewise, collectors of Rembrandt's etchings met at sales, competed for rarities, swapped duplicates, and shared discoveries.[77] Connoisseurs of both shells and prints marked up the pages of their catalogues with handwritten notes evaluating their own examples against others, tracking sales, describing novelties, and correcting published descriptions on the basis of firsthand observation. These annotated volumes became desirable reference sources and collectibles in their own right, often offered at auction as part of a collector's estate.[78]

**FIG. 85.** Wenzel Hollar, *Conus imperialis*, c. 1646. Etching, 9.8 × 13.5 cm. Rijksmuseum, Amsterdam [RP-P-1920-2766].

In his 1736 essay promoting the virtues of shell collecting, Gersaint expresses regret that his catalogue lacks engraved illustrations recording from life "the different faces and profiles" of shells and "the diverse characters and ornaments that Nature has imprinted *(imprimés)* on their surfaces."[79] Encouraging print connoisseurs to add shells to their repertoire, he recommends the 1705 edition of Rumphius's *Ambonese Curiosity Cabinet* over that of 1711, noting that its illustrations are superior in beauty, craftsmanship, and resemblance to their models.[80] Here, then, Gersaint values prints *of* shells for their evidentiary and aesthetic properties, while invoking the technology of printing to describe nature's ingenious inscription of colors and patterns *on* the shells themselves.

Rembrandt's etching anticipates the descriptive clarity of illustrations in conchology handbooks by Rumphius and others; it even provided a model for the *Conus marmoreus* illustrated by Lister.[81] However, in 1650, there were few precedents for independent prints featuring shells. The most frequently cited antecedent is the series of thirty-eight etchings of shells completed by Wenzel Hollar around 1646.[82] Unsigned and exceptionally rare, they were never formally published, but Hollar's name is mentioned in Rembrandt's inventory of 1656, so he may have been aware of them.[83] Hollar visited the Netherlands in 1634 and copied several of Rembrandt's etchings.[84] Like Rembrandt, Hollar must have studied shells firsthand. In 1745, George Vertue stated that Hollar worked from specimens owned by Thomas Howard, Earl of Arundel (1586–1646). However, it has recently been suggested that a reference by Schijnvoet in Rumphius's *Ambonese Curiosity Cabinet* links Hollar's project with shell prints commissioned by Jan Volkertsz. He is the Amsterdam merchant we met earlier as the owner of a precious wentletrap; Schijnvoet describes him as one of the greatest shell collectors of his time.[85] Volkertsz's collection was inherited by his two sons, one of whom, Volckert Jansz (d. 1681), is represented in Rembrandt's imposing group portrait *The Sampling Officials of the Amsterdam Drapers' Guild, known as 'The Syndics'* (1662, Amsterdam, Rijksmuseum). Rembrandt made preparatory drawings of several sitters, including Volkert Jansz. Thus he may well have seen this esteemed collection of *naturalia*.[86]

The *Conus marmoreus* does not appear in Hollar's series, so there was no exact model to follow: Hollar depicts the rarer *Conus imperialis* (fig. 85).[87] If, indeed, Rembrandt took Hollar's prints as his starting point, his pictorial strategy was markedly different. Hollar's motif floats against a plain white ground, its form effectively reduced to a schematic ideal. Rembrandt presents his shell as a weighty, tangible object, casting a shadow on the surface it occupies. Hollar's cone shell is flattened by its placement parallel to the picture plane, while Rembrandt's is angled in space, exposing its intricate spiral and emphasizing its solid three-dimensionality. The shadowy context added in the second state (see fig. 77) further transforms the specimen into a singular possession, tucked away in a cabinet to await the return of its owner's inquiring touch.

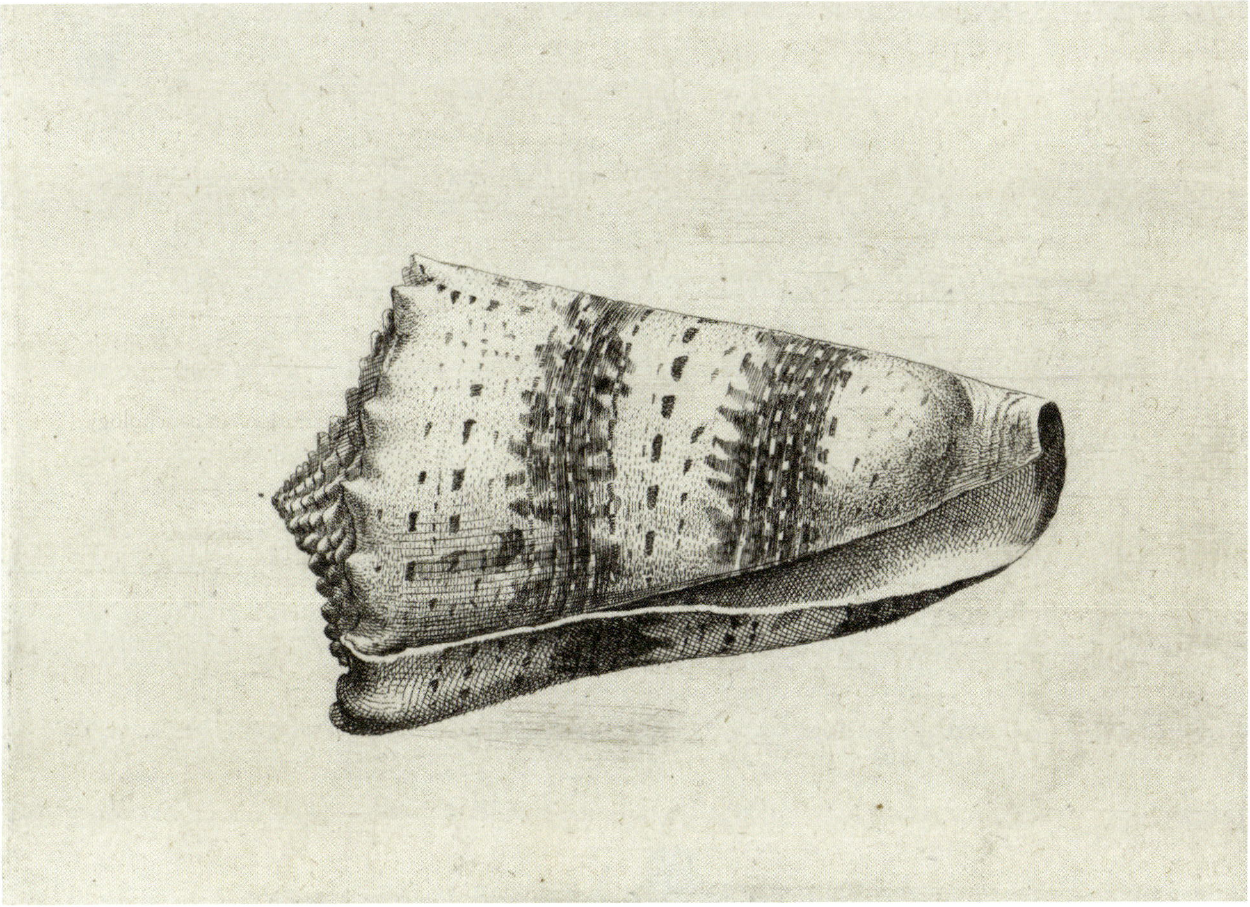

This dichotomy between the disembodied ideal and the individuated material object is central to the development of natural history illustration in the early modern period. The earliest book plates, rendered in the blunt medium of woodcut, did not allow for fine detail, but by the seventeenth century skilled engravers could capture the complexities of natural form with greater accuracy. Faced with this enticing option, illustrators were torn between accurately depicting individual specimens and synthesizing observations to create an ideal type. Even after the advent of photography, the relative efficacy of these strategies for scientific communication remains a matter of debate.[88] For Rembrandt, however, there was no question: the principle of *naar het leven,* or observation from life, was central to his art. *The Shell*, perhaps because of its inanimate subject matter, escaped the attention of classicist critics, among them Arnold Houbraken, who argued that artists should improve on nature (hence favoring the type over the specimen). Those who condemned Rembrandt's commitment to representing observed reality, with all its imperfections, took offense primarily at his rendering of a more charged motif, the female nude.[89] Yet perhaps Gersaint echoes this critique when he questions Rembrandt's choice to expend his talents on a "common" and "trifling" cone shell. Rembrandt's *Conus marmoreus,* by its pictorial insistence on individuality and material substance, makes a case for direct observation as the artist's mandate.

If Rembrandt's aim was faithful representation, it may seem surprising that he committed an obvious mistake: like most snail shells, *Conus marmoreus* curls to the right, but the dextral spiral drawn on the plate is reversed in the printing process, so that on paper the aperture opens to the left. Many early illustrators made the same mistake when depicting mollusks; the plates in Buonanni's *Ricreatione dell'occhio* are one example.[90] In Hollar's etching, too, the spiral is reversed. However, Rembrandt inscribed his signature backward, so that it prints correctly even though the shell does not, suggesting that he knew what he was doing. In fact, image reversal was common in prints of Rembrandt's time not only for natural history illustrations but also for cityscapes, reproductions of paintings, and other motifs where one might expect more fidelity to the model.[91] My theory is that print connoisseurs understood that the printing process produces a mirror image. Once accepted as such, reversal becomes a guarantee of veracity because it records the artist's unmediated encounter with his motif. Thus, like Rembrandt's etched portraits of patrons and friends, *The Shell* captures the appearance of an individual, observed from life.[92]

Whether or not it records an object from Rembrandt's own cabinet, *The Shell* advertises its maker as a member of the community of amateurs who valued both art and *naturalia,* prints and shells. As a printmaker and a collector, Rembrandt must have been alert to the novelty of etching a still life: the portrait of a distinctive object, possessed by the artist's gaze and multiplied through the replicative technology of print. Turning the sleek little cone to catch the light, was he conscious of holding the exoskeleton of a venomous carnivore, an alien relic from a distant realm? More likely, he was captivated by its asymmetrical geometry, an engaging challenge for the draftsman, and by its distinctive reticulated pattern. There is visual wit in applying the black-and-white medium of etching to an object in which nature plays so boldly with the contrast of dark and light. By making it the subject of his own artistry, Rembrandt engaged in a rivalry familiar to art theory of his time: the creative encounter between human and divine artifice, between art and nature. By calling our attention to its strangeness and beauty, his print invites us to consider the central question that has driven this book. What makes shells different from other collectible and representable things? By offering his viewers one shell out of many to collect and to contemplate, the artist leaves the answer in our hands.

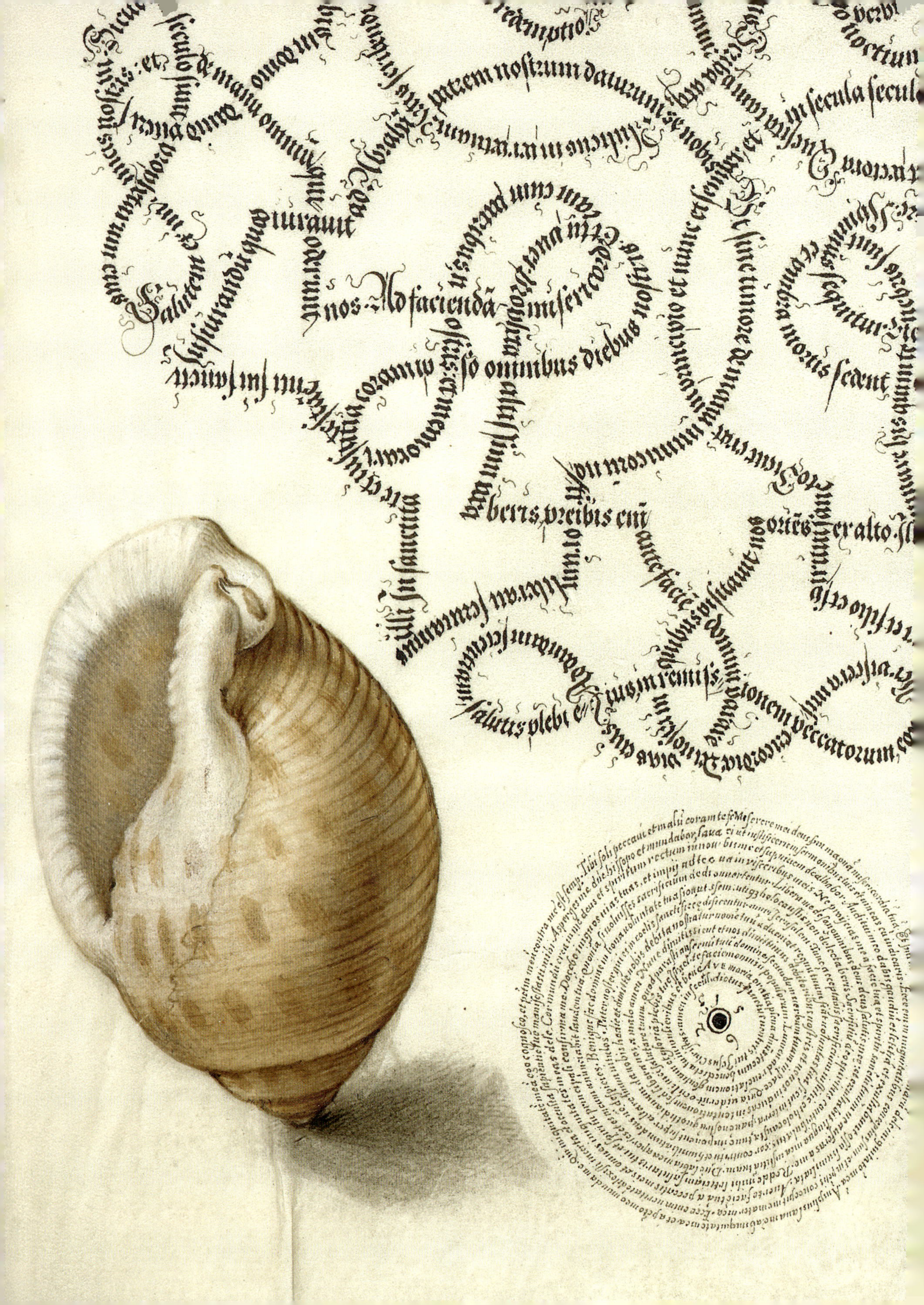

# ACKNOWLEDGMENTS

This book began over a *borrel* at De Engelbewaarder in Amsterdam, when we discovered that the four of us share a love of shells. We came to this topic as scholars of early modern art and history invested in the study of collecting, commerce, representation, and thinking with material things. Over the years that our shared passion developed and grew, we experienced what so often happens with a rich topic of research: shells were everywhere once we started to look for them. We may have conceived the word "conchophilia" in that café on the Kloveniersburgwal, but as it turned out, we had simply given a name to something that had always been there—a phenomenon that shapes the engagement with shells in any time and place.

Our first thanks must go to the Netherlands Institute for Advanced Study, which brought us together in spring 2017 when we were each working on other projects. In the spring of the following year, we organized two sessions at the Renaissance Society of America in New Orleans in which we embarked on our exploration of the subject in earnest and were joined by Krista de Jonge and our three fellow contributors to the volume: Stephanie Dickey, Anna Grasskamp, and Róisín Watson. We were inspired both by their wonderful papers and by the warm response of our audience in New Orleans.

We are beyond grateful to Michelle Komie, our editor at Princeton University Press, for her tremendous support and for her incomparable team of colleagues including Kenneth Guay, Steven Sears, and our brilliant copyeditor Lauren Lepow. We also thank Jenny Chan for the beautiful design of the volume, Tom Broughton-Willett for his skillful indexing, and the anonymous readers for their insightful feedback on the manuscript at an earlier stage. A grant from the Frederick W. Hilles Publication Fund of Yale University generously supported the production of this book, as did funds from Yale University's History of Art Department and the Warnock Publication Fund of the Department of Art History at Northwestern University.

Lastly, it means a great deal to us that this volume is the result of a collaboration among a cast of all-female scholars who span generations, and that this collaboration happened naturally rather than by design. We dedicate *Conchophilia* to the many women in early modern Europe who expressed their love for shells long before us in words, pictures, and collections of their own making.

Marisa Anne Bass, Anne Goldgar, Hanneke Grootenboer, and Claudia Swan
22 October 2020

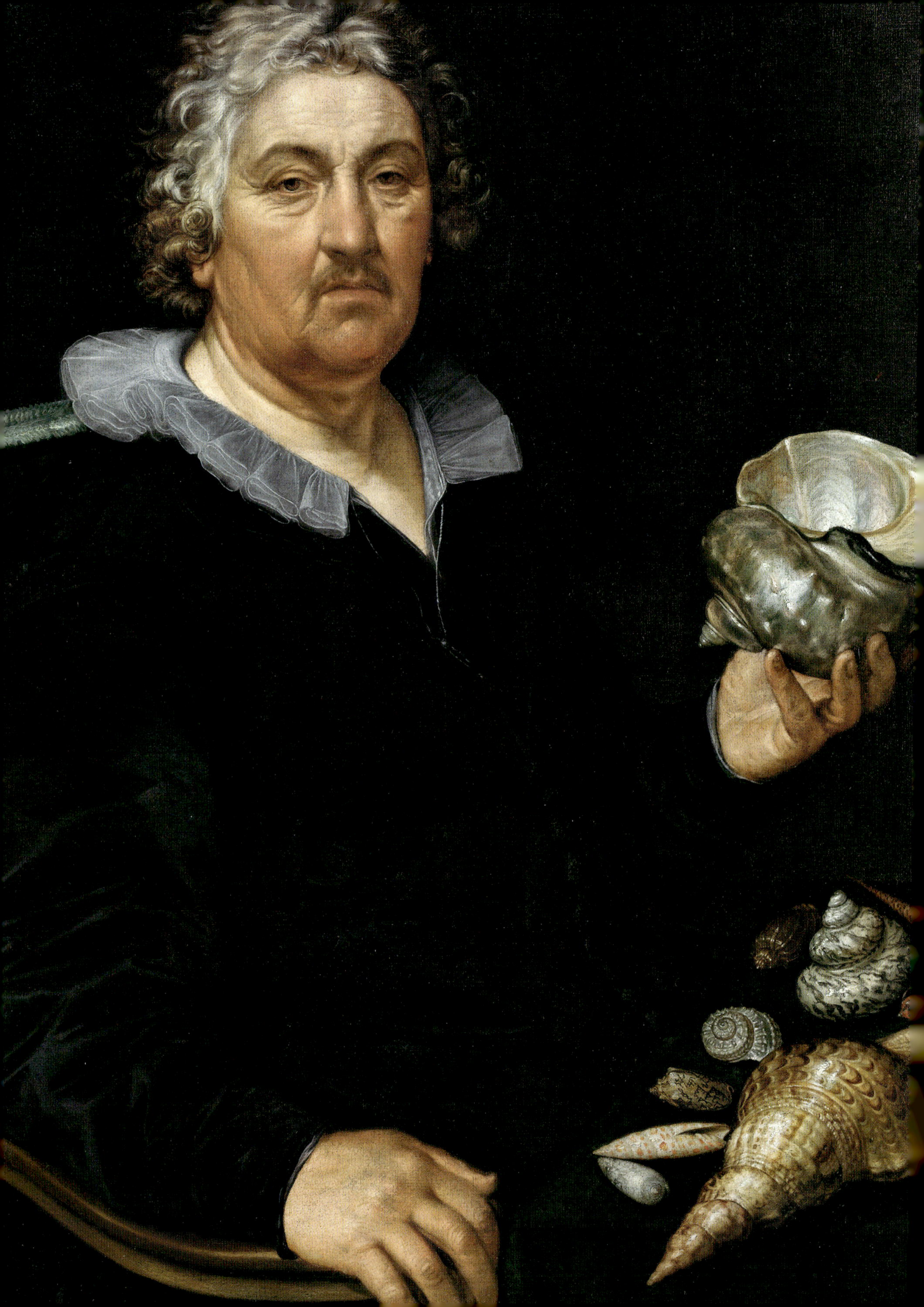

# NOTES

## INTRODUCTION. FOR THE LOVE OF SHELLS

I am grateful to my fellow authors and to the anonymous reviewers of the manuscript for their comments and assistance on earlier versions of this piece.

1. As Emma Spary has powerfully argued about shells in her "Scientific Symmetries," aesthetics cannot easily be divorced from science in this period.
2. Valentyn, *Verhandeling der Zee-Horenkens*, 47.
3. On *liefhebbers*, see Skogh et al, *The Varied Role of the Amateur*.
4. See Rumphius, *The Ambonese Curiosity Cabinet*, and chapter 1 of this book.
5. Blok and Molhuysen, *Nieuw Nederlandsch Biografisch Woordenboek*, 5:989–90.
6. Valentyn, *Verhandeling der Zee-Horenkens*, 44–70.
7. Ibid., 30.
8. Appadurai, *The Social Life of Things*; Gerritsen and Riello, *The Global Lives of Things*; Findlen, *Early Modern Things*; and on method: Hannan and Longair, *History through Material Culture*, Gerritsen and Riello, *Writing Material Culture History*; and Harvey, *History and Material Culture*.
9. There is a vast literature on this topic, but see, for example, Ambrose, "Archaeology and Shell Middens," and Álvarez et al., "Shell Middens as Archives of Past Environments."
10. D'Errico and Blackwell, "Earliest Evidence of Personal Ornaments"; Steele, Álvarez-Fernández, and Hallett-Desguez, "A Review of Shells as Personal Ornamentation."
11. Again, there is a wide body of literature on this, but for some examples, see Trubitt, "The Production and Exchange of Marine Shell Prestige Goods"; Sonak, "Shell Art and Shell Craft of Goa"; Bradley, "Re-visiting Wampum."
12. On the paper nautilus, see Leonhard, "Shell Collecting," 209; Leonhard and Leuker, "Who Commissioned Hollar's Shells?" 228–32; Leuker, "Knowledge Transfer," 153–54. On nautilus cups, see esp. Kemp, "'Wrought by No Artist's Hand'"; Mette, *Der Nautiluspokal*; and chapter 2 of this book.
13. Pliny, *Natural History* 9.52; Cicero, *De oratore* 2.6.23; Suetonius, *The Lives of the Twelve Caesars*, Life of Caligula, par. 46. On this, and on what follows, see esp. Prosperetti, "'Conchas legere,'" 393, 395–96.
14. Suetonius, *The Lives of the Twelve Caesars*, Life of Nero, par. 31, refers to the *domus aurea* as "adorned with precious stones and shells"; with thanks to Barbara Graziosi for this reference.
15. Szafranska, "The Philosophy of Nature and the Grotto"; Dernie, "The Use and Meaning of Materials," 69; Minor, "G. B. Piranesi's *Diverse Maniere*." On both ancient and early modern grottoes, see chapter 5 of this book.
16. Prosperetti, "'Conchas legere,'" 395.
17. Leonhard, "Shell Collecting"; Prosperetti, "'Conchas legere'"; Spary, "Scientific Symmetries"; Tobin, *The Duchess's Shells*; Roos, *Martin Lister and His Remarkable Daughters*.
18. Stadsarchief Amsterdam 5072/353/44–44v, Desolate Boedelkamer, inventory of Francisco Gomes da Costa, 16 September 1646.
19. On exotic goods, see further chapters 1 and 2 of this book as well as (among many others) Gerritsen and Riello, *The Global Lives of Things*; Findlen, *Early Modern Things*; Schmidt, *Inventing Exoticism*; Smith and Findlen, *Merchants and Marvels*.
20. Van der Veen, "Dit klain Vertrek," and Coomans, "Schelpenverzamelingen."
21. Visscher, *Sinnepoppen*, bk. 1, no. 4.
22. I owe this point to Claudia Swan.

23. Green, *A Fistful of Shells*, 17–18.
24. Hogendorp and Johnson, *The Shell Money of the Slave Trade*.
25. See esp. Emmer, *The Dutch Slave Trade*.
26. See chapter 2 of this volume.
27. I owe this observation to Hanneke Grootenboer, as well as, of course, to Anna Grasskamp and Claudia Swan in this volume.
28. See chapter 6 of this volume.
29. Cited in Leonhard, "Shell Collecting," 188: Lonitzer, *Naturalis historiae opus novum* (1551–55); Gesner, *Historia animalium liber IV qui est de piscium et aquatilium animantium natura* (1558); Belon, *De aquatilibus libri duo* (1553); Guillaume Rondelet, *Universae aquatilium historiae pars altera* (1555).
30. Buonanni, *Ricreatione dell'occhio*. On the Duchess of Portland, see Tobin, *The Duchess's Shells*.
31. Van de Roemer, "Neat Nature"; Spary, "Scientific Symmetries"; Dietz, "Mobile Objects."
32. Beekman, "Introduction," lxvii–lxviii.
33. Ibid., civ.
34. On Lister, see Roos, *Martin Lister and His Remarkable Daughters*; Tobin also observes that most conchologists were not interested in local shells: Tobin, *The Duchess's Shells*, 51.
35. I have previously discussed this in Goldgar, *Tulipmania*, 86–88.
36. Van Borsselen, *Strande*.
37. Valentyn, *Verhandeling der Zee-Horenkens*, 54.
38. See Van Berkel, *Citaten uit het Boek van Natuur*; Jorink, *Reading the Book of Nature*. *Cedo Nulli* was also the name of a type of Dutch tulip in the 1630s and 1640s.
39. Rumphius, *The Ambonese Curiosity Cabinet*, 88.
40. On natural theology, see esp. Jorink, *Reading the Book of Nature*.
41. Vincent, *Wondertoneel der Nature*, 23.
42. Seba, *Locupletissimi rerum naturalium thesauri*, bk. 1, preface.
43. For an overview, see Daston, "Nature by Design," and also Goldgar, "Nature as Art."
44. See Findlen, "Jokes of Nature"; Daston, "Nature by Design"; Kemp, "'Wrought by No Artist's Hand'"; Daston and Park, *Wonders and the Order of Nature*, 277ff.
45. Seba, *Locupletissimi rerum naturalium thesauri*, 1: plate 87.
46. Findlen, "Jokes of Nature," 318; on Johanna van Breda's embroidery, see François Halma, "Op de weêrgadelooze Natuur Kabinet van den Heere Levinus Vincent, en zyne Ega Juffr. Johanna Breda," in Vincent, *Korte Beschryving*, 4.
47. Dance, *Shells and Shell Collecting*, one of the main resources until recently for the history of shell collecting, attacked early modern collections as having "a complete lack of scientific arrangement," saying that "it is evident that pleasing arrangements meant more to them than the shells themselves." Dance, *Shells and Shell Collecting*, 65. For a more nuanced view, see Spary, "Scientific Symmetries"; Van de Roemer, "Neat Nature"; Dietz, "Mobile Objects."
48. Spary, "Scientific Symmetries"; and see also Van de Roemer, "Neat Nature," on Simon Schijnvoet's collection drawers preserved at the University of Amsterdam.
49. Dietz and Nutz, "Collections Curieuses."
50. This type of sociability is also highlighted in the context of auctions by Montias, *Art at Auction*; in the context of drawings, see Warwick, *The Arts of Collecting*.
51. On solitude, see Shapin, "'The Mind'"; Findlen, "The Museum."
52. Leonhard, "Shell Collecting," 181–82.
53. Grootenboer, "How to Become a Picture," 329–30.
54. Cummings, "Damian Hirst."

## CHAPTER I. THE NATURE OF EXOTIC SHELLS

Doerr, "The Shell Collector," 14. Profound thanks to Patrick and Monique de Koster-van Rijckevorsel for their loving support, and for introducing me to Rumphius's work long ago. For access to critical resources, I am grateful to Niels de Boer and Hans Mulder. For feedback and inspiration, I thank Stephen Campbell, Lieke van Deinsen, Stephanie Dickey, Olivia Dill, Christine Göttler, Robert Hariman, Antien Knaap, Giancarla Periti, Ulrich Pfisterer, Arianna Ray, Alessandra Russo, and especially Jon Swan; the readers for Princeton University Press; and my fellow authors. Unless otherwise indicated, translations are my own.

1. Kolfin, "African Man with a Conch," discusses adaptations in other works by Bloemaert, and attributes a painting slightly larger than the drawing, which it echoes precisely, to the artist (São Paulo, Collection Beatriz and Mário Pimenta Camargo).
2. The artist's attention to the play of light on skin and shell, the folds of cloth, and pearlescent conch participate in an aestheticization of Black bodies described in early modern art theory, a feature of numerous seventeenth-century works of art. See Hochstrasser, *Still Life and Trade*, 204–27; Chadwick, "'This deepe and perfect glosse of Blacknesse'"; and Kolfin and Runia, *Black in Rembrandt's Time*.
3. *D'Amboinsche rariteitkamer* was published in Amsterdam in 1705; the herbal was compiled and published in Amsterdam decades later (1741–50). For an English translation of *D'Amboinsche rariteitkamer*, with annotations and an extensive introduction by E. M. Beekman, see Rumphius, *The Ambonese Curiosity Cabinet*.
4. Rumphius intended his work to comprise a natural history of Ambon and nearby regions. The Amsterdam collector Simon Schijnvoet added annotations to *D'Amboinsche rariteitkamer* in which he cites shells in prominent collections in the Netherlands, including specimens from the Atlantic. Schijnvoet assembled the sixty engraved plates for the book as well; the original drawings are in the Koninklijke Bibliotheek, The Hague. Maria Sibylla Merian is sometimes credited with preparing the plates; on this and on her working relationship with Schijnvoet, see van de Roemer, "Merians netwerk," 21–23.
5. Rumphius, *D'Amboinsche rariteitkamer*, 1, on which see Dance, *A History of Shell Collecting*, 26–31; Van 't Zelfde, "'O seldsaem dierken,'" 60–61; Arens and Kießling, "Knowledge and Power"; Leuker, "Knowledge Transfer"; Leonhard, "Shell Collecting," 203–12; and Leonhard and Leuker, "Who Commissioned Hollar's Shells?" On *liefhebbers*, see the introduction to this book.
6. On curiosity, see inter alia Daston and Park, *Wonders and the Order of Nature*, 303–38.
7. Rumphius, *D'Amboinsche rariteitkamer*, 1. On the role of collectors in the publication of the book, and on the book as a manual for amateur shell collectors and as a collector's item in and of itself, see Leuker, "Knowledge Transfer," 165. Simon Schijnvoet included references to and images of shells from all over the world, thus distending Rumphius's original intention to write a local natural history.
8. Rumphius, *D'Amboinsche rariteitkamer*, printer's preface, 3r. See also Rumphius, *The Ambonese Curiosity Cabinet*, 9. In addition to standing as a record of Rumphius's malacological and conchological explorations, the text of *The Ambonese Curiosity Cabinet* evokes and commemorates Rumphius's own collection of shells—a collection he sold under financial duress.

9. Rumphius, *D'Amboinsche rariteitkamer*, 163.
10. On the sale, see ibid., bk. 2, chapter 39, and Beekman, "Introduction," lxxii, civ–cv, who describes it as having preyed on Rumphius throughout his work.
11. On Cospi's collection, see esp. Findlen, *Possessing Nature*, passim. On receipt of the shells, Cospi had them painted; a volume of watercolors survives in the Biblioteca universitaria di Bologna (Manoscritti, Ms. 4312, Jacopo Tosì, *Testacei, cioe nicchi chioccioe [sic] e conchiglie di piu spezie*).
12. Dance, *A History of Shell Collecting*, 35.
13. We know that he relied on clerks and illustrators and his son Paulus Augustus; in the Ambonese Herbal, he speaks of working with "borrowed pen and eyes." See Beekman, "Introduction," lxxii–lxxiii; Rumphius, *The Poison Tree*, 19; and Leuker, "Knowledge Transfer," 148 and passim.
14. Beekman, *Paradijzen van weleer*, 110; Beekman, *The Poison Tree*, 21–22.
15. Aristotle, *History of Animals* LCL 438:152–53.
16. Shell, "Casting Life," 22. See also Leonhard, "Shell Collecting," 180–82.
17. Van Borsselen, *Strande*, C6v.
18. Letter to Clusius, Leiden University Library VUL 101 (GOV_J001). The existence of a portrait drawing of Govertsz dated 1614 by Goltzius (Kunstsammlungen der Veste Coburg, Coburg) recommends a date of death later than 1612, proposed by Nichols, "Jan Govertsz. van der Aar," passim, and Nichols, *The Paintings of Hendrick Goltzius*, 169–71.
19. Van Mander, *Schilder-Boeck*, fol. 286.
20. I owe the identifications to Nichols, *The Paintings of Hendrick Goltzius*, 170; see also Dance, *A History of Shell Collecting*, 11.
21. On the association of the man and his shells staged here, see Grootenboer, "How to Become a Picture," 329–30.
22. Cornelis Cornelisz van Haarlem, *Allegory of the Arts in a Time of Peace*, 1607, oil on canvas, 70 × 92 cm, private collection of Lord Sackville, Knole House, on loan to the National Trust UK. Van der Aar is depicted seated at a table with an array of shells before him, passing a *Hippopus hippopus* (bear paw clam, horse's hoof clam, or strawberry clam shell) to the painter at an easel behind him. See Dance, *A History of Shell Collecting*, 11 and plate III.
23. On his association with the VOC, see Nichols, "Jan Govertsz. van der Aar," 247, 253–54.
24. On the wreck of the *Witte Leeuw*, see Sténuit, "De 'Witte Leeuw'"; on its shells, Bergvelt and Kistemaker, *De wereld binnen handbereik. Catalogus*, cat. no. 61; and, on painterly interest in the sorts of shells and the porcelain on board, Wallert, "Balthasar van der Ast," 81–82, 89.
25. Brinck, *Adversaria*, OAH 2059, 81v.
26. Ibid.
27. Visscher, *Sinnepoppen*, bk. 1, nos. 4–5.
28. Brinck, *Adversaria*, OAH 2059, 82r.
29. See Kisluk-Grosheide, "Dirck van Rijswijck," and Corrigan, van Campen, and Diercks, *Asia in Amsterdam*, cat. nos. 86a–b, 299–301.
30. On Bellekin's work, see van Seters, "Oud-Nederlandse Parelmoerkunst," and on his many abilities, including the repair of East Indian rarities, see van der Veen, "Dit klain Vertrek," esp. 244.
31. For a fuller discussion of this sale and the gift, see Swan, *Rarities of These Lands*, 117, 152.
32. Brinck, *Adversaria*, OAH 2059, 82v. Van Borsselen, *Strande*, A6v, mentions what he calls "Angolan gold" and describes women gathering them on the beach.
33. Anonymous ("Written by a Dutch Gentleman"), *A Voyage to the Island of Ceylong & Etc* (London, 1754), quoted by Rumphius, *The Ambonese Curiosity Cabinet*, 449. See Hogendorp and Johnson, *The Shell Money of the Slave Trade*, and Green, *A Fistful of Shells*. See also Rumphius, *D'Amboinsche rariteitkamer*, 2:24.
34. On Porret, see Swan, *Rarities of These Lands*, esp. 100–103, with prior literature.
35. On Paludanus's collection of shells and other maritime creatures, see ibid., esp. 94–95, with prior literature.
36. Jorink, *Reading the Book of Nature*, 277.
37. Cornelisz's painting is related to the awkwardly sensual depiction of Neptune and Amphitrite (or Venus) and Cupid, by Jacques de Gheyn II (see fig. 25 of this book). Extracted from far-flung origins and transported overseas, the shells depicted bear witness to a single individual's dominion over them, their forms. To hold them is to rule over them.
38. On such shells in various stages of polish, see Koeppe, "*Artificialia* and Their Meanings in *The Paston Treasure*," 123.
39. The monograph devoted to shell vessels made of nautilus shells—Mette, *Der Nautiluspokal: wie Kunst und Natur miteinander spielen*—takes its subtitle from a text by the art agent and impresario Philip Hainhofer that makes reference to "how art and nature play with one another" (*wie ars und natura miteinander spielen*). On the intertwining of art and nature in *Wunderkammern* and a concomitant distinction between art and nature in the study of nature, see Daston and Park, *Wonders and the Order of Nature*, 255–301. "As a habit of the understanding," they write, "the Aristotelian opposition between art and nature still framed the mental world of early modern Europeans" (265). On shell collections, aesthetics, and natural history, see van de Roemer, "Neat Nature" and Spary, "Scientific Symmetries." See Kehoe, "The Nautilus Cup between Foreign and Domestic," on art and nature and the extension of this paradigm to domestic and foreign goods.
40. On figured stones, see Daston and Park, *Wonders and the Order of Nature*, 286.
41. Beekman, "Introduction," civ, xcviii, and passim.
42. Rumphius, *D'Amboinsche rariteitkamer*, 70.
43. Ibid., 69; cf. Rumphius, *The Ambonese Curiosity Cabinet*, 99. Beekman translates *doodverf* as "the color of a corpse" but it is a painter's color, usually used to prepare canvases. "Spanish green" is the pigment verdigris, as Beekman notes. Beekman is rightly appreciative of Rumphius's poetic language, especially when writing in Dutch; see Beekman, "Introduction," cix–cx, and Leonhard, "Shell Collecting," 209–12.
44. Rumphius, *D'Amboinsche rariteitkamer*, 70. See, for annotations, Rumphius, *The Ambonese Curiosity Cabinet*, 99–101. Referring to its iridescence, he calls it a "moon shell." Rumphius does not associate the forms of the shell with a turban, as in Linnaean nomenclature. Turbans were worn by Melanesian natives, but they were not the elaborate white Ottoman or Persian style; see, for instance, fig. 6 of this book.
45. Rumphius, *D'Amboinsche rariteitkamer*, 163; Rumphius, *The Ambonese Curiosity Cabinet*, 223.
46. Rumphius, *D'Amboinsche rariteitkamer*, 163–66; Rumphius, *The Ambonese Curiosity Cabinet*, 225–26. See Leonhard, "Shell Collecting," 203–4.

47. Kemp, "'Wrought by No Artist's Hand.'"
48. See Mette, *Der Nautiluspokal*; Kehoe, "The Nautilus Cup between Foreign and Domestic"; Grasskamp, "Spirals and Shells."
49. Kemp, "'Wrought by No Artist's Hand,'" 179.
50. On slaves who procured specimens acquired by Hans Sloane and James Petiver, see Murphy, "Collecting Slave Traders"; Delbourgo, *Collecting the World*, 97–100 and passim. On the pictorial trope of "fantasized availability of the world's bounty" to European Whites, see Chadwick, "'This deepe and perfect glosse of Blacknesse,'" 103 and passim.
51. For her analysis of the frontispiece as it relates to the production of scientific knowledge, see Bleichmar, *Visible Empire*, 155–58.

## CHAPTER 2. SHELLS, BODIES, AND THE COLLECTOR'S CABINET

I would like to thank Marisa Anne Bass and her coauthors for advice and comments.

1. Inventory of the *Kunstkammer* by Archduke Ferdinand II (1596) cited from List, "Wiener Goldschmiede und ihre Beziehungen zum kaiserlichen Hofe," 293: "Ein scheener indianischer, mit vergultem beschlecht, schneggen."
2. List, "Wiener Goldschmiede und ihre Beziehungen zum kaiserlichen Hofe," 292.
3. On Kaltemarckt, see Gutfleisch and Menzhausen, "'How a *Kunstkammer* Should Be Formed.'"
4. Kaltemarckt, *Bedenken wie eine Kunst-cammer*, translation cited from Gutfleisch and Menzhausen, "'How a *Kunstkammer* Should Be Formed,'" 11: "So uff der Erden, in der Erden / in Wassern und Meer gefunden wird . . . was durch Natur und Kunst von solchen / Gewechsen zu trinck und andern Geschirrn / formirt und gemacht . . . Sintemal derer ding hin und wider / in Teutschland und Italia vil zubekommen."
5. Nautilus cups formed part of the 1520 church treasury of Halle. An example of a merchant collection that held a nautilus cup is the one kept by Octavian Fugger (1549–1600) in Augsburg. On relationships between merchant and court collections in sixteenth-century Germany, see Meadow, "Merchants and Marvels."
6. Mette, *Der Nautiluspokal*. For a recent interpretation of nautilus cups in a seventeenth-century Dutch context, see Kehoe, "The Nautilus Cup between Foreign and Domestic." For an eighteenth-century specialist's approach to sixteenth-century and seventeenth-century vessels, see Zuroski, "Nautilus Cups."
7. See Tlusty, *Bacchus and the Civic Order*. Sexual activities related to drinking are an established trope in the representation of peasant feasts (see, for example, Hans Sebald Beham, *The Peasants' Feast or the Twelve Months*, 1546–47), but also appear on the surfaces of shell cups made for elite consumers. See, for example, Cornelis Bellekin, *Nautilus Cup*, seventeenth century, Bayerisches Nationalmuseum, inv. no. R 2504.
8. See, for example, Kaufmann, "Remarks on the Collections of Rudolf II"; Grote, *Macrocosmos in Microcosmo*.
9. Meadow, "Merchants and Marvels," 195; Meadow, "Quiccheberg and the Copious Object," 53.
10. Roberts, "A World of Wonder," 403.
11. Grasskamp, "Spirals and Shells."
12. Mette, *Der Nautiluspokal*, 68.
13. On the case of oyster shells in the visualization of proverbs, see Cheney, "The Oyster in Dutch Genre Paintings."
14. Rothstein, "Jan van Hemessen's Anatomy of Parody," 478.
15. While the displayed collectibles change throughout different workshop versions of the same mythological theme, the central scene of Lycomedes's daughter pointing at shells appears in at least three of them: the one illustrated in this chapter, the one in the Pushkin Museum in Moscow, and Workshop of Frans Francken the Younger, *Daughters of Lycomedes*, seventeenth century, oil on panel, sold by Antiquité Promenade. https://www.anticstore.art/73198P.
16. The story is briefly narrated by Ulysses in Ovid, *Metamorphoses* 13.123–381. http://ovid.lib.virginia.edu/trans/Metamorph13.htm#Bkthirteen123. The *Achilleid* features a longer version of the tale with further details such as the use of the trumpet: Statius, *Achilleid* 1.709, 874. https://www.theoi.com/Text/StatiusAchilleid1B.html.
17. Ovid, *Metamorphoses* 13.123–381. http://ovid.lib.virginia.edu/trans/Ovhome.htm; Statius, *Achilleid* 1.841.
18. Rosenthal, "Frans Francken the Younger's Discovery of Achilles," 727. Similarly, paintings of early modern collections by another seventeenth-century Antwerp painter, Guillam van Haecht (1593–1637), have been interpreted as reflections on the desire to collect art as similar to and entangled with amorous desires. Schwartz, "Love in the Kunstkamer," 46–48.
19. Rosenthal, "Frans Francken the Younger's Discovery of Achilles," 720.
20. See, for example, Frans Francken II, *An Art and Curio Cabinet*, 1620–25, oil on wood, 74 × 78 cm, Kunsthistorisches Museum Wien, inv. no. GG 1048, and Frans Francken II, *A Collector's Cabinet*, 1625, oil on panel, 53.2 cm × 73 cm, sold at art fair, May 2000, London.
21. On general connections among Christ, coral, and blood, Pointon, *Brilliant Effects*, 135ff.; on the motif in Netherlandish art, Rijks, "'Unusual Excrescences of Nature,'" 12; on examples in German *Kunstkammer* collecting, Grasskamp, "Metamorphose in Rot," 19.
22. Finlay, *The Pilgrim Art*, 70.
23. Mish, *The Merriam-Webster New Book of Word Histories*, 371.
24. Belon, *L'histoire naturelle*, 53.
25. Julius Scaliger, *Exotericarum exercitationum* (1557), quoted in Lightbown, "Oriental Art and the Orient," 231.
26. Weststeijn, "Cultural Reflections," 217, 223.
27. Syndram and Minning, *Das Inventar von 1619*, fol. 15.
28. Ibid., fol. 12r: "An italienischen trinck- und andern geschirren, welche anno 1590 von dem hertzog von Florentz sind verehret worden."
29. Kopytoff, "The Cultural Biography of Things."
30. Grasskamp, "Frames of Reflection," 72ff.
31. Aldrovandi, *De reliquis animalibus*, 266. Aldrovandi identified the nautilus as coming from "the Indies" (*ab indis*). At times the treatise specifically refers to the "West Indies" (*Occidentalia India/India Occidentalis*), but overall uses the generic term "India" to denote Asia as well as the Americas.
32. Mette, *Der Nautiluspokal*, 82ff.
33. Ibid., 84ff.
34. Grasskamp, "Frames of Reflection," 79ff.
35. For an example of a shell incised with the image of a dragon, see London, British Museum, inv. no. WB.114. Examples of nautilus shells with bird-and-flower patterns are discussed in Mette, *Der Nautiluspokal*, 82.
36. Grasskamp, *Objects in Frames*, 25–51.
37. Hay, *Sensuous Surfaces*, 67.

38. Ibid., 13.
39. Grasskamp, *Objects in Frames*.
40. Manning Stevens, "New World Contacts."
41. Scot, *The discoverie of witchcraft*, 8.
42. See, for example, Albrecht Dürer, *Witch Riding Backwards on a Goat*, engraving, 1501–2. Hans Baldung (called Hans Baldung Grien), *The Witches*, chiaroscuro woodcut in two blocks, printed in gray and black, 1510.
43. For a representation of an octopus-like creature in a nautilus shell, see Aldrovandi, *De reliquis animalibus*, 260; for the image of a large foreign sea snail opposite depictions of local specimens, see Aldrovandi, *De reliquis animalibus*, 390f. The motif is copied in later treatises such as Buonanni, *Ricreatione dell' occhio*, fig. 32.
44. Albertinus, *Der Welt Tummel- und Schaw-Platz*, cited from Grasskamp, "Spirals and Shells," 152ff.
45. Grasskamp, *Objects in Frames*, 121–23.
46. Mason, *Deconstructing America*, 173.
47. Scholars hold different opinions on the letters' authorship, the authenticity of their contents, and other aspects of their interpretation. See, for example, Roukema, "The Mythical 'First Voyage.'"
48. Daly and Rengel, *Greek and Roman Mythology*, 59.
49. For period references to poetic associations between female skin and white ivory, see Vickers, "Diana Described," 266; Yandell, "Iconography and Iconoclasm," 549. The German poet Christian Hoffmann von Hoffmannswaldau's "Description of Immaculate Beauty" (*Beschreibung vollkommener Schönheit*) published in 1670 employs the metaphor of alabaster to refer to the skin of a dangerously seductive woman.
50. Ovid, *Metamorphoses* 13.791–804.
51. Grasskamp, "Spirals and Shells," 151.
52. Pedersen, *Mermaids*, 27.
53. Edward Weston, "Diary Entry, July 7, 1927," in Weston, *The Daybooks of Edward Weston*, 2:32.
54. For a historical overview on sexual aspects of the symbolism of shells, see Eliade, *Images and Symbols*, 125–50.
55. Burke, "Nakedness and Other Peoples."

## CHAPTER 3. SHELL LIFE, OR THE UNSTILL LIFE OF SHELLS

In writing this chapter, I am grateful for assistance, advice, comments, and inspiration from Jim Clifton, Ned Cooke, Milette Gaifman, Denise Gill, Rafeeq Hasan, Pamela Lee, Rhodri Lewis, Renata Nagy, Verity Platt, and Carolyn Yerkes. I also thank my fellow authors, and especially Hanneke Grootenboer. All translations, unless otherwise indicated, are my own.

1. See Mund et al., *The Mayer van den Bergh Museum*, 378–401.
2. For further representations of the saint tormented by sea monsters and marine creatures, dating back to as early as the thirteenth century, see Benker, *Christophorus*, 106–9. See also Stahl, *Die Legende*, passim.
3. The first to associate this triptych with Bosch's legacy was Unverfehrt, *Hieronymus Bosch*, 192, 266, no. 63. On Bosch's inventive strategies in the use of nature study, see also Koerner, "Impossible Objects."
4. Bass and Wyckoff, *Beyond Bosch*, 96–101, no. 5, and Pokorny, "Alart du Hameel," 269.
5. Pieters et al., "Een 15de-eeuwse sector," 227, 229, fig. 17. An exotic trumpet shell also appears in another closely contemporary Netherlandish painting, Jan Gossart's so-called *Neptune and Amphitrite* of 1516 (Staatliche Museen zu Berlin, Gemäldegalerie, inv. no. 648), where it is worn by the sea god as a precarious codpiece. Although identified in recent literature as a *Charonia tritonis* (Ainsworth, *Jan Gossart's Renaissance*, 220), this shell is more likely a *Charonia lampas* given its stocky and knobbed form.
6. See most recently Ekserdjian, *Still Life*, and also Kemperdick and Sander, "The Still Life."
7. Take Sandrart's reference to the still-life painter Cornelis de Heem, whom he describes as a specialist in representing "all sorts of still-standing natural things" (*allerley stillstehende Natürlichkeiten*). See Von Sandrart, *Teutsche Academie*, CCXVIII. See further Hochstrasser, "*Stil-staende dingen*," 106–7; Fehrenbach, "Cut Flowers," 584–85; Chong, "Contained under the Name"; and Woodall, "Laying the Table," esp. 983–84.
8. Ribouillault, "Seeing Christ, Reading Nature," esp. 534–40, for an example of how shells signify within another innovative devotional painting of the period: Jan Swart van Groningen's *Christ in a Landscape* (c. 1530–40) in the Montreal Museum of Fine Arts (inv. no. 1996.15).
9. For an excellent account of shell formation, see Vermeij, *A Natural History*.
10. On shell vessels, see further chapter 2 of this book.
11. See Goldgar, *Tulipmania*, esp. 62–130; Leonhard, "Shell Collecting"; Bass, "Florilegium"; and the introduction to this book.
12. Vergara, *The Art of Clara Peeters*, 106–9, no. 11, with prior literature.
13. Identified (from left to right) as a *Hexaplex rosarium*, *Harpa harpa*, *Harpa doris*, and *Cittarium pica*. See ibid., 108.
14. See Brusati, "Stilled Lives," esp. 172–74. For compelling discussion of still-life painting as a site of "ungendered imaginary" for early modern women artists, see also Garrard, "The Not-So-Still Lifes," esp. 73.
15. Ayooghi, Böhmer, and Trümper, *Die Stillleben*, 117–19, no. 8.
16. On the destructive role of insects in early modern still life, see further Berger, *Caterpillage*, passim.
17. Pace Segal, "Shell Still Lifes," 88, and passim. See further the introduction to this book.
18. See discussion of the painting's technical examination in Wallert, *Still Lifes*, 52–55, no. 4, esp. 53–54, fig. 1.
19. See the examples illustrated in Ayooghi, Böhmer, and Trümper, *Die Stillleben*, 132–36, nos. 13–14, 143–45, no. 17, and 164–65, no. 24.
20. Zuidervaart, "'Scientia' in Middelburg," esp. 57–59.
21. Van Borsselen, *Strande*, A5r: "Maer moeten al te gaer der Schelpen schoonheydt wijcken, / Geen Tulipa sagh m'oyt hier by te vergelijcken." For further discussion of this passage, see also Prosperetti, "'Conchas legere,'" 401–2; Goldgar, *Tulipmania*, 81–82; and Dance, *A History of Shell Collecting*, 11.
22. Van Borsselen, *Strande*, A5r: "op dese sietmen leven / De woeste water-plas, hier d'ebbe, daer den vloed / D'wijl Son, Maen, Sterr' end Zee de Schelpen groeyen doet."
23. Keyes et al., *Masters of Dutch Painting*, 238–39, no. 99, with additional literature.
24. Ibid., 239, fig. 2, and n5. The current location of this pendant is unknown.
25. It is interesting to note that shells were associated with resuscitation and vitality in other cultures as well—in particular, among the indigenous inhabitants of North America. See

Hamell, "Trading in Metaphors," esp. 6, with thanks to Ruth Phillips for this reference.

26. Pliny the Elder, *Naturalis historia* 9.52: "concharum genera, in quibus magna ludentis naturae varietas: tot colorum differentiae, tot figurae." Translation adapted from Pliny, *Natural History*, 231–33.
27. Ibid., 9.52: "iam distinctione virgulata, crinita, crispa, canaliculatim, pectinatim divisa, imbricatim undata, cancellatim reticulata, in obliquum, in rectum expansa, densata, porrecta, sinuata; brevi nodo ligatis, toto latere conexis, ad plausum apertis, ad bucinam recurvis."
28. See Hendrix and Vignau-Wilberg, *Mira Calligraphiae Monumenta*, passim, and Vignau-Wilberg, *Joris and Jacob Hoefnagel*, 82–91, no. A4.
29. J. Paul Getty Museum, MS 20 (86.MV.527), fol. 118r. For further discussion of Hoefnagel's understanding of nature, see Bass, *Insect Artifice*.
30. Bachelard, *The Poetics of Space*, 126.
31. See Clifton et al., *Pleasure and Piety*, 140–42, no. 30, with prior literature.
32. As compellingly observed in Woodall, "Wtewael's *Perseus and Andromeda*," esp. 41–44.
33. For identification of the other shells in the painting, see Le Chanu, *Joachim Wtewael*, 30–31.
34. Shakespeare, *Richard III*, 1.4.30–34.
35. See, for example, Shakespeare, *The Tempest*, 1.2.474–82.
36. On Shakespeare's larger interest in visual effects and deceits of the eye, see Elan, *Shakespeare's Pictures*, esp. 260–64. On the playwright's use of shell imagery, see also Kelley, "Shells."
37. I borrow the phrase "uncanny effect" from Freud, *The Uncanny*, esp. 150.
38. See Brooks, "*Richard III*."
39. Martirano describes the grotto in his own verses as "fatta è sol di cocchiglie / di color bianco, nero, giallo e misco." His friend and fellow poet Luigi Tansillo also described the grotto in verses dedicated to the delights of Leucopetra. See Martirano, *Il pianto*, 76; Tansillo, *L'egloga*, 257 ("Stanze a Bernardino Martirano"); and also Milburn, *Luigi Tansillo*, 55–57. For further discussion of shell grottoes, see chapter 5 of this volume.
40. See esp. Smith, "The Genre and Critical Reception." See also Brown, "Underworld Sailors," passim, and Piepho, "Phineas Fletcher."
41. Sannazaro, *Eclogae*, I: 38–41: "En tibi caerulei muscum aequoris, en tibi conchas / purpureas, nec non toto quaesita profundo, / et vix ex imis evulsa corallia saxis / adferimus." Text and translation from Sannazaro, *Latin Poetry*, 104–6.
42. Sannazaro, *Eclogae*, I: 72–75: "Nunc iuvat immensi fines lustrare profundi, / perque procellosas errare licentius undas / Tritonum immistum turbis, scopulosaque cete / inter et informes horrenti corpore phocas." Sannazaro, *Latin Poetry*, 108.
43. Eliasson et al., *Italian Paintings*, 187–88, no. 71.
44. For a particularly penetrating discussion of this painting, see Grootenboer, "Sublime Still Life," esp. 20, including illustrations of Coorte's other paintings of shells. See also Buvelot, *The Still Lifes*, 48–54.
45. Hooke, *Lectures and Discourses*, 335. See also Rossi, *The Dark Abyss of Time*, 15–16. For a parallel connection between antiquarianism and the investigation of fossilized shells in seventeenth-century Italy, see Findlen, "Agostino Scilla," esp. 131–32.
46. See esp. Langereis, "De verzameling," and also chapter 1 of this book.
47. Gorlaeus, *Dactyliotheca*, unpaginated front matter (from Gorlaeus's dedicatory letter to the Leiden doctor and botanist Aelius Everhardus Vorstius): "Non facile dixerim quam inter reliqua mihi fuerit volupe uberrimam illam rerum antiquarum apud te conflugem summo cum animi mei stupore intueri, singulaque minutim perlustrare, ac digitis contrectare."
48. On Linard and his career, see esp. Szanto, "Pour Jacques Linard." See also Nusbaumer, *Jacques Linard*, 13–16; Coatalem et al., *La nature morte*, 240–58; Salvi, *D'après nature*, 21–27; and Rosenberg, "Jacques Linard."
49. Schnapper, *Le géant*, esp. 189–90.
50. See Wellington, *Antiquarianism*, esp. 80–81; Kenny, *The Uses of Curiosity*, esp. 224–25; and on the relation between the collecting of shells and antiquities in eighteenth-century France, see also Pomian, *Collectors and Curiosities*, 121–38.
51. Belon, *L'histoire naturelle*.
52. For an overview of Palissy's work and practice, see Amico, *Bernard Palissy*, and Shell, "Casting Life," See also Newman, *Promethean Ambitions*, 153, 156–57, for Palissy's particular interest in shells and the origins of fossils.
53. Palissy, *Recepte veritable*, 213: "Penses-tu que ces petites concavitez et nervures, qui sont desdites coquilles, soyent faites seulement par ornement, et beauté? Non, non: il y a quelque chose d'avantage: cela augmente en telle sorte la force de ladite forteresse, comme feroyent certains arbotans appuyez contre une muraille, pour la consolider."
54. Nusbaumer, *Jacques Linard*, 76, no. 23.
55. Callimachus, *Epigrams* 6. See also Prescott, "Callimachus' Epigram," and Gutzwiller, "The Nautilus."
56. I am especially grateful to Verity Platt for her assistance with this translation.
57. Translation from Valéry, *Man and the Sea Shell*, 134. Originally published as Valéry, *L'homme et la coquille*.
58. Nitze-Ertz, Kleinmann, and Brakensiek, *Das flämische Stilleben*, 106–9, no. 32, with prior literature.
59. Van Mander, *Den grondt*, 1:248–49, fols. 45v–46r (2.11.8): "Exempel de bespraeckte Papegaeyen, / Voghels, schelpen, en meer dinghen gheschapen, / Hoe alle verwen malcander verknapen, / Dus Natuere, die ons alles maeckt vroeder, / is van het schilderen voester en moeder."

### CHAPTER 4. THINKING WITH SHELLS IN PETRONELLA OORTMAN'S DOLLHOUSE

1. The argument that dollhouses are female cabinets of curiosity is made by, among others, Pijzel-Dommisse, *Het Hollandse pronkpoppenhuis*, 13–18; and Broomhall and Spinks, "Imagining Domesticity."
2. Stewart, *On Longing*, 55.
3. Valéry, "Man and the Sea Shell," 117.
4. Bachelard, *The Poetics of Space*, esp. chapters 5, 7, and 9.
5. Pijzel-Dommisse, *Het Hollandse pronkpoppenhuis*, esp. 247–347; Broomhall and Spinks, "Imagining Domesticity"; van Eeghen, "Het poppenhuis van Petronella Oortman," 113–17.
6. Few documents that give insight into the self-images of early modern Dutchwomen are extant. Broomhall and Spinks, "Imagining Domesticity," 122, suggest expanding Rudolf Dekker's

notion of ego-document to include dollhouses, given the investment and energy usually spent on them.

7. Appadurai, *The Social Life of Things*; Smith, "Material Christianity in the Early Medieval Household."
8. Carruthers, *The Craft of Thought*.
9. Jaap van der Veen examined ninety seventeenth-century Amsterdam collections and observed that the core of an assemblage of *naturalia* was formed by a series of shells. Van der Veen, "Dit klain Vertrek," 233.
10. Ibid.
11. The porcelain vases adorning the cabinet are carefully placed to contrast with the porcelain-like luster of the shells' insides. In still life painting, the resonance between porcelain and mother-of-pearl is deliberately sought.
12. Ter Molen, *Plezierreis*, 268. Pijzel-Dommisse, *Het Hollandse pronkpoppenhuis*, 249.
13. It is tempting to think that the shells may have found their way to Europe via Georgius Rumphius, who catered to many European collectors. Another possible candidate is François Valentyn (1656–1727), a preacher who spent considerable time in Ambon in the service of the East India Company, though not as long as the forty years cited by Von Uffenbach. Van der Roemer, "Neat Nature" 71.
14. "voorts zyn de laadjes gestoffeerd met allerhande natuurlyke Orientaalse Hoortntjes en Schelpen in Cierelyke Compartementen en Lofsgewys geschikt, alles zeer kleyn en egt." *Beschryving* (eighteenth century, no date), quoted in Pijzel-Dommisse, *Het Hollandse pronkpoppenhuis*, 329.
15. Pijzel-Dommisse, *Het Hollandse pronkpoppenhuis*, 331.
16. Von Uffenbach writes that he admired "ein Stuck curieux" with the face of a bearded man. Ter Molen reads "ein Stuck" to indicate a single shell, and though this is possible, on the basis of the layout used for other shell collections, I assume, with Pijzel-Dommisse, that it refers to an overall design consisting of several shells.
17. Examples of small-scale cabinets can be found in paintings, for instance Cornelis de Man, *Merchant in an Interior*, present location unknown (see Bergvelt and Kistemaker, *De wereld binnen handbereik*, 77), and in various still lifes by Jan van der Heyden (1637–1712).
18. Von Uffenbach, *Merkwürdige Reisen*, 659: "Auf dem Tische stund ein artig schwartz Cabinetgen, so seiner Frau, wie er sagte, gehorte. In selbigem waren fast alle Sorten vond Conchylien ins kleine, wier er sie ins grosse hatten. Sie waren all gar shon und sher zierlich gelegt." See also Pijzel-Dommisse, *Het Hollandse pronkpoppenhuis*, 331.
19. Pijzel-Dommisse, *Het Hollandse pronkpoppenuis*, 251.
20. Broomhall and Spinks, "Imagining Domesticity," 114–15.
21. Like most of the paintings, this one is signed by the well-known Amsterdam master painter Johannes Voorhout (1647–1717).
22. Pijzel-Dommisse, *Het Hollandse pronkpoppenhuis*, 17, and Schotel, *Letter en Oudheidkundige Avondstonde*, 83. See also Leonhard and Leuker, "Who Commissioned Hollar's Shells?"
23. I am very grateful to Dror Wahrman for sharing this unpublished information with me.
24. "Voorts, dat yder Meubel enz: heeft zyne volslagene eygenschap, niet alleen in stof, maar bewerking; zynde niets gelapt, gelymt, of uyt een stuk te voorschyn gebragt; maar, 't geen gespykert moet weezen, is gespykerdt; 't gelymde, gelymdt; . . . gesmeedt, gesmeedt . . . als zulk in het groot, immer wordt gevonden." Pijzel-Dommisse, *Het Hollandse pronkpoppenhuis*, 249–50.
25. "Efin dit wonder konststuk met zo veel vernuft, oordeelmoeyte enorme kosten en arbeydt te zamen gesteld! is het volmaaktste middel dat bedacht kan worden, om aan vreemdelingen een net denkbeeld te geven van de alom beruchte Hollandsche zindelyke en proper Oeconomie of Huishouding." Van de Eeuw 1600–1700. *Beschryving*, w.d., p. II. (Bibliotheek Rijksmuseum, Amsterdam). Quoted in Pijzel-Dommisse, *Het Hollandse pronkpoppenhuis*, 250.
26. Zacharias von Uffenbach made two trips through the Low Countries, one with his brother Friedrich between 1709 and 1711, and a shorter journey with various family members in 1718, during which he visited Oortman's dollhouse. For more on Von Uffenbach's detailed account of this last trip, see Ter Molen, *Plezierreis*.
27. Ibid., 268.
28. Ter Molen, "Een bezichtiging," and Ter Molen, *Plezierreis*.
29. Cats, *Houwelick*, "Kinder Spel: Dit spel, al schijntet sonder sin, Dat heeft een kleyne vverelt in; De vveerelt en haer gans gestel / En is maer enckel kinder-spel."
30. "Al wat men hier op Aerden siet / Is poppe goed en anders niet/ De mensch, al wat hij daar van vint / Die speelter mee, gelijk een Kint / Hij heeft het Lief een Korten tijt / Dat hij daar naar Licht van hem smijt / Zoo is the mensch dans als men vindt / Niet tweemals, maar altijds een Kindt." De Brune, *Emblemata of Sinne-werck, no. LII (first published in 1624)*. See also Willemsen, "Kinder-Spel en Poppe-Goet"; and Broomhall and Spinks, "Imagining Domesticity," 118.
31. Van Eeghen, "Het poppenhuis."
32. "Blijft binnen u schelpen, dat zal u helpen" Cats, *Spiegel*, appendix. Also cited in translation in Hollander, *An Entrance for the Eyes*, 162.
33. "'t Waer wel soo goet dat de Mensch in zijn eiygen Schelp kroope, ende Aldus by hem selven dachte: Wat was ick, eer ick gebooren was?" Quoted in Prosperetti, "'Conchas legere,'" 409.
34. The snail as a metaphor for the housing of the soul is a rich iconographic motif, which had already been cited by Pietro Valeriano in his *Hieroglyphia* (1567). See also Arasse, "The Snail's Gaze."
35. Erasmus, "The New Mother," 592, 594.
36. Ibid., 600.
37. Ibid.
38. Jorink, *Nederlandse geleerden*, 140.
39. Findlen, *Possessing Nature*, esp. pt. 1.
40. Cicero, *De oratore* 2.22, 213–15. See also Prosperetti, "'Conchas legere,'" 395.
41. Prosperetti, "'Conchas legere,'" passim.
42. "Their [Scipio and son] triumphal habit put off, strolling along at leisure, and unconscious of depressing cares, over the hills and along the shore, and often picking up little shells and sea pebbles, of different sorts, both black and white." Petrarch, *The Life of Solitude*, 288.
43. Cicero, *De oratore* 2.22.
44. Dance, *A History of Shell Collecting*, 43. This was the inventory of the collection of Bonnier de la Masson comprised by Edmé-François Gersaint. For my discussion of that collection and its philosophical potential, see Grootenboer, "A Clock Picture."
45. Jutting, *Brief History*, 199; Leonard, "Shell Collecting," 183. Joan Coenraad Brandt's shell collection, sold in 1792, lists "a very fine

and bright Amboina Wentletrap" put down for fifty guilders. Dance, *A History of Shell Collecting*, 59.

46. Rumphius, *The Ambonese Curiosity Cabinet*, 220–21.
47. Jutting, *Brief History*, 199. See also Dance, *A History of Shell Collecting*, 12.
48. The mathematical term *limaçon de Pascal* refers to a bicircular algebraic curve. The term was coined in 1650 by Gilles de Roberval (1602–75) in reference to the earliest research on the curve by Étienne Pascal (father of Blaise).
49. Beudeker, *De Sprekende Konstkamer*, 160–61.
50. Van Borsselen, *Strande*, D2r: "Laet dyne Schepsel zijn een cromme wendel-trap."
51. Montaigne, *The Complete Essays*, 250–63.
52. Derrida, *Margins*, 253.

## CHAPTER 5. SHELLS AND GROTTOES IN EARLY MODERN GERMANY

I would like to thank in particular Marisa Bass, Hanneke Grootenboer, and Anne Goldgar for their comments on this piece.

1. On Duchess Magdalena Sibylla von Württemberg, see Kremer, *Magdalena Sibylla von Württemberg*.
2. Weismann, *Christliche Betrachtung*, 24–42.
3. Bepler, "Lektüre, Schrift und Gebet," 316–20.
4. On early modern automata, see Marr, "Gentille curiosité," and Keating, *Animating Empire*.
5. For work on shells and the aesthetic of the early modern grotto, see Brunon and Mosser, *L'imaginaire des grottes*, 170–78; Rincón, "The Aesthetics of the Early Modern Grotto"; Franke, "Natürliche Kunst und künstliche Natur."
6. Fleischhauer, *Renaissance im Herzogtum Württemberg*, 304–13.
7. Hauptstaatsarchiv Stuttgart (henceforth HStAS) A62 Bü 7. "Grottenbau" (1609–28), unfoliated; Frischlin, "Schöne lustige Antiquitäten und denckwürdige Historien von Ursprung, alt Herkommen und Erbawung des fürstlichen Hauses Württemberg" (1618). Württembergische Landesbibliothek, Cod.Hist.2°73, 237. An inventory from 1634 described the space as adorned with "a large number of all kinds of stones, shells of snails and others." See Fleischhauer, *Renaissance im Herzogtum Württemberg*, 310. All translations are my own unless otherwise indicated.
8. For literature on early modern grottoes, see Miller, *Heavenly Caves*; Morel, *Les grottes maniéristes*; Brunon and Mosser, *L'imaginaire des grottes*.
9. Quoted in Morgan, *The Monster in the Garden*, 1.
10. On early modern curiosity, see Marr, "Introduction," esp. 6; Benedict, *Curiosity*.
11. For more on shells as objects to think with, see chapter 4 of this book.
12. Elderkin, "The Natural and the Artificial Grotto."
13. Ovid, *Metamorphoses* 11.20. On early modern interaction with classical ideas about grottoes, see Rincón, "The Aesthetics of the Early Modern Grotto," esp. 15–18.
14. The seminal text on Italian gardens remains Lazzaro, *The Italian Renaissance Garden*.
15. HStAS A62 Bü 6. "Grottenbau" (1609–28), unfoliated.
16. Maxwell, "The Pursuit of Art."
17. Morgan, *Nature as Model*, 152–97.
18. Broomhall and Van Gent, *Gender, Power and Identity*, 227–28.
19. Fleischhauer, "Zur Tätigkeit des Salomon de Caus," 377–79.
20. Quoted in Rincón, "The Aesthetics of the Early Modern Grotto," 24.
21. Furttenbach, *Newes Itinerarium Italiae*, 105.
22. Furttenbach, *Architectura civilis*, 35.
23. On Genoese grottoes, see Hanke, "The Splendour."
24. On Furttenbach's life, see Stemshorn, *Der Kunst-Garten*; Furttenbach, *Lebenslauff*.
25. Furttenbach, *Lebenslauff*, 137, 144.
26. On Furttenbach's grotto, see Siebenhüner, "Entwerfen."
27. Instructions on grotto building also appear in *Architectura recreationis* (1640), *Feriae architectonicae* (1662), and *Mannhaffter Kunst-Spiegel* (1663).
28. Detail on the construction of the grotto structure itself appears in *Architectura recreationis* and *Mannhaffter Kunst-Spiegel*.
29. Furttenbach, *Architectura Privata*, 53. The later publication *Mannhaffter Kunst-Spiegel* details how to build a grotto.
30. Fleischhauer, *Renaissance im Herzogtum Württemberg*, 308.
31. HStAS A 62 Bü 6. "Grottenbau" (1609–28), unfoliated.
32. Furttenbach, *Architectura privata*, 53.
33. Furttenbach, *Architectura civilis*, 48.
34. Fleischhauer, *Renaissance im Herzogtum Württemberg*, 307–8.
35. HStAS A256 Bd 102. "Landschreiberei Rechnung" (1615/16), fols. 364v, 357r; A 256 Bd 101; "Landschreiberei Rechnung" (1614/15), fol. 370v.
36. Maxwell, "The Pursuit of Art," 428.
37. HStAS A25 Bü 115. "Das Grottenwerk im herzoglichen Lustgarten zu Stuttgart" (1742–48), unfoliated.
38. Ibid.
39. Lesser, *Testaceo-theologia*, 66.
40. Furttenbach, *Newes Itinerarium Italiae*, 102.
41. On the cost and labor associated with shell preparation, see also chapter 1 of this book.
42. Furttenbach, *Architectura Privata*, 55.
43. Savage, "Natural History," 36.
44. Fleischhauer, *Renaissance im Herzogtum Württemberg*, 306.
45. Ibid., 310.
46. A similar form of viewing is found in Arcimboldo's composite portraits. See Kaufmann, *Arcimboldo*, 10. On grotesques in grottoes, see Lazzaro, *The Italian Renaissance Garden*, 140–42.
47. Furttenbach, *Architectura civilis*, 38.
48. Von Oechelhäuser, "Philipp Hainhofers Bericht," 270.
49. Bredekamp, *The Lure of Antiquity*, 49.
50. Findlen, "Jokes of Nature," 296.
51. Maxwell, "The Pursuit of Art," 417; Franke, "Natürliche Kunst und künstliche Natur," 1080.
52. Kraus, *Description*, 7.
53. On early modern ingenuity, see Marr et al., *Logodaedalus*.
54. Furttenbach, *Architectura privata*, 61.
55. Furttenbach, *Architectura civilis*, 37.
56. Fleischhauer, "Zur Tätigkeit des Salomon de Caus," 378.
57. Furttenbach, *Architectura Privata*, 68.
58. Lesser, *Testaceo-theologia*, 20.
59. Kelley, "Arno River Floods," 751. On terror and early modern gardens, see Morgan, *The Monster in the Garden*; Lazzaro, *The Italian Renaissance Garden*, 206.
60. On the shell and its potential for reanimation, see also chapter 3 of this book.
61. Lesser, *Testaceo-theologia*, 9.
62. Miller, *Heavenly Caves*, 29–34.

63. Furttenbach, *Architectura civilis*, 35; Furttenbach, *Mannhaffter Kunst-Spiegel*, 149–50.
64. Leonhard, "Shell Collecting," 177; Lesser, *Testaceo-theologia*, 99–101.
65. Lesser, *Testaceo-theologia*, 9.
66. Furttenbach, *Architectura privata*, 60.
67. Landesmuseum Württemberg, *Die Kunstkammer der Herzöge von Württemberg*, 2:458–59; Koeppe, *Making Marvels*, 73, no. 15.
68. Findlen, "Jokes of Nature," 297–98.
69. On the relationship between the *Kunstkammer* and knowledge production, see Bredekamp, *The Lure of Antiquity*, passim.
70. Furttenbach, *Architectura privata*, 24.
71. On the emblem, see Graham, "*Emblema Multiplex*"; Wesseling, "Testing Modern Emblem Theory."

### CHAPTER 6. SHELLS, PRINTS, AND THE DISCERNING EYE

I am grateful to the editors and peer reviewers of this volume for their stimulating comments. Preliminary versions of this essay were presented in 2018 at the Renaissance Society of America Conference and at the Center for Reformation and Renaissance Studies, University of Toronto. Research funding was provided by Queen's University (Kingston, Canada) and the Social Sciences and Humanities Research Council of Canada.

1. On Rembrandt as a collector, see esp. Scheller, "Rembrandt en de encyclopedische Kunstkamer"; Van den Boogert, et al., *Rembrandt's Treasures*; Chapman, "Curiosity and Desire." He bought paintings, prints, and drawings for inspiration, investment, resale, and exchange; *naturalia* more likely for his own curiosity.
2. Hinterding and Rutgers, New Hollstein *Rembrandt*, cat. 247. For analysis, see esp. Hinterding, *Rembrandt Etchings from the Frits Lugt Collection*, cat. 129, 300–2; Luijten, "*The Shell* 1650"; Busch, "Rembrandts *Muschel*"; Leonhard, "Die Muschel"; Rife, "The Exotic Gift," 34–72.
3. On religion, art, and *naturalia*, see Van de Roemer, "God en het rariteitenkabinet"; Jorink, *Reading the Book of Nature*; Prosperetti, "'Conchas Legere.'" On the rivalry of art and nature in Rembrandt's milieu, see Dickey, "Saskia as Glycera," with additional references.
4. See Tomasi et al., *L'art multiplié*, on the rise of multiples and mass production. See Scholten, "The Larson Workshop," on small-scale sculpture in the Dutch Republic. On shells and porcelain, see chapter 2 of this book.
5. On the contemporary market for collectibles, see, inter alia, Rosier and Rogers, "Profits and Perils"; Boltanski and Esquerre, "The Economic Life of Things."
6. For networks of collectors, see also, inter alia, Bergvelt and Kistemaker, *De wereld binnen handbereik*; Montias, *Art at Auction*; Jorink, *Reading the Book of Nature*; Dietz, "Mobile Objects"; Tobin, *The Duchess's Shells*.
7. Van Borsselen, *Strande*, A5R. See further the introduction to this book and Goldgar, *Tulipmania*, 87 and passim, who notes that "marbled" patterns in nature were widely admired.
8. Bosschaert's motif is a *Polymita picta* while Van der Ast depicts *Liguus fasciatus*, native to Florida. On land snails, see Breure and De Heere, "From a 'domestic commodity' to a 'secret of trade.'" On van der Ast's watercolor (formerly attributed to Bartholomeus Assteyn), see also Luijten, "*The Shell* 1650," 260–61; Busch, "Rembrandts *Muschel*," 94, 97, fig. 7; on still life, Segal, "Shell Still Lifes"; Leonhard, "Shell Collecting"; and chapter 3 of this book.
9. For related works by Francken, see Rijks, "A Painter, a Collector, and a Horseshoe Crab"; Busch, "Rembrandts *Muschel*," 97, fig. 9.
10. On Rumphius, see esp. Beekman, "Introduction," xxxv–cx; Van de Roemer, "The Serious Naturalist"; and chapter 1 of this book. On the cleaning and polishing of shells to enhance their beauty, see also, inter alia, Gersaint, *Catalogue raisonné de Coquilles*, 14–19; Grasskamp, "The Frames of Reflection."
11. Bergvelt and Kistemaker, *De wereld binnen handbereik*, 201–2, 232–58, 313–34. For an Antwerp collection, see Rijks, "A Painter, a Collector, and a Horseshoe Crab," 344–45.
12. Strauss, *Rembrandt Documents*, doc. 1637/2.
13. Crenshaw, *Rembrandt's Bankruptcy*, 94.
14. Rumphius, *D'Amboinsche rariteitenkamer*, 222; Leonard and Leuker, "Who Commissioned Hollar's Shells?," 236–37; Dietz, "Mobile Objects," 381. On Volkertsz, see further below. On the wentletrap, see also chapter 4 of this book.
15. Van Hoogstraten, *Inleyding*, 212.
16. Purchased by Rembrandt's student Leendert van Beyeren; whether for himself or for Rembrandt is a matter of debate; Strauss, *Rembrandt Documents*, doc. 1637/2.
17. Hinterding and Rutgers, New Hollstein *Rembrandt*, cat. 239; see recently Golahny, "The *One Guilder Print*," with additional references.
18. On the social status of collecting in the Dutch Republic, see Bergvelt and Kistemaker, *De wereld binnen handbereik*, 241–47; Montias, *Art at Auction*; Van de Roemer, "Neat Nature."
19. See Crenshaw, *Rembrandt's Bankruptcy*.
20. Strauss, *Rembrandt Documents*, doc. 1656/12; marine specimens, items 175, 176, and 179.
21. Hinterding, *Rembrandt Etchings from the Frits Lugt Collection*, 301n2; on Röver, see Slive, *Rembrandt and His Critics*, 56, 172–76.
22. Blanc, *Rembrandt*, 372, cat. 353. See also Rife, "The Exotic Gift," 62–71. It is possible Rembrandt's etching depicts a shell owned by an acquaintance, but most scholars presume it came from his own collection (see references in nn1 and 2 above).
23. *Portrait of Johannes Wtenbogaert (The Goldweigher)*, 1639; see Dickey, *Portraits in Print*, 66–88, with further references.
24. Smit, Sanders, and Van der Veer, *Hendrick Engel's Alphabetical List*, 95, nos. 527–29, 309, no. 1720; Dudok van Heel, "Mr. Johannes Wtenbogaert," esp. 152, 162; Leonhard and Leuker, "Who Commissioned Hollar's Shells?," 232.
25. Dudok van Heel, "Mr. Johannes Wtenbogaert," 162.
26. Dance, *A History of Shell Collecting*, 35; Beekman, "Introduction," lxxii, cvi; and chapter 1 in this book. On Cosimo's travels in the Netherlands, see also Van Veen, "Cosimo de' Medici's reis"; Jorink, *Reading the Book of Nature*, 247.
27. Beekman, "Introduction," cvii. On Sloane, see Delbourgo, *Collecting the World*; on Seba, Spary, "Scientific Symmetries," and the introduction to this book.
28. On Sloane as a shell collector, see Wilkins, "A Catalogue and Historical Account"; on his print collection, Griffiths, "Sir Hans Sloane." On Courten, see Gibson-Wood, "Classification and Value"; Kusukawa, "William Courten's Lists."
29. British Library, Sloane MS 3961, fol. 28v; Wilkins, "A Catalogue and Historical Account," 5–7; Kusukawa, "William Courten's Lists," 2, fig. 1. See also Tobin, *The Duchess's Shells*, 205–8; Roos, *Martin Lister and His Remarkable Daughters*, esp. 9, 115–16.
30. Lippincott, *Selling Art in Georgian London*, 29, 36–37, 118–22, 146.

31. Griffiths, "The Reverend Clayton Mordaunt Cracherode"; Wilkins, "The Cracherode Shell Collection."
32. Tobin, *The Duchess's Shells*, 27, 233; Pelling, "Selling the Duchess."
33. Pelling, "Selling the Duchess," 16.
34. Inv. F,5.107 (1973,u.1035), bequeathed by Cracherode in 1799, with Remy's signature and mark, Lugt 2136.
35. Dietz, "Mobile Objects," 377–79; on Remy, see also Bleichmar, "Learning to Look"; Marandet, "Pierre Remy."
36. See esp. Glorieux, *À l'Enseigne de Gersaint*; McClellan, "Watteau's Dealer"; McClellan, "Edmé Gersaint and the Marketing of Art"; Spary, "Scientific Symmetries."
37. Gersaint, *Catalogue raisonné de Coquilles*, 20; Gersaint, *Catalogue raisonné des différents Effets*, 230–31; Glorieux, *À l'Enseigne de Gersaint*, 309; Spary, "Scientific Symmetries," 13.
38. Glorieux, *À l'Enseigne de Gersaint*, 386; Dietz, "Mobile Objects," 375–379. On Vincent, see Spary, "Scientific Symmetries," 1–12, and the introduction to this book.
39. *Mercure de France*, November 1735, 2459–61; Spary, "Scientific Symmetries," 13.
40. Gersaint, *Catalogue raisonné de Coquilles*; Glorieux, *À l'Enseigne de Gersaint*, 385; McClellan, "Watteau's Dealer," 445–48.
41. On Boucher, see Priebe, "The Artist as Collector."
42. Buonanni, *Ricreatione dell'occhio*; Gersaint, *Catalogue raisonné de Coquilles*, vii, 6, 1–22. See also Dietz, "Mobile Objects," 374–75; Spary, "Scientific Symmetries," 12–19; Bleichmar, "Learning to Look."
43. Glorieux, *À l'Enseigne de Gersaint*, 420.
44. Gersaint, *Catalogue raisonné de toutes les Pièces*; Gersaint, *A Catalogue and Description*. See Preston, "Gersaint on Rembrandt"; Dickey, "'The Finest Possible State,'" with further references.
45. Gersaint, *A Catalogue and Description*, 80, cat. 154. The nickname *Damier*, or checkerboard, derives from the shell's distinctive pattern. The English translation slightly condenses the French of 1751, in which the print is deemed one of Rembrandt's "most rare" and its subject "a fairly common shell . . . of no interest"; Gersaint, *Catalogue raisonné de toutes les Pièces*, 129, cat. 154.
46. Gersaint, *Catalogue raisonné de toutes les Pièces*, "Avertissement des Éditeurs," ix–x.
47. On Gersaint and "conchological aesthetic principles," see Spary, "Scientific Symmetries," 13; Spary, "Rococo Readings"; Leonhard, "Die Muschel."
48. Gersaint, *Catalogue raisonné de Diverses Curiosités*, 38–42; Glorieux, *À l'Enseigne de Gersaint*, 416.
49. Gersaint, *Catalogue raisonné de Diverses Curiosités*, 226, lots 203 (17 prints) and 204 (27 prints). Cf. 222, lot 182, comprising twenty-four portraits, including rare prints by Rembrandt and Cornelis Visscher.
50. Ibid., iv–v, 49–130; Glorieux, *L'Enseigne de Gersaint*, 420–26.
51. Gersaint, *A Catalogue and Description*, t.p., n.p.
52. Gersaint, *Catalogue raisonné de Diverses Curiosités*, 281.
53. Ibid., prints, 38–260, shells, 281–94. On descriptive language, see also Dietz, "Mobile Objects," 367–71.
54. *Catalogue of the Portland Museum*, 121, lot 2704; it fetched £1.8.0 (annotated copy, Washington, Smithsonian Institution), about average for the lots of prints, which were sold on 23 May 1786.
55. "A very rare species of Turbo, *undescribed*," brought £1.11.6 at the Duchess of Portland's sale, while most lots of shells fetched only a few shillings; *Catalogue of the Portland Museum*, 36, lot 874 (3 May 1786).
56. Houbraken, *De Groote Schouburgh*, 1:270–71; Slive, *Rembrandt and His Critics*, 190.
57. Goldgar, *Tulipmania*, 79–98.
58. On rarity in nature, see Gaston, *Rarity*; Violle et al., "Functional Rarity"; Courchamp et al., "Rarity Value and Species Extinction"; on the rarity principle in market economics, Kapferer, Valette-Florence, and Lucatello, "The Impact of Brand Penetration."
59. See, e.g., Houbraken, *De Grote Schouburgh*, 1:271; Slive, *Rembrandt and His Critics*, 189–91; and recently Fucci, *Rembrandt's Changing Impressions*.
60. Gersaint, *Catalogue raisonné de Coquilles*, 14–16.
61. Spary, "Rococo Readings," 266, notes that Dezallier d'Argenville considered the Dutch notorious for this.
62. It appears extant impressions of *The Shell* were printed by Rembrandt himself, but many of his plates were reprinted by others. For a copy of *The Shell* by George Bickham the Younger (1721–71), see London, British Museum, inv. 1868,0822.671; Hinterding and Rutgers, New Hollstein *Rembrandt*, cat. 247 and copy a; see also Luijten, "*The Shell* 1650," 261–62.
63. Sotheby's London, 6 December 2001, £16,025. Impressions in fine condition sold at Christie's London, 18 June 1992, for £66,000 and Christie's New York, 13 May 1997, for $118,000.
64. On shell cabinets, see Vincent, *Wondertoneel der Nature*; Spary, "Scientific Symmetries"; and chapters 1 and 4 of this book. Prints and albums could also be stored in cabinets, like one built to house the nine-volume *Atlas Maior* published by Joan Blaeu in 1662 (Amsterdam, Rijksmuseum, inv. RP-D-BR-2012–1).
65. Van de Waals, *De prentschat van Michiel Hinloopen*, 10, 39n72, 127. On Schijnvoet and Rumphius, see Beekman, "Introduction"; Van de Roemer, "The Serious Naturalist."
66. A rare example is a decorative collage of his own prints by the Haarlem printmaker Cornelis van Noorde (1731–95), c. 1750–70 (Amsterdam, Rijksmuseum); Hinterding, Leeflang, and Van der Mullen, "Acquisitions," 364–65.
67. Gersaint, *Catalogue raisonné de Coquilles*, 19–20.
68. Glorieux, *À l'Enseigne de Gersaint*, 394; Tobin, *The Duchess's Shells*, 143.
69. Wilkins, "The Cracherode Shell Collection," 124–26, 134; Tobin, *The Duchess's Shells*, 241–42.
70. See Griffiths, *Landmarks in Print Collecting*.
71. Glorieux, *À l'Enseigne de Gersaint*, 309, 427.
72. Inv. NG-1977–210-W, NG-1977–214-W, and NG-1977–206-W. I am grateful to Claudia Swan for discussions about *De Witte Leeuw*; see further chapter 1 of this book, with an illustration (see fig. 15) of one shell retrieved from the shipwreck.
73. In "A Catalogue and Historical Account," and "The Cracherode Shell Collection," Wilkins traces the histories of specific Sloane and Cracherode shells; see also Dietz, "Mobile Objects," 380–81; Tobin, *The Duchess's Shells*, 178–99; Daston, "Type Specimens."
74. Gersaint, *Catalogue raisonné des Diverses Curiosités*, xiii; Glorieux, *À l'Enseigne de Gersaint*, 390.
75. For collecting, sociability, and exchange, see Leonhard, "Shell Collecting"; Prosperetti, "'Conchas legere'"; Dietz, "Mobile Objects"; Lippincott, *Selling Art in Georgian London*; Tobin, *The Duchess's Shells*. On the social capital of Rembrandt's collection, see Scheller, "Rembrandt en de encyclopedische Kunstkamer."
76. See Beekman, "Introduction," cvii; Coomans, "Schelpenverzamelingen," 199; and chapter 1 of this book.
77. For this practice in Britain, see Dickey, "'The Finest Possible State.'"

78. For instance, one copy of Gersaint's 1736 *Catalogue raisonné de Coquilles*, now in the British Library, passed through the hands of the naturalists Emanuel Mendes da Costa and Joseph Banks; McClellan, "Watteau's Dealer," 445n37; Glorieux, *L'Enseigne de Gersaint*, 396. On shell books, see, inter alia, Spary, "Scientific Symmetries," 14–16; Spary, "Rococo Readings." On annotated copies of the Portland sale catalogue, see Pelling, "Selling the Duchess," 23–27; Tobin, *The Duchess's Shells*, 142. On annotated copies of Rembrandt catalogues raisonnés, see Dickey, "'The Finest Possible State.'"
79. Gersaint, *Catalogue raisonné de Coquilles*, 2, 4–5, 17. A companion volume of illustrations announced in *Mercure de France*, October 1736, 2321, was never published; Glorieux, *À l'Enseigne de Gersaint*, 394–95.
80. Gersaint, *Catalogue raisonné de Coquilles*, 49–50.
81. Leonhard, "Shell Collecting," 192–93, 198–201; Leonhard, "Die Muschel," 136–39.
82. Turner, New Hollstein *Hollar*, cat. 1273–11; see, e.g., Luijten, "*The Shell* 1650," 260.
83. Rembrandt's inventory mentions unnamed prints by Hollar stored in an "East Indian basket" with prints by Hieronymus Cocq and Rembrandt himself; Strauss, *Rembrandt Documents*, doc. 1656/12, no. 235.
84. Turner, "Hollar in Holland," 77–78.
85. Also implicated is Salomon Savery, an Amsterdam printmaker with ties to Rembrandt. Rumphius, *The Ambonese Curiosity Cabinet*, 97; Leonhard and Leuker, "Who Commissioned Hollar's Shells?," 227–32; see also Busch, "Rembrandts *Muschel*," 94–97; Grasskamp, "The Frames of Reflection," 81–83.
86. Rembrandt, *Study for the Syndics: Volkert Jansz*, c. 1662, pen and ink, Rotterdam, Museum Boijmans van Beuningen, inv. R133. Van Eeghen, "De Staalmeesters"; Leonard and Leuker, "Who Commissioned Hollar's Shells?," 233.
87. Turner, New Hollstein *Hollar*, cat. 1280; Luijten, "*The Shell* 1650," 260; Leonard and Leuker, "Who Commissioned Hollar's Shells?," 235. On cone shell varieties, see Puillandre et al., "One, Four or 100 Genera?"
88. See, inter alia, Parshall, "Imago Contrafacta"; Daston and Galison, *Objectivity*; Egmond, *Eye for Detail*; Van de Roemer, "Neat Nature"; Kusukawa, "Illustrating Nature," 105–8; Spary, "Rococo Readings."
89. Houbraken, *De Groote Schouburgh*, 1:265–68; see esp. Sluijter, *Rembrandt and the Female Nude*; Dickey, "Disgust and Desire."
90. Dance, *A History of Shell Collecting*, 17, 24; Wilkins, "A Catalogue and Historical Account," 34; Luijten, "*The Shell* 1650," 260; Leonhard, "Shell Collecting," 192. For perspectives on reversal and dextrality, see also Leonhard, "Die Muschel"; Allmon, "The Evolution of Accuracy"; Breure and De Heer, "From a 'domestic commodity' to a 'secret of trade,' 81–82, 93, 95.
91. On reversal in printmaking, see Parshall, "Dürer and the Axis of Meaning"; Posèq, "Left and Right in Rembrandt's *Defeat of Goliath*," with further references (esp. 8n1).
92. White, *Rembrandt as an Etcher*, 1:169; Grasskamp, "The Frames of Reflection," 82; on Rembrandt's portrait prints, see Dickey, *Portraits in Print*.

# BIBLIOGRAPHY


### ARCHIVAL SOURCES

Brinck, Ernst. *Adversaria*. Streekarchivariaat Noordwest-Veluwe (SANwV), Archief Stadsbestuur Harderwijk (1231–1813), OAH 2013–2061.

Frischlin, Jakob. "Schöne lustige Antiquitäten und denckwürdige Historien von Ursprung, alt Herkommen und Erbawung des fürstlichen Hauses Württemberg" (1618). Cod.Hist.2°73, 237, Württembergische Landesbibliothek.

"Grottenbau" (1609–28). A62 Bü 6, Hauptstaatsarchiv Stuttgart.

"Grottenbau" (1609–28). A62 Bü 7, Hauptstaatsarchiv Stuttgart.

"Das Grottenwerk im herzoglichen Lustgarten zu Stuttgart" (1742–48). A25 Bü 115, Haupstaatsarchiv Stuttgart.

"Landschreiberei Rechnung" (1614/15). A256 Bd 101, Hauptstaatsarchiv Stuttgart.

"Landschreiberei Rechnung" (1615/16). A256 Bd 102, Hauptstaatsarchiv Stuttgart.

Stadsarchief Amsterdam 5072/353/44–44v, Desolate Boedelkamer, inventory of Francisco Gomes da Costa, 16 September 1646.

### PRIMARY AND SECONDARY SOURCES

Ainsworth, Maryan, ed. *Man, Myth, and Sensual Pleasures: Jan Gossart's Renaissance*. New York: Metropolitan Museum of Art, 2010.

Albertinus, Aegidius. *Der Welt Tummel- und Schaw-Platz*. Augsburg: Johann Krüger, 1612.

Aldrovandi, Ulisse. *De reliquis animalibus*. Bologna: Ioannem Baptistam Bellagambam, 1606.

Allmon, W. D. "The Evolution of Accuracy in Natural History Illustration: Reversal of Printed Illustrations of Snails and Crabs in Linnaean Works Suggest Indifference to Morphological Detail." *Archives of Natural History* 34 (2007): 174–91.

Álvarez, Myriam, et al. "Shell Middens as Archives of Past Environments, Human Dispersal and Specialized Resource Management." *Quaternary International* 239.1–2 (2011): 1–7.

Ambrose, W. R. "Archaeology and Shell Middens." *Archaeology and Physical Anthropology in Oceania* 2.3 (1967): 169–87.

Amico, Leonard. *Bernard Palissy: In Search of Earthly Paradise*. Paris: Flammarion, 1996.

Appadurai, Arjun. *The Social Life of Things: Commodities in Cultural Perspective*. Cambridge: Cambridge University Press, 1986.

Arasse, Daniel. "The Snail's Gaze." In *Take a Closer Look*, 17–38. Princeton, NJ: Princeton University Press, 2013.

Arens, Esther Helena, and Charlotte Kießling. "Knowledge and Power: Rumphius' Ambonese Herbal and Ambonese Curiosity Cabinet as Colonial Contact Zones." *European Review* 26 (2018): 461–70.

Ayooghi, Sarvenaz, Sylvia Böhmer, and Timo Trümper, eds. *Die Stillleben des Balthasar van der Ast (1593/94–1657)*. Petersberg: Michael Imhof Verlag, 2016.

Bachelard, Gaston. *The Poetics of Space*. Translated by Richard Kearney. New York: Penguin Books, 2014.

Barten, Sigrid. *Die Muschel in der Kunst: von der Renaissance bis zur Gegenwart*. Zurich: Museum Bellerive, 1985.

Bass, Marisa Anne. "*Florilegium*: The Origins of the Flower Still Life in the Early Modern Netherlands." In *Festschrift for David Freedberg*, edited by Claudia Swan, 11–25. Turnhout: Brepols, 2019.

———. *Insect Artifice: Nature and Art in the Dutch Revolt*. Princeton, NJ: Princeton University Press, 2019.

Bass, Marisa, and Elizabeth Wyckoff, eds. *Beyond Bosch: The Afterlife of a Renaissance Master in Print*. St. Louis, MO: St. Louis Art Museum, 2015.

Beekman, E. M. "Introduction: Rumphius' Life and Work." In Georgius Rumphius, *The Ambonese Curiosity Cabinet*, edited and translated by E. M. Beekman, xxxv–cxii. New Haven, CT: Yale University Press, 1999.

———. *Paradijzen van weleer: koloniale literatuur uit Nederlands-Indië, 1600–1950*. Amsterdam: Prometheus, 1998.

Belon, Pierre. *L'histoire naturelle des estranges poissons marins, avec la vraie peincture & description du Daulphin, & de plusiers autres de son espece*. Paris: Regnaud Chaudiere, 1551.

Benedict, Barbara M. *Curiosity: A Cultural History of Early Modern Inquiry*. Chicago: University of Chicago Press, 2002.

Benker, Gertrud. *Christophorus: Patron der Schiffer, Fuhrleute und Kraftfahrer. Legende, Verehrung, Symbol*. Munich: Verlag Georg D. W. Callwey, 1975.

Bepler, Jill. "'zu meinem und aller dehrer die sichs gebrauchen wollen, nutzen, trost undt frommen': Lektüre, Schrift und Gebet im Leben der fürstlichen Witwen in der Frühen Neuzeit." In *Witwenschaft in der Frühen Neuzeit: Fürstliche und adlige Witwen zwischen Fremd- und Selbstbestimmung*, edited by Martina Schattkowsky, 303–19. Leipzig: Leipziger Universitätsverlag, 2003.

Berger, Harry. *Caterpillage: Reflections on Seventeenth-Century Dutch Still-Life Painting*. New York: Fordham University Press, 2011.

Bergvelt, Ellinoor, and Renée Kistemaker, eds. *De wereld binnen handbereik: Nederlandse kunst- en rareitenverzamelingen, 1585–1735*. Zwolle: Waanders; Amsterdam: Amsterdams Historisch Museum, 1992.

———, eds. *De wereld binnen handbereik. Nederlandse kunst- en rareitenverzamelingen, 1585–1735. Catalogus*. Zwolle: Waanders; Amsterdam: Amsterdams Historisch Museum, 1992.

Van Berkel, Klaas. *Citaten uit het boek der natuur*. Amsterdam: Bert Bakker, 1998.

Beudeker, Christoffel. *De Sprekende Konstkamer. Vertoonende het regte Gebruik der Boeken, Konsten en natuurkundige Wetenschappen in zedige Gedachten*. By the Author, 1748.

Blanc, Charles. *L'oeuvre complet de Rembrandt*. 2 vols. Paris: Gide, 1859–61.

Bleichmar, Daniela. "Learning to Look: Visual Expertise across Art and Science in Eighteenth-Century France." *Eighteenth Century Studies* 46 (2012): 85–111.

———. *Visible Empire: Botanical Expeditions and Visual Culture in the Hispanic Enlightenment*. Chicago: University of Chicago Press, 2012.

Boltanski, Luc, and Arnaud Esquerre. "The Economic Life of Things: Commodities, Collectibles, Assets." *New Left Review* 98 (2016): 31–54.

Van den Boogert, Bob, et al., eds. *Rembrandt's Treasures*. Amsterdam: Museum het Rembrandthuis, 1999.

Van Borsselen, Philibert. *Strande oft ghedichte van de schelpen, kinck-hornen, ende andere wonderlicke zee-schepselen, tot lof van den Schepper aller dinghen*. Haarlem: by Adriaen Rooman, 1611.

Bradley, James W. "Re-visiting Wampum and Other Seventeenth-Century Shell Games." *Archaeology of Eastern North America* 39 (2011): 25–51.

Bredekamp, Horst. *The Lure of Antiquity and the Cult of the Machine: The Kunstkammer and the Evolution of Nature, Art and Technology*. Translated by Allison Brown. Princeton, NJ: M. Wiener Publishers, 1995.

Breure, Abraham S. H., and Susanne R. de Heer. "From a 'domestic commodity' to a 'secret of trade': Snails and Shells of Land Molluscs in Early (mainly 16th and 17th century) Visual Arts." *Basteria. Tijdschrift van de Nederlandse Malacologische Vereniging* 79.4–6 (December 2015): 81–97.

Brooks, Harold F. "*Richard III*: Antecedents to Clarence's Dream." *Shakespeare Survey* 32 (1980): 145–50.

Broomhall, Susan, and Jennifer Spinks. "Imagining Domesticity in Early Modern Dutch Dolls' Houses." In *Early Modern Women in the Low Countries: Feminizing Sources and Interpretations of the Past*, 99–122. Farnham: Ashgate, 2011.

Broomhall, Susan, and Jacqueline Van Gent. *Gender, Power and Identity in the Early Modern House of Orange-Nassau*, London: Routledge, 2016.

Brown, Eric C. "Underworld Sailors in Milton's *Lycidas* and Virgil's *Aeneid*." *Milton Quarterly* 36.1 (2002): 34–45.

De Brune, Johan. *Emblemata of Zinne-werck*. Edited by P. J. Meertens. Soest: Davaco, 1971.

Brunon, Hervé, and Monique Mosser. *L'imaginaire des grottes dans les jardins européens*. Paris: Hazan, 2014.

Brusati, Celeste. "Stilled Lives: Self-Portraiture and Self-Reflection in Seventeenth-Century Netherlandish Still-Life Painting." *Simiolus* 20.2/3 (1990–91): 168–82.

Buonanni, Filippo. *Ricreatione dell'occhio e della mente nell'osseruation delle chiocciole*. Rome: il Varese, a spesi di Felice Cesaretti, 1681.

Burke, Jill. "Nakedness and Other Peoples: Rethinking the Italian Renaissance Nude." *Art History* 4 (2013): 714–39.

Busch, Werner. "Rembrandts *Muschel*—Nachahmung der Natur? Ein methodisches Lehrstück." In *Vom Objekt zum Bild. Piktorale Prozesse in Kunst und Wissenschaft, 1600–2000*, edited by Bettina Göckel, with Julia Häckl and Miriam Volkmert, 93–122. Berlin: Akademie Verlag, 2011.

Buvelot, Quentin. *The Still Lifes of Adriaen Coorte (active c. 1683–1707)*. Zwolle: Waanders, 2008.

Carruthers, Mary. *The Craft of Thought: Meditation, Rhetoric, and the Making of Images, 400–1200*. Cambridge: Cambridge University Press, 1998.

*A Catalogue of the Portland Museum, Lately the Property of the Duchess Dowager of Portland Deceased which will be sold by Auction, by Mr. Skinner and Co., on Monday the 24th of April, 1786, and the thirty-seven following days*. London, 1786.

Cats, Jacob. *Houwelick: Dat is het gansche gelegenheyt des echten-staets*. Middelburg: Jan Pieterss van der Venne, 1625.

Chadwick, Esther. "'This deepe and perfect glosse of Blacknesse': Colour, Colonialism, and *The Paston Treasure*'s Period Eye." In *The Paston Treasure: Microcosm of the Known World*, edited by Andrew Moore, Nathan Flis, and Francesca Vanke, 102–9. New Haven, CT: Yale University Press, 2018.

Le Chanu, Patrick. *Joachim Wtewael: Persée et Andromède*. Paris: Musée du Louvre, 1999.

Chapman, H. Perry. "Curiosity and Desire: Rembrandt's Collection as Historiographic Barometer." In *Rembrandt and His Circle: Insights and Discoveries*, edited by Stephanie S. Dickey, 99–121. Amsterdam: Amsterdam University Press, 2017.

Cheney, Liana De Girolami. "The Oyster in Dutch Genre Paintings: Moral or Erotic Symbolism." *Artibus et Historiae* 8.15 (1987): 135–58.

Chong, Alan. "Contained under the Name of Still Life: The Associations of Still Life Painting." In *Still Life Paintings from the Netherlands 1550–1720*, edited by Alan Chong and Wouter Kloek, 11–38. Zwolle: Waanders; Amsterdam: Rijksmuseum, 1999.

Cicero, *De oratore*. Cambridge, MA: Harvard University Press, 2014.

Clifton, James, et al. *Pleasure and Piety: The Art of Joachim Wtewael*. Washington, DC: National Gallery of Art, 2015.

Coatalem, Éric, et al. *La nature morte française au XVIIe siècle. 17th-Century Still-Life Painting in France*. Dijon: Éditions Faton, 2014.

Coomans, H. E. "Schelpenverzamelingen." In *De wereld binnen handbereik. Nederlandse kunst- en rareitenverzamelingen, 1585–1735*, edited by Ellinoor Bergvelt and Renée Kistemaker, 192–203. Zwolle: Waanders; Amsterdam: Amsterdams Historisch Museum, 1992.

Corrigan, Karina, Jan van Campen, and Femke Diercks, eds. *Asia in Amsterdam: The Culture of Luxury in the Dutch Golden Age*. Salem, MA: Peabody Essex Museum; Amsterdam: Rijksmuseum, 2015.

Courchamp, Franck, Elena Angulo, Philippe Ravalan, Richard J. Hall, Laetitia Signoret, Leigh Bull, Yves Meinard, and Georgina Mace. "Rarity Value and Species Extinction: The Anthropogenic Allee Effect." *Public Library of Science: Biology*, 4.12 (2006–11), e415. DOI 10.1371/journal.pbio.0040415.

Crenshaw, Paul. *Rembrandt's Bankruptcy: The Artist, His Patrons, and the Art Market in Seventeenth-Century Netherlands*. Cambridge: Cambridge University Press, 2006.

Cummings, Laura. "Damian Hirst: Treasures from the Wreck of the Unbelievable Review—Beautiful and Monstrous," *Observer*, 16 April 2017.

Daly, Kathleen, and Marian Rengel, eds. *Greek and Roman Mythology A to Z*. 1992. New York: Chelsea House, 2009.

Dance, S. Peter. *A History of Shell Collecting*. Leiden: Brill, 1986.

———. *Shells and Shell Collecting*. London: Hamlyn, 1972.

Daston, Lorraine. "Nature by Design." In *Picturing Science, Producing Art*, edited by Caroline A. Jones and Peter Galison, 232–52. New York: Routledge, 1998.

———. "Type Specimens and Scientific Memory." *Critical Inquiry* 31 (Autumn 2004): 153–82.

Daston, Lorraine, and Peter Galison. *Objectivity*. New York: Zone Books, 2007.

Daston, Lorraine, and Katharine Park. *Wonders and the Order of Nature*. New York: Zone Books, 1998.

Delbourgo, James. *Collecting the World: Hans Sloane and the Origins of the British Museum*. Cambridge, MA: Harvard University Press, 2017.

Dernie, David. "The Use and Meaning of Materials in the Garden of the Villa d'Este at Tivoli." *Architectural Research Quarterly* 2.3 (1997): 64–73.

D'Errico, Francisco, and Lucinda Blackwell. "Earliest Evidence of Personal Ornaments Associated with Burial: The *Conus* Shells from Border Cave." *Journal of Human Evolution* 93 (2016): 91–108.

Derrida, Jacques. *Margins of Philosophy*. Chicago: University of Chicago Press, 1985.

Dickey, Stephanie S. "Disgust and Desire: Responses to Rembrandt's Nudes." In *The Primacy of the Image in Northern European Art 1400–1700: Essays in Honor of Larry Silver*, edited by Debra Cashion, Ashley West, and Henry Luttikhuizen, 447–62. Leiden: Brill, 2017.

———. "'The Finest Possible State'. Cataloguing and Collecting Rembrandt's Prints, 1700–1840." In *Rembrandt: Britain's Discovery of the Master*, edited by Christian Tico Seifert, 69–80. Edinburgh: National Galleries of Scotland, 2018.

———. *Rembrandt. Portraits in Print*. Amsterdam: John Benjamins, 2004.

———. "Saskia as Glycera. Rembrandt's Emulation of an Antique Prototype." In *Aemulatio: Emulation, Imitation, and Invention in Netherlandish Art, Essays in Honor of Eric Jan Sluijter*, edited by Anton Boschlo et al., 233–47. Zwolle: Waanders, 2011.

Dietz, Bettina. "Mobile Objects. The Space of Shells in 18th-Century France." *British Journal for the History of Science* 39.3 (2006): 363–82.

Dietz, Bettina, and Thomas Nutz. "Collections Curieuses: The Aesthetics of Curiosity and Elite Lifestyle in Eighteenth-Century Paris." *Eighteenth-Century Life* 29.3 (2005): 44–75.

Doerr, Anthony. "The Shell Collector." In *The Shell Collector*, 9–39. New York: Scribner, 2002.

Dudok van Heel, S.A.C. "Mr. Johannes Wtenbogaert (1608–1680), een man uit remonstrants milieu en Rembrandt van Rijn." *Jaarboek Amstelodamum* 70.2 (1978): 146–69.

Van Eeghen, I. H. "De Staalmeesters." *Jaarboek Amstelodamum* 49 (1957): 65–80.

———. "Het poppenhuis van Petronella Oortman." *Amstelodamum* 40 (1953): 113–17.

Egmond, Florike. *Eye for Detail: Images of Plants and Animals in Art and Science, 1500–1630*. London: Reaktion Books, 2016.

Ekserdjian, David. *Still Life before Still Life*. New Haven, CT: Yale University Press, 2018.

Elan, Keir. *Shakespeare's Pictures: Visual Objects in the Drama*. London: Bloomsbury Arden Shakespeare, 2017.

Elderkin, G. W. "The Natural and the Artificial Grotto." *Hesperia: The Journal of the American School of Classical Studies at Athens* 10.2 (1941): 125–37.

Eliade, Mircea. *Images and Symbols: Studies in Religious Symbols*. Princeton, NJ: Princeton University Press, 1991.

Eliasson, Sabrina Norlander, et al. *Italian Paintings: Three Centuries of Collecting. Nationalmuseum, Stockholm*. Vol. 1. Stockholm: Nationalmuseum, 2015.

Emmer, P. C. *The Dutch Slave Trade, 1500–1800*. Translated by Chris Emery. New York: Berghahn Books, 2006.

Erasmus, Desiderius. "The New Mother." In *The Collected Works of Erasmus: Colloquies*, vol. 39. Translated and annotated by Craig R. Thompson. Toronto: University of Toronto Press, 1997.

Fehrenbach, Frank. "Cut Flowers." *Nuncius* 32 (2017): 583–614.

Findlen, Paula. "Agostino Scilla: A Baroque Painter in Pursuit of Science." In *Science in the Age of Baroque*, edited by Ofer Gal and Raz Chen-Morris, 119–59. Dordrecht: Springer, 2013.

———, ed. *Early Modern Things: Objects and Their Histories, 1500–1800*. 2nd edition. Abingdon; New York: Routledge, 2021.

———. "Jokes of Nature and Jokes of Knowledge: The Playfulness of Scientific Discourse in Early Modern Europe." *Renaissance Quarterly* 43.2 (1990): 292–331.

———. "The Museum: Its Classical Etymology and Renaissance Genealogy." *Journal of the History of Collections* 1.1 (1989): 59–78.

———. *Possessing Nature: Museums, Collecting, and Scientific Culture in Early Modern Italy*. Berkeley: University of California Press, 1994.

Finlay, Robert. *The Pilgrim Art: Cultures of Porcelain in World History*. Berkeley: University of California Press, 2010.

Fleischhauer, Werner. *Renaissance im Herzogtum Württemberg*. Stuttgart: Kohlhammer, 1971.

———. "Zur Tätigkeit des Salomon de Caus an den Grottenwerken zu Brüssel, Heidelberg und Stuttgart." In *Landesgeschichte und Geistesgeschichte, Festschrift für Otto Herding zum 65. Geburtstag*, edited by K. Elm, E. Gönner, and E. Hillenbrand, 372–81. Stuttgart: Kohlhammer, 1977.

Franke, Birgit. "Natürliche Kunst und künstliche Natur: Ein Beitrag zur Grottenkunst des 16. und frühen 17. Jahrhunderts." In *Künste und Natur in Diskursen der Frühen Neuzeit*, edited by Hartmut Laufhütte, 2:1075–94. Wiesbaden: Harrassowitz, 2000.

Freud, Sigmund. *The Uncanny*. Translated by David Mclintock. New York: Penguin Books, 2003.

Fucci, Robert. *Rembrandt's Changing Impressions*. New York: Wallach Art Gallery, Columbia University; Cologne: Walther König, 2016.

Furttenbach the elder, Joseph. *Architectura civilis: Das ist: Eigentlich Beschreibung wie ma[n] nach bester form / und gerechter Regul / Furs Erste: Pallast / mit dero Lust: und Thiergarten / darbey auch Grotten [ . . . ] erbawen soll*. Ulm: Jonam Saurn, 1628.

———. *Architectura privata; das ist: Gründtliche Beschreibung, neben conterfetischer Vorstellung, inn was Form und Manier, ein gar Irregular, Burgerliches Wohn-Hauss . . . erbawet . . . unnd also schon zu gutem Ende ist gebracht worden*. Augsburg: Johann Schultes, 1641.

———. *Architectura recreationis. Das ist von allerhand nützlich und erfrewlichen civilischen Gebäwen*. Augsburg: Johann Schultes, 1640.

———. *Feriae architectonicae: der siebende Theil, das seynd die nur noch übrige Früe- und Spattstunden, mit liebreichen Delectationen der drey Arten hochnützlichen Gebäwen*. Augsburg: n.p., 1662.

———. *Lebenslauff 1652–1664*, edited by Kaspar von Greyerz, Kim Siebenhüner, and Roberto Zaugg. Cologne: Böhlau, 2013.

———. *Mannhaffter Kunst-Spiegel, oder, Continuatio, und Fortsetzung allerhand mathematisch- und mechanisch-hochnutzlich-so wol auch sehr erfröhlichen Delectationen, und respective im Werck selbsten experimentirten freyen Künsten*. Augsburg: Johann Schultes, 1663.

———. *Newes Itinerarium Italiae in welchem der Reisende nicht allein gründtlichen Bericht, durch die herrlichste namhaffteste örter Italiae sein Reiß wol zubestellen*. Ulm: Jonam Saurn, 1637.

Garrard, Mary D. "The Not-So-Still Lifes of Giovanna Garzoni." In *'The Immensity of the Universe' in the Art of Giovanna Garzoni*, edited by Sheila Barker, 62–77. Livorno: Sillabe, 2020.

Gaston, Kevin J. *Rarity*. London: Chapman & Hall, 1994.

Van Gemert, Leo. "Rumphius over het verzamelen en schoonmaken van schelpen." *De Kreukel* 41 (2005): 59–65.

Gerritsen, Anne, and Giorgio Riello, eds. *The Global Lives of Things: The Material Culture of Connections in the Early Modern World*. London: Routledge, 2015.

———. *Writing Material Culture History*. London: Bloomsbury, 2015.

Gersaint, Edmé-François. *A Catalogue and Description of the Etchings of Rembrandt Van-Rhyn, With some Account of his Life, To which is added, A List of the best pieces of this Master for the Use of those who would make a Select Collection of his Works. Written originally by the late M. Gersaint, and published by Mess. Helle and Glomy, with considerable Additions and Improvements*. London: T. Jeffrys, 1752.

———. *Catalogue raisonné de Coquilles, Insects, Plantes marines, et autres Curiosités naturelles*. Paris: Flahault and Prault, 1736.

———. *Catalogue raisonné des différents Effets curieux et rares contenus dans le Cabinet de feu M. le Chevalier de La Roque*. Paris: Jacques Barois, 1745.

———. *Catalogue raisonné des Diverses Curiosités du Cabinet de feu M. Quentin de L'Orangère, Composé de Tableaux originaux de meilleurs Maîtres de Flanders; d'une très-nombreuse Collection de Desseins & d'Estampes de toutes les Écoles; de plusiers Atlas & suites de Cartes; de quantité de Morceaux de Topographie, & d'un Coquillier fait avec Choix* . . . Paris: Jacques Barois, 1744.

———. *Catalogue raisonné de toutes les Pièces qui forment l'Oeuvre de Rembrandt, composé par feu M. Gersaint, & mis au jour, avec les Augmentations nécessaires, par les Sieurs Helle & Glomy*. Paris: Hochereau l'Aîné, 1751.

Gibson-Wood, Carol. "Classification and Value in a Seventeenth-Century Museum. William Courten's Collection." *Journal of the History of Collections* 9 (1997): 61–77.

Glorieux, Guillaume. *À L'Enseigne de Gersaint. Edme-François Gersaint, Marchand d'Art sur le Pont Notre-Dame (1694–1750)*. Preface by Daniel Roche. Seyssel: Champ Vallon, 2002.

Golahny, Amy. "Rembrandt's *One Guilder Print*. Value and Invention in 'the most beautiful [print] that ever came from the burin of this Master.'" In *Rembrandt and His Circle: Insights and Discoveries*, edited by Stephanie S. Dickey, 230–251. Amsterdam: Amsterdam University Press, 2017.

Goldgar, Anne. "Nature as Art: The Case of the Tulip." In *Merchants and Marvels: Commerce and the Representation of Nature in Early Modern Europe*, edited by Pamela Smith and Paula Findlen, 324–46. New York: Routledge, 2002.

———. *Tulipmania: Money, Honor, and Knowledge in the Dutch Golden Age*. Chicago: University of Chicago Press, 2007.

Gorlaeus, Abraham. *Dactyliotheca seu annulorum sigillarium quorum apud priscos tam Graecos quam Romanos usus*. [Antwerp]: [publisher not identified], 1601.

Graham, David. "*Emblema Multiplex*: Towards a Typology of Emblematic Forms, Structures and Functions." In *Emblem Scholarship: Directions and Developments. A Tribute to Gabriel Hornstein*, edited by Peter M. Daly, 131–58. Turnout: Brepols, 2005.

Grasskamp, Anna. "The Frames of Reflection: 'Indian' Shell Surfaces and European Collecting, 1550–1650." In *Objets frontière*, edited by Sabine du Crest, 69–83. Paris: Boccard, 2018.

———. "Metamorphose in Rot: Die Inszenierung von Korallenfragmenten in Kunstkammern des 16. und 17. Jahrhunderts." *Tierstudien* 4 (2013): 13–24.

———. *Objects in Frames: Displaying Foreign Collectibles in Early Modern China and Europe*. Berlin: Reimer, 2019.

———. "Spirals and Shells: Breasted Vessels in Sixteenth-Century Nuremberg." *Res: Anthropology and Aesthetics* 67–68 (2016/17): 146–63.

Green, Toby. *A Fistful of Shells: West Africa from the Rise of the Slave Trade to the Age of Revolution*. Chicago: University of Chicago Press, 2019.

Griffiths, Antony, ed. *Landmarks in Print Collecting: Connoisseurs and Donors at the British Museum since 1753*. London: British Museum Press, 1996.

———. "The Reverend Clayton Mordaunt Cracherode (1730–99)." In *Landmarks in Print Collecting: Connoisseurs and Donors at the British Museum since 1753*, edited by Antony Griffiths, 43–64. London: British Museum Press, 1996.

———. "Sir Hans Sloane (1660–1753)." In *Landmarks in Print Collecting: Connoisseurs and Donors at the British Museum since 1753*, edited by Antony Griffiths, 21–42. London: British Museum Press, 1996.

Grootenboer, Hanneke. "A Clock Picture as a Philosophical Experiment: The Tableau Mécanique in the Physics Cabinet of Bonnier de La Mosson." In *Art and Technology in Early Modern Europe*, edited by Richard Taws and Genevieve Warwick, special issue of *Art History* 39.2 (2016): 340–55.

———. "How to Become a Picture: Theatricality as Strategy in Seventeenth-Century Dutch Portraits." *Art History* 33.2 (April 2010): 320–33.

———. *The Pensive Image: Art as a Form of Thinking*. Chicago: University of Chicago Press, 2021.

———. "Sublime Still Life: On Adriaen Coorte, Elias van den Broeck, and the *Je ne sais quoi* of Painting." Special issue edited by Stijn Bussels and Bram van Oostveldt of the *Journal of Historians of Netherlandish Art* 8.2 (2016): https://jhna.org/articles/sublime-still-life-adriaen-coorte-elias-van-den-broeck-je-ne-sais-quoi-painting/.

Grote, Andreas, ed. *Macrocosmos in Microcosmo. Die Welt in der Stube Zur Geschichte des Sammelns 1450 bis 1800*. Opladen: Leske + Budrich, 1994.

Gutfleisch, Barbara, and Joachim Menzhausen, "'How a Kunstkammer Should Be Formed': Gabriel Kaltemarckt's Advice to Christian I of Saxony on the Formation of an Art Collection, 1587." *Journal of the History of Collections* 1 (1989): 3–32.

Gutzwiller, Kathryn. "The Nautilus, the Halycon, and Selenaia: Callimachus's 'Epigram' 5 Pf. = 14 G.–P." *Classical Antiquity* 11.2 (1992): 194–209.

Hamell, George R. "Trading in Metaphors: The Magic of Beads. Another Perspective upon Indian-European Contact in Northeastern North America." In *Proceedings of the 1982 Glass Trade Bead Conference*, edited by Charles F. Hayes III, 5–28. Rochester, NY: Rochester Museum & Science Center, 1983.

Hanke, Stephanie. "The Splendour of Bankers and Merchants: Genoese Garden Grottoes of the Sixteenth Century." *Urban History* 37 (2010): 399–418.

Hannan, Leonie, and Sarah Longair, eds. *History through Material Culture*. Manchester: Manchester University Press, 2017.

Harvey, Karen, ed. *History and Material Culture*. Abingdon: Routledge, 2009.

Hay, Jonathan. *Sensuous Surfaces: The Decorative Object in Early Modern China*. London: Reaktion Books, 2010.

Hendrix, Lee, and Thea Vignau-Wilberg, eds. *Mira Calligraphiae Monumenta: A Sixteenth-Century Calligraphic Manuscript Inscribed by Georg Bocskay and Illuminated by Joris Hoefnagel*. Malibu, CA: J. Paul Getty Museum, 1992.

Hinterding, Erik. *Rembrandt Etchings from the Frits Lugt Collection*. Bussum: Toth, 2008.

Hinterding, Erik, Huigen Leeflang, and Manon van der Mullen. "Acquisitions from the F. G. Waller-Fonds." *Rijksmuseum Bulletin* 67.4 (2019): 340–73.

Hinterding, Erik, and Jaco Rutgers. *Rembrandt: New Hollstein Dutch & Flemish Etchings, Engravings and Woodcuts 1450–1700*. 7 vols. Oudekerk aan den Ijssel: Sound and Vision Publishers, 2013.

Hochstrasser, Julie. "*Stil-staende dingen*: Picturing Objects in the Dutch Golden Age." In *Early Modern Things: Objects and Their Histories, 1500–1800*, edited by Paula Findlen, 103–23. 2nd edition. Abingdon; New York: Routledge, 2021.

Hochstrasser, Julie Berger. *Still Life and Trade in the Dutch Golden Age*. New Haven, CT: Yale University Press, 2007.

Hogendorp, Jan, and Marion Johnson. *The Shell Money of the Slave Trade*. Cambridge: Cambridge University Press, 1986.

Hollander, Martha. *An Entrance for the Eyes: Space and Meaning in Seventeenth-Century Dutch Art*. Berkeley: University of California Press, 2002.

Van Hoogstraten, Samuel. *Inleyding tot de Hooge Schoole der Schilderkonst, anders de Zichtbaere Werelt*. Rotterdam: François van Hoogstraten, 1678.

Hooke, Robert. *Lectures and Discourses of Earthquakes and Subterraneous Eruptions*. New York: Arno Press, 1978.

Houbraken, Arnold. *De Groote Schouburgh der Nederlantsche Konstschilders en Schilderessen*. 2nd ed. 2 vols. The Hague: J. Swart, C. Boucquet, and M. Gaillard, 1753 (first ed. 1718–21). Facsimile edition, Amsterdam: B.M. Israël, 1976.

Jorink, Eric. *Het 'Boeck der Natuere': Nederlandse geleerden en de wonderen van Gods schepping 1575–1715*. Leiden: Primavera Pers, 2006.

———. *Reading the Book of Nature in the Dutch Golden Age, 1575–1715*. Leiden: Brill, 2010.

Jutting, W.S.S., van B. "A Brief History of the Conchological Collections at the Zoological Museum of Amsterdam." *Bijdragen tot de Dierkunde* 27.1 (1939): 167–246.

Kaltemarckt, Gabriel. *Bedenken wie eine Kunst-cammer aufzurichten seyn möchte* (1587). In Barbara Gutfleisch and Joachim Menzhausen, "'How a Kunstkammer Should Be Formed': Gabriel Kaltemarckt's Advice to Christian I of Saxony on the Formation of an Art Collection, 1587." *Journal of the History of Collections* 1 (1989): 7–32.

Kapferer, Jean-Noel, Pierre Valette-Florence, and Coralie Lucatello. "The Impact of Brand Penetration and Awareness on Luxury Brand Desirability: A Cross Country Analysis of the Relevance of the Rarity Principle." *Journal of Business Research* 83.2 (2018): 38–50.

Kaufmann, Thomas DaCosta. *Arcimboldo: Visual Jokes, Natural History, and Still-Life Painting*. Chicago: University of Chicago Press, 2009.

———. "Remarks on the Collections of Rudolf II: The *Kunstkammer* as a Form of Representatio." *Art Journal* 38.1 (1978): 22–28.

Kehoe, Marsely L. "The Nautilus Cup between Foreign and Domestic in the Dutch Golden Age." *Dutch Crossing* 35.3 (2011): 275–85.

Kelley, Shannon. "Arno River Floods and the Cinquecento Grotto at the Boboli Garden." *Renaissance Studies* 30.5 (2016): 729–51.

———. "Shells, *Pericles*, and the Fantasy of Shell-Dwelling." In *The Shakespearean International Yearbook. 15: Special Section: Shakespeare and the Human*, edited by Tiffany Jo Werth, 167–83. Farnham: Ashgate, 2015.

Kemp, Martin. "'Wrought by No Artist's Hand': The Natural, Artificial and Exotic in Some Artefacts from the Sixteenth Century." In *Reframing the Renaissance: Visual Culture in Europe and Latin America, 1450–1650*, edited by Claire Farago, 177–96. New Haven, CT: Yale University Press, 1996.

Kemperdick, Stephan, and Jochen Sander. "The Still Life before Still Life." In *The Magic of Things: Still-Life Painting 1500–1800*, edited by Jochen Sander, 21–27. Frankfurt am Main: Städel Museum, 2008.

Kenny, Neil. *The Uses of Curiosity in Early Modern France and Germany*. New York: Oxford University Press, 2004.

Keyes, George S., et al. *Masters of Dutch Painting: The Detroit Institute of Arts*. Detroit: Detroit Institute of Arts, 2004.

Kisluk-Grosheide, Daniëlle. "Dirck van Rijswijck (1596–1679), a Master of Mother-of-Pearl." *Oud Holland* 111.2 (1997): 77–94.

Koeppe, Wolfram. "*Artificialia* and Their Meanings in *The Paston Treasure*." In *The Paston Treasure: Microcosm of the Known World*, edited by Andrew Moore, Nathan Flis, and Francesca Vanke, 120–25. New Haven, CT: Yale University Press, 2018.

———, ed. *Making Marvels: Science and Splendor at the Courts of Europe*. New Haven, CT: Yale University Press, 2019.

Koerner, Joseph. "Impossible Objects: Bosch's Realism." *Res: Anthropology and Aesthetics* 46 (2004): 73–97.

Kolfin, Elmer. "African Man with a Conch, ca. 1620–1625: A New Attribution to Abraham Bloemaert." *Oud Holland* 131.3/4 (2018): 123–32.

Kolfin, Elmer, and Epco Runia, eds. *Black in Rembrandt's Time*. Zwolle: WBooks; Amsterdam: Museum Het Rembrandthuis, 2020.

Kopytoff, Igor. "The Cultural Biography of Things: Commoditization as a Process." In *The Social Life of Things: Commodities in Cultural Perspective*, ed. Arjun Appadurai, 64–91. Cambridge: Cambridge University Press, 1986.

Kraus, Johann Ulrich. *Description de la Grotte de Versailles oder: Beschreibung der Grotten zu Versailles*. Augsburg, c. 1700.

Kremer, Joachim, ed. *Magdalena Sibylla von Württemberg: politisches und kulturelles Handeln einer Herzogswitwe im Zeichen des frühen Pietismus*. Ostfildern: Jan Thorbecke Verlag, 2017.

Kusukawa, Sachiko. "Illustrating Nature." In *Books and the Sciences in History*, edited by Marina Frasca-Spada and Nick Jardine, 90–113. Cambridge: Cambridge University Press, 2000.

———. "William Courten's Lists of 'Things Bought' from the Late Seventeenth Century." *Journal of the History of Collections* 29 (2017): 1–17.

Landesmusem Württemberg, ed. *Die Kunstkammer der Herzöge von Württemberg*. Vol. 2. Ostfildern: Jan Thorbecke Verlag, 2017.

Langereis, Sandra. "De verzameling munten, gemmen en schelpen van Abraham Gorlaeus te Delft." In *Schatten in Delft: burgers verzamelen, 1600–1750*, edited by Ellinoor Bergvelt, Michiel Jonker, and Agnes Wiechmann, 91–99. Delft: Stedelijk Museum Het Prinsenhof, 2002.

Lazzaro, Claudia. *The Italian Renaissance Garden: From the Conventions of Planting, Design, and Ornament to the Grand Gardens of Sixteenth-Century Central Italy*. New Haven, CT: Yale University Press, 1990.

Leonhard, Karin. "Die Muschel als symbolische Form, oder: Wie Rembrandts *Conus marmoreus* nach Oxford kam." In *Vom Objekt zum Bild. Piktorale Prozesse in Kunst und Wissenschaft, 1600–2000*, edited by Bettina Göckel, with Julia Häckl and Miriam Volkmert. 123–56. Berlin: Akademie Verlag, 2011.

———. "Shell Collecting: On 17th-Century Conchology, Curiosity Cabinets and Still Life Painting." In *Early Modern Zoology: The Construction of Animals in Science, Literature and the Visual Arts*, edited by Karl A. E. Enenkel and Paul J. Smith, 177–214. Leiden: Brill, 2007.

Leonhard, Karin, and Maria-Theresia Leuker. "Who Commissioned Hollar's Shells?" *Simiolus: Netherlands Quarterly for the History of Art* 37.3/4 (2013): 227–39.

Lesser, Friedrich Christian, *Testaceo-theologia, oder, gründlicher Beweis des Daseyns und der vollkommensten Eigenschaften eines göttlichen Wesens: aus natürlicher und geistlicher Betrachtung der Schnecken und Muscheln zur gebührender Verherrlichung des grossen Gottes und Beförderung des ihm schuldigen Dienstes ausgefertiget mit Kupfern*. Ulm, 1756.

Leuker, Maria-Theresia. "Knowledge Transfer and Cultural Appropriation: Georg Everhard Rumphius's 'D'amboinsche rariteitkamer' (1705)." In *The Dutch Trading Companies as Knowledge Networks*, edited by Siegfried Huigen, Jan de Jong, and Elmer Kolfin, 145–70. Leiden: Brill, 2010.

Lightbown, Ronald W. "Oriental Art and the Orient in Late Renaissance and Baroque Italy." *Journal of the Warburg and Courtauld Institutes* 32 (1969): 228–79.

Lippincott, Louise. *Selling Art in Georgian London: The Rise of Arthur Pond*. New Haven, CT: Yale University Press, 1983.

List, Camillo. "Wiener Goldschmiede und ihre Beziehungen zum kaiserlichen Hofe." In *Jahrbuch der Kunsthistorischen Sammlungen des Allerhöchsten Kaiserhauses* 17 (1896): 291–306.

Luijten, Ger. "*The Shell* 1650." In Erik Hinterding, Ger Luijten, and Martin Royalton-Kisch, *Rembrandt the Printmaker*, 258–62. London, British Museum; Amsterdam, Rijksmuseum; Zwolle: Waanders 2000.

Van Mander, Karel. *Den grondt der edel vry schilder-const*. Edited by Hessel Miedema. 2 vols. Utrecht: Haentjens Dekker & Gumbert, 1973.

———. *Het Schilder-Boeck*. Haarlem: Paschier van Wesbusch, 1604.

Manning Stevens, Scott. "New World Contacts and the Trope of the 'Naked Savage.'" In *Sensible Flesh: On Touch in Early Modern Culture*, edited by Elizabeth D. Harvey, 125–40. Philadelphia: University of Pennsylvania Press, 2003.

Marandet, François. "Pierre Remy (1715–97). The Parisian Art Market in the Mid-Eighteenth Century." *Apollo* 158.498 (2003): 32–42.

Marr, Alexander. "Gentille curiosité: Wonder-Working and the Culture of Automata in the Late Renaissance." In *Curiosity and Wonder from the Renaissance to the Enlightenment*, edited by Robert J. W. Evans and Alexander Marr, 149–70. Aldershot: Ashgate, 2006.

———. "Introduction." In *Curiosity and Wonder from the Renaissance to the Enlightenment*, edited by Robert J. W. Evans and Alexander Marr, 1–20. Aldershot: Ashgate, 2006.

Marr, Alexander, Raphaële Garrod, José Ramón Marcaida and Richard J. Oosterhoff. *Logodaedalus: Word Histories of Ingenuity in Early Modern Europe*. Pittsburgh, PA: University of Pittsburgh Press, 2019.

Martirano, Bernardino. *Il pianto d'Aretusa*. Edited by Tobia R. Toscano. Naples: Paolo Loffredo, 1993.

Mason, Peter. *Deconstructing America: Representations of the Other*. London: Routledge, 1990.

Maxwell, Susan. "The Pursuit of Art and Pleasure in the Secret Grotto of Wilhelm V of Bavaria." *Renaissance Quarterly* 61.2 (2008): 414–62.

McClellan, Andrew. "Edmé Gersaint and the Marketing of Art in Eighteenth-Century Paris." *Eighteenth-Century Studies* 29.2 (1995/96): 218–22.

———. "Watteau's Dealer: Gersaint and the Marketing of Art in Eighteenth-Century Paris." *Art Bulletin* 78.3 (1996): 439–53.

Meadow, Mark A. "Merchants and Marvels: Hans Jacob Fugger and the Origins of the Wunderkammer." In *Merchants and Marvels: Commerce, Science, and Art in Early Modern Europe*, edited by Pamela H. Smith and Paula Findlen, 182–200. London: Routledge, 2001.

———. "Quiccheberg and the Copious Object: Wenzel Jamnitzer's Silver Writing Box." In *The Lure of the Object*, edited by Stephen Melville, 39–58. New Haven, CT: Yale University Press, 2005.

Mette, Hanns-Ulrich. *Der Nautiluspokal: Wie Kunst und Natur miteinander spielen*. Munich: Klinkhardt & Biermann, 1995.

Milburn, Erika. *Luigi Tansillo and Lyric Poetry in Sixteenth-Century Naples*. Leeds: Maney Publishing, 2003.

Miller, Naomi. *Heavenly Caves: Reflections on the Garden Grotto*. Boston: Allen & Unwin, 1982.

Minor, Heather Hyde. "G. B. Piranesi's *Diverse Maniere* and the Natural History of Ancient Art." *Memoirs of the American Academy in Rome* 56/57 (2011/2012): 323–51.

Mish, Frederick C., ed. *The Merriam-Webster New Book of Word Histories*. Springfield, MA: Merriam-Webster, 1991.

Ter Molen J. R. "Een bezichtiging van het poppenhuis van Petronella Brandt-Oortman in de zomer van 1718," *Bulletin van het Rijksmuseum* 42 (1994): 120–36.

———. *Een plezierreis in de zomer van 1718: De familie Von Uffenbach in de Nederlanden*. Zwolle: Waanders, 2017.

Möller, Karin Annette, ed. *Schimmern aus der Tiefe: Muscheln, Perlen, Nautilus*. Petersberg: Imhof, 2013.

Montaigne, Michel de. *The Complete Essays*. Translated by M. A. Screech. London: Penguin Books, 2003.

Montias, John Michael. *Art at Auction in Seventeenth-Century Amsterdam*. Amsterdam: Amsterdam University Press, 2002.

Morel, Philippe. *Les grottes maniéristes en Italie au XVIe siècle. Théâtre et alchimie de la nature*. Paris: Macula, 1998.

Morgan, Luke. *The Monster in the Garden: The Grotesque and the Gigantic in Renaissance Landscape Design*. Philadelphia: University of Pennsylvania Press, 2016.

———. *Nature as Model: Salomon de Caus and Early Seventeenth-Century Landscape Design*. Philadelphia: University of Pennsylvania Press, 2007.

Mund, Hélène, et al. *The Mayer van den Bergh Museum, Antwerp*. Brussels: Centre d'étude de la peinture du quinzième siècle, 2003.

Murphy, Kathleen. "Collecting Slave Traders: James Petiver, Natural History, and the British Slave Trade." *William and Mary Quarterly* 70.4 (2013): 637–70.

Newman, William R. *Promethean Ambitions: Alchemy and the Quest to Perfect Nature*. Chicago: University of Chicago Press, 2004.

Nichols, Lawrence W. "Jan Govertsz. van der Aar: On the Identification of Goltzius's Patron." *Nederlands Kunsthistorisch Jaarboek / Netherlands Yearbook for History of Art* 38 (1987): 241–55.

———. *The Paintings of Hendrick Goltzius (1558–1617): A Monograph and Catalogue Raisonné*. Aetas Aurea, vol. 22. Doornspijk: Davaco Publishers 2013.

Nitze-Ertz, Christa, Ute Kleinmann, and Stephan Brakensiek, eds. *Das flämische Stillleben, 1550–1680*. Vienna: Kunsthistorisches Museum, 2002.

Nusbaumer, Philippe. *Jacques Linard, 1597–1645: Catalogue de l'oeuvre peint*. Le Pecq-sur-Seine: Nusbaumer, 2006.

Von Oechelhäuser, Adolf. "Philipp Hainhofers Bericht über die Stuttgarter Kindtaufe im Jahre 1616." *Neue Heidelberger Jahrbücher* 1 (1891): 254–335.

Ovid. *Ovid's Metamorphoses*. Translated by Anthony S. Kline. Cambridge, MA: Harvard University Press, 2000.

Palissy, Bernard. *Recepte veritable*. Edited by Keith Cameron. Geneva: Libraire Droz, 1988.

Parshall, Peter. "Albrecht Dürer and the Axis of Meaning." *Allen Memorial Art Museum Bulletin* 50 (1997): 4–31.

———. "Imago Contrafacta: Images and Facts in the Northern Renaissance." *Art History* 16.4 (1993): 554–79.

Pedersen, Tara. *Mermaids and the Production of Knowledge in Early Modern England*. Farnham: Ashgate, 2015.

Pelling, Madeleine. "Selling the Duchess: Narratives of Celebrity in *A Catalogue of the Portland Museum* (1786)." *Early Modern Women* 13.2 (2019): 3–32.

Petrarch, Francesco. *The Life of Solitude*. Translated by Jacob Zeitlin. Urbana: University of Illinois Press, 1924.

Piepho, Lee. "The Latin and English Eclogues of Phineas Fletcher: Sannazaro's *Piscatoria* among the Britons." *Studies in Philology* 81.4 (1984): 461–72.

Pieters, Marnix, et al. "Een 15de-eeuwse sector van het verdwenen vissersdorp te Raversijde (stad Oostende, prov. West-Vlaanderen)." *Archeologie in Vlaanderen* 4 (1994): 219–36.

Pijzel-Dommisse, J., *Het Hollandse pronkpoppenhuis: Interieur en huishouden in de 17e en 18e eeuw*. Zwolle: Waanders; Amsterdam: Rijksmuseum, 2000.

Pliny the Elder. *Natural History*. Volume 3. Translated by H. Rackham. Loeb Classical Library. Cambridge, MA: Harvard University Press, 1983.

Pointon, Marcia. *Brilliant Effects: A Cultural History of Gem Stones and Jewellery*. New Haven, CT: Yale University Press, 2010.

Pokorny, Erwin. "Alart du Hameel and Jheronimus Bosch: Artistic Relations and Chronologies." In *Jheronimus Bosch: His Life and His Work*, edited by Jo Timmermans, 264–76. 's-Hertogenbosch: Jheronimus Bosch Art Center, 2016.

Pomian, Krzysztof. *Collectors and Curiosities: Paris and Venice, 1500–1800*. Translated by Elizabeth Wiles-Portier. Cambridge, UK: Polity Press, 1990.

Posèq, Avigdor W. G. "Left and Right in Rembrandt's *Defeat of Goliath*." *Studia Rosenthaliana* 23.1 (1989): 8–27.

Prescott, Henry W. "Callimachus' Epigram on the Nautilus." *Classical Philology* 16.4 (1921): 327–37.

Preston, Frances L. "Gersaint on Rembrandt: À Son Seul Génie." *Dutch Crossing* 25.2 (2001): 200–220.

Priebe, Jessica. "The Artist as Collector: François Boucher (1703–1770)." *Journal of the History of Collections* 28.1 (2016): 27–42.

Prosperetti, Leopoldine van Hogendorp. "'Conchas legere': Shells as Trophies of Repose in Northern European Humanism." *Art History* 29 (2006): 387–413.

Puillandre, N., T. F. Duda, C. Meyer, B. M. Olivera, and P. Bouchet. "One, Four or 100 Genera? A New Classification of Cone Snails." *Journal of Molluscan Studies* 81.1 (2015): 1–23.

Ribouillault, Denis. "Seeing Christ, Reading Nature: Jan Swart van Groningen's *Christ in a Landscape* in the Montreal Museum of Fine Arts." *Nuncius* 32 (2017): 514–55.

Rife, Ellen O'Neil. "The Exotic Gift and the Art of the Seventeenth-Century Dutch Republic." PhD diss., University of Kansas, 2013.

Rijks, Marlise. "A Painter, a Collector, and a Horseshoe Crab: Connoisseurs of Art and Nature in Early Modern Antwerp." *Journal of the History of Collections* 31.2 (2019): 343–61.

———. "'Unusual Excrescences of Nature': Collected Coral and the Study of Petrified Luxury in Seventeenth-Century Antwerp." *Dutch Crossing* 41.2 (2017): 1–29.

Rincón, Luis Rodríguez. "The Aesthetics of the Early Modern Grotto and the Advent of an Empirical Nature." In *The Aesthetics of the Undersea*, edited by Margaret Cohen and Killian Quigley, 14–27. London: Routledge, 2018.

Ritchie, Carson I. A. *Shell Carving: History and Techniques*. South Brunswick, NJ: A. S. Barnes, 1974.

Roberts, Lissa. "A World of Wonder, a World of One." In *Merchants and Marvels: Commerce, Science, and Art in Early Modern Europe*, edited by Pamela H. Smith and Paula Findlen, 399–411. London: Routledge, 2001.

Van de Roemer, Bert. "God en het rariteitenkabinet: het religieuze motif van Noord-Nederlandse rariteitenverzamelaars eind zeventiende en begin achtieende eeuw." *Theoretische geschiedenis* 25 (1998): 242–55.

———. "Merians netwerk van verzamelende natuurliefhebbers / Merian's Network of Collector-Naturalists." In *Maria*

*Sibylla Merian Metamorphosis insectorum Surinamensium 1705*, edited by Marieke van Delft and Hans Mulder, 19–28. Tielt: Lannoo, 2016.

———. "Neat Nature. The Relation between Nature and Art in a Dutch Cabinet of Curiosities from the Early Eighteenth Century." *History of Science* 42 (2004): 47–84.

Van de Roemer, Gijsbert M. "The Serious Naturalist and the Frivolous Collector: Convergent and Divergent Approaches to Nature in *D'Amboinsche Rariteitkamer*." *Early Modern Low Countries* 3.2 (2019): 208–33.

Roos, Anna Marie. *Martin Lister and His Remarkable Daughters: The Art of Science in the Seventeenth Century*. Oxford: The Bodleian Library, 2019.

Rosenberg, Pierre. "Jacques Linard, 'peintre de coquilles.'" In *Napoli, l'Europa: Ricerche di Storia dell'Arte in onore di Ferdinando Bologna*, edited by Francesco Abbate and Fiorella Sricchia Santoro, 231–34. Catanzaro: Meridiana Libri, 1995.

Rosenthal, Lisa. "Frans Francken the Younger's *Discovery of Achilles*: Desire, Deception, and Inalienable Possession." In *Ut pictura amor. The Reflexive Imagery of Love in Artistic Theory and Practice, 1500–1700*, edited by Walter S. Melion, Joanna Woodall, and Michael Zell, 704–29. Leiden: Brill, 2017.

Rosier, Catherine A., and Mary Margaret Rogers. "Profits and Perils in the Antiques and Collectibles Market: The Influence of Product Knowledge Structures on Dealer Exchange Outcomes." *International Review of Retail, Distribution and Consumer Research* 26.1 (2016): 1–16.

Rossi, Paolo. *The Dark Abyss of Time: The History of the Earth and the History of Nations from Hooke to Vico*. Translated by Lydia G. Cochrane. Chicago: University of Chicago Press, 1984.

Rothstein, Bret. "Jan van Hemessen's Anatomy of Parody." In *The Anthropomorphic Lens: Anthropomorphism, Microcosmism and Analogy in Early Modern Thought and Visual Arts*, edited by Walter S. Melion, Bret Rothstein, and Michel Weemans, 455–79. Leiden: Brill, 2015.

Roukema, Edzer. "The Mythical 'First Voyage' of the Soderini Letter." *Imago Mundi* 16 (1962): 70–75.

Rumphius, Georgius. *D'Amboinsche rariteitenkamer*. Amsterdam: François Halma, 1705. 2nd ed., 1711.

———. *The Ambonese Curiosity Cabinet*. Edited and translated by E. M. Beekman. New Haven, CT: Yale University Press, 1999.

———. *The Poison Tree: Selected Writings of Rumphius on the Natural History of the Indies*. Edited and translated by E. M. Beekman. Amherst: University of Massachusetts Press, 1981.

Salvi, Claudia. *D'après nature: la nature morte en France au XVIIe siècle*. Tournai: La Renaissance du Livre, 2000.

Von Sandrart, Joachim. *Teutsche Academie der Bau-, Bild- und Mahlerey-Künste, Nuremberg 1675–1680*. Edited by T. Kirchner, A. Nova, C. Blüm, A. Schreurs, and T. Wübbena. 2008–12: http://ta.sandrart.net/en/.

Sannazaro, Jacopo. *Latin Poetry*. Translated by Michael C. J. Putnam. The I Tatti Renaissance Library. Cambridge, MA: Harvard University Press, 2009.

Savage, Robert J. G. "Natural History of the Goldney Garden Grotto, Clifton, Bristol." *Garden History* 17.1 (1989): 1–40.

Scheller, Robert. "Rembrandt en de encyclopedische kunstkamer." *Oud Holland* 84 (1969): 81–147.

Schmidt, Benjamin. *Inventing Exoticism: Geography, Globalism, and Europe's Early Modern World*. Philadelphia: University of Pennsylvania Press, 2015.

Schnapper, Antoine. *Le géant, la licorne et la tulipe: collections et collectionneurs dans la France du XVIIe siècle*. Paris: Flammarion, 1988.

Scholten, Frits. "The Larson Workshop: Reproducing Sculpture in Seventeenth-Century Holland." In *Collecting Sculpture in Early Modern Europe*, edited by Nicholas Penny and Eike D. Schmidt, 291–300. New Haven, CT: Yale University Press, 2008.

Schotel, G.D.J. *Letter en oudheidkundige avondstonden*. Dordrecht: Blussé en van Braam, 1841.

Schwartz, Gary. "Love in the Kunstkamer: Additions to the Work of Guillam van Haecht (1593–1637)." *Tableau* 18 (1996): 43–52.

Scot, Reginald. *The discoverie of witchcraft*. 1584, 1651, 1665. London: Eliot Stock, 1886.

Seba, Albertus. *Locupletissimi rerum naturalium thesauri accurata descriptio*. 4 vols. Amsterdam: apud J. Wetstenium, & Gul. Smith, & Janssonio-Waesbergios, 1734–65.

Segal, Sam. "Shell Still Lifes." In *A Prosperous Past: The Sumptuous Still Life in the Netherlands, 1600–1700*, edited by William B. Jordan, 77–92. The Hague: SDU Publishers, 1988.

van Seters, Wouter Hendrik. "Oud-Nederlandse parelmoerkunst: het werk van leden der familie Belquin, parelmoergraveurs en schilders in de 17e eeuw." *Nederlands Kunsthistorisch Jaarboek / Netherlands Yearbook for History of Art* 9 (1958): 173–238.

Shapin, Steven. "'The Mind Is Its Own Place': Science and Solitude in Seventeenth-Century England." *Science in Context* 4.1 (1991): 191–218.

Shell, Hanna Rose. "Casting Life, Recasting Experience: Bernard Palissy's Occupation between Maker and Nature." *Configurations* 12.1 (2004): 1–40.

Siebenhüner, Kim. "Entwerfen, Modelle bauen, ausstellen. Joseph Furttenbach und seine Rüst- und Kunstkammer." In *Lebenslauff 1652–1664*, edited by Kaspar von Greyerz, Kim Siebenhüner, and Roberto Zaugg, 45–65. Cologne: Böhlau, 2013.

Skogh, Lisa, et al. *The Varied Role of the Amateur*. Special issue of *Nuncius* 31.3 (2016).

Slive, Seymour. *Rembrandt and His Critics 1630–1730*. The Hague: Martinus Nijhoff, 1953.

Sluijter, Eric Jan. *Rembrandt and the Female Nude*. Amsterdam: Amsterdam University Press, 2006.

Smit, Pieter, with A.P.M. Sanders and J.P.F. van der Veer. *Hendrik Engel's Alphabetical List of Dutch Zoological Cabinets and Menageries*. 2nd, enlarged ed. Nieuwe Nederlandse Bijdragen tot de Geschiedenis der Geneeskunde en der Natuurwetenschappen, No. 19. Amsterdam: Rodopi, 1986.

Smith, Julia. "Material Christianity in the Early Medieval Household." In *Religion and the Household*, edited by J. Doran, C. Methuen, and A. Walsham, 23–46. Woodbridge: Boydell and Brewer, 2014.

Smith, Nicholas. "The Genre and Critical Reception of Jacopo Sannazaro's 'Eclogae piscatoriae' (Naples, 1526)." *Humanistica Lovaniensia* 50 (2001): 199–219.

Smith, Pamela, and Paula Findlen, eds. *Merchants and Marvels: Commerce and the Representation of Nature in Early Modern Europe*. New York: Routledge, 2002.

Sonak, Sangeeta M. "Shell Art and Shell Craft of Goa." In *Marine Shells of Goa*, 185–217. N.p.: Springer, 2017.

Spary, Emma C. "Rococo Readings of the Book of Nature." In *Books and the Sciences in History*, edited by Marina Frasca-Spada and Nick Jardine, 255–75. Cambridge: Cambridge University Press, 2000.

———. "Scientific Symmetries." *History of Science* 62 (2004): 1–46.

Stahl, Ernst Konrad. *Die Legende vom Heil. Riesen Christophorus in der Graphik des 15. und 16. Jahrhunderts*. Munich: J. J. Lentner'sche Buchhandlung, 1920.

Statius, Publius Papinius. *Achilleid*. Translated by John H. Mozley. Cambridge, MA: Harvard University Press; London, William Heinemann, 1928.

Steele, Teresa E., Estevan Álvarez-Fernández, and Emily Hallett-Desguez. "A Review of Shells as Personal Ornamentation during the African Middle Stone Age." *PaleoAnthropology* (2019): 24–51.

Stemshorn, Max, ed. *Der Kunst-Garten. Gartenentwürfe von Joseph Furttenbach 1591–1667*. Ulm: Süddeutsche Verlagsgesellschaft, 1999.

Sténuit, Robert. "De 'Witte Leeuw'. De schipbreuk van een schip van de V.O.C. in 1613 en het onderwateronderzoek naar het wrak in 1976." *Bulletin van het Rijksmuseum* 25.4 (1977): 165–78.

Stewart, Susan. *On Longing: Narratives of the Miniature, the Gigantic, the Souvenir, the Collection*. Durham: Duke University Press, 1993.

Strauss, Walter, et al. *The Rembrandt Documents*. New York: Abaris Books, 1979.

Swan, Claudia. *Rarities of These Lands: Art, Trade, and Diplomacy in the Dutch Republic*. Princeton, NJ: Princeton University Press, 2021.

Syndram, Dirk, and Martina Minning, eds. *Das Inventar von 1619*. In *Die Inventare der kurfürstlich-sächsischen Kunstkammer in Dresden*, vol. 3. Dresden: Sandstein, 2010.

Szafranska, Malgorzata. "The Philosophy of Nature and the Grotto in the Renaissance Garden." *Journal of Garden History* 9.2 (1989): 76–85.

Szanto, Mickaël. "Pour Jacques Linard, peintre de natures mortes (Troyes 1597–Paris, 1645)." *Bulletin de la Société de l'Histoire de l'Art français*, 2001/2, 25–62.

Tansillo, Luigi. *L'egloga e i poemetti*. Edited by Tobia R. Toscano. Commentary by Carmine Boccia and Rossano Pestarino. Naples: Paolo Loffredo, 2017.

Tlusty, Ann. *Bacchus and the Civic Order: The Culture of Drink in Early Modern Germany*. Charlottesville: University of Virginia Press, 2001.

Tobin, Beth Fowkes. *The Duchess's Shells: Natural History Collecting in the Age of Cook's Voyages*. New Haven, CT: Yale University Press, 2014.

Tomasi, Michele, et al. *L'art multiplié. Production de masse, en série, pour le marché dans les arts entre Moyen Âge et Renaissance*. Rome: Viella, 2011.

Trubitt, Mary Beth. "The Production and Exchange of Marine Shell Prestige Goods." *Journal of Archaeological Research* 11.3 (2003): 243–77.

Turner, Simon. "Hollar in Holland: Drawings from the Artist's Visit to the Dutch Republic in 1634." *Master Drawings* 48.1 (Spring 2010): 73–104.

———. *Wenceslaus Hollar: New Hollstein German Engravings, Etchings, and Woodcuts 1400–1700*. 2 vols. Oudekerk aan den Ijssel: Sound and Vision Publishers, 2010.

Von Uffenbach, Zacharias. *Herrn Zacharias Conrad von Uffenbach Merckwürdige Reisen durch Niedersachsen Holland und Engelland*. Vol. 3. Frankfurt; Leipzig: Christian Ulrich Wagner, 1754.

Unverfehrt, Gerd. *Hieronymus Bosch: die Rezeption seiner Kunst im frühen 16. Jahrhundert*. Berlin: Gebr. Mann Verlag, 1980.

Valentyn, François. *Verhandeling der Zee-Horenkens en Zee-Gewassen in en omtrent Amboina en de Nabygelegen Eilanden*. Amsterdam: Joannes van Keulen, 1754.

Valéry, Paul. *L'homme et la coquille*. Paris: Gallimard, 1937.

———. *"Man and the Sea Shell."* In *Paul Valéry: An Anthology*, selected, with an introduction by James R. Lawler, 108–35. Princeton, NJ: Princeton University Press, 1977.

Van Veen, H. Th. "Cosimo de' Medici's reis naar de Republiek in een nieuw perspectief." *Bijdragen en mededelingen betreffende de geschiedenis der Nederlanden* 102 (1987): 44–52.

Van der Veen, Jaap. "Dit klain Vertrek bevat een Wereld vol gewoel. Negentig Amsterdammers en hun kabinetten." In *De wereld binnen handbereik: Nederlandse kunst- en rareitenverzamelingen, 1585–1735*, edited by Ellinoor Bergvelt and Renée Kistemaker, 232–58. Zwolle: Waanders; Amsterdam: Amsterdams Historisch Museum, 1992.

Vergara, Alejandro, ed. *The Art of Clara Peeters*. Antwerp: Koninklijk Museum voor Schone Kunsten; Madrid: Museo Nacional del Prado, 2016.

Vermeij, Geerat J. *A Natural History of Shells*. Princeton, NJ: Princeton University Press, 1993.

Vickers, Nancy. "Diana Described: Scattered Woman and Scattered Rhyme." *Critical Inquiry* 8.2 (1981): 265–79.

Vignau-Wilberg, Thea. *Joris and Jacob Hoefnagel: Art and Science around 1600*. Berlin: Hatje Cantz Verlag, 2017.

Vincent, Levinus. *Korte Beschryving van den inhout der Cabinetten begreepen in de Rariteit Kamer of Wonder Toneel der Natuur van Levinus Vincent. Tegenwoordig geplaatst in de Nieuwe Doelen, in 's Gravenhage*. The Hague: for Vincent, 1727.

———. *Wondertoneel der Nature, geopent in eene korte Beschrijvinge der Hoofdelen van de Byzondere Zeldsaamheden daar in begrepen*. Amsterdam: François Halma, 1706.

Violle, Cyrille, et al. "Functional Rarity: The Ecology of Outliers." *Trends in Ecology and Evolution* 32.5 (May 2017): 356–67.

Visscher, Roemer. *Sinnepoppen*. Amsterdam: by Willem Jansz., 1614.

Van der Waals, Jan. *De prentschat van Michiel Hinloopen. Een reconstructie van de eerste openbare papierkunstverzameling in Nederland*. The Hague: SDU; Amsterdam: Rijksmuseum, 1988.

Wallert, Arie. "Balthasar van der Ast. Materialien und Techniken." In *Die Stillleben des Balthasar van der Ast (1593/94–1657)*, edited by Sarvenaz Ayooghi, Sylvia Böhmer, and Timo Trümper, 81–93. Petersberg: Michael Imhof Verlag, 2016.

———, ed. *Still Lifes: Techniques and Style. An Examination of Paintings from the Rijksmuseum*. Amsterdam: Rijksmuseum, 1999.

Warwick, Genevieve. *The Arts of Collecting: Padre Sebastiano Resta and the Market for Drawings in Early Modern Europe*. Cambridge: Cambridge University Press, 2000.

Weismann, Ehrenreich. *Christliche Betrachtung der betrübten Zeit und freuden-vollen Ewigkeit*. Nuremberg: Christian Sigismund Froberger, 1680.

Wellington, Robert. *Antiquarianism and the Visual Histories of Louis XIV: Artifacts for a Future Past*. Farnham: Ashgate, 2015.

Wesseling, Ari. "Testing Modern Emblem Theory: The Earliest Views of the Genre (1564–1566)." In *The Emblem Tradition and the Low Countries: Selected Papers of the Leuven International Emblem Conference 18–23 August, 1996*, edited by John Manning, Karel Porteman, and Mare Vaeck, 4–22. Turnout: Brepols, 1999.

Weston, Edward. *The Daybooks of Edward Weston*. Edited by Nancy Newhall. New York: Aperture, 1961.

Weststeijn, Thijs. "Cultural Reflections on Porcelain in the 17th-Century Netherlands." In *Chinese and Japanese Porcelain for the Dutch Golden Age*, edited by Jan van Campen and Titus Eliëns, 213–29. Zwolle: Waanders, 2014.

White, Christopher. *Rembrandt as an Etcher: A Study of the Artist at Work*. 2 vols. London: A. Zwemmer, 1969.

Wilkins, Guy L. "A Catalogue and Historical Account of the Sloane Shell Collection." *Bulletin of the British Museum (Natural History), Historical Series* 1.1 (1953): 1–47.

———. "The Cracherode Shell Collection." *Bulletin of the British Museum (Natural History), Historical Series* 1.4 (1957): 121–84.

Willemsen, Annemarieke. "Kinder-Spel en Poppe-Goet: 17e eeuwse miniatuur-gebruiksvoorwerpen en hun function," *Antiek* 28 (1993–94): 392–99.

Woodall, Joanna. "Laying the Table: The Procedures of Still Life." *Art History* 35.5 (2012): 976–1003.

———. "Wtewael's *Perseus and Andromeda*: Looking for Love in Early Seventeenth-Century Dutch Painting." In *Manifestations of Venus: Art and Sexuality*, edited by Caroline Arscott and Katie Scott, 39–68. Manchester: Manchester University Press, 2000.

Yandell, Cathy. "Iconography and Iconoclasm: The Female Breast in French Renaissance Culture." *French Review* 83.3 (2010): 540–58.

Van 't Zelfde, Reinier. 2001. "'O seldsaem dierken dat so schoon paleys bewoonet!': Exotische schelpen in de Noord- en Zuid-Nederlandse schilderkunst van de zestiende en de zeventiende eeuw." *Standplaats: RKD / Standplaats: Academie*, edited by Ingrid Bons and Suzanne Laemers, 41–76. Zwolle: Waanders, 2001.

Zuidervaart, Huib J. "'Scientia' in Middelburg rond 1600: De wisselwerking tussen academici, kooplieden en handwerkers in het vroegmoderne Middelburg: essentieel voor de creatie van 's werelds eerste verrekijker?" *Archief: Mededelingen van het Koninklijk Zeeuwsch Genootschap der Wetenschappen* (2017): 43–100.

Zuroski, Eugenia. "Nautilus Cups and Unstill Life." *Journal18: A Journal of Eighteenth-Century Art and Culture* 3 (2017): http://www.journal18.org/1493.

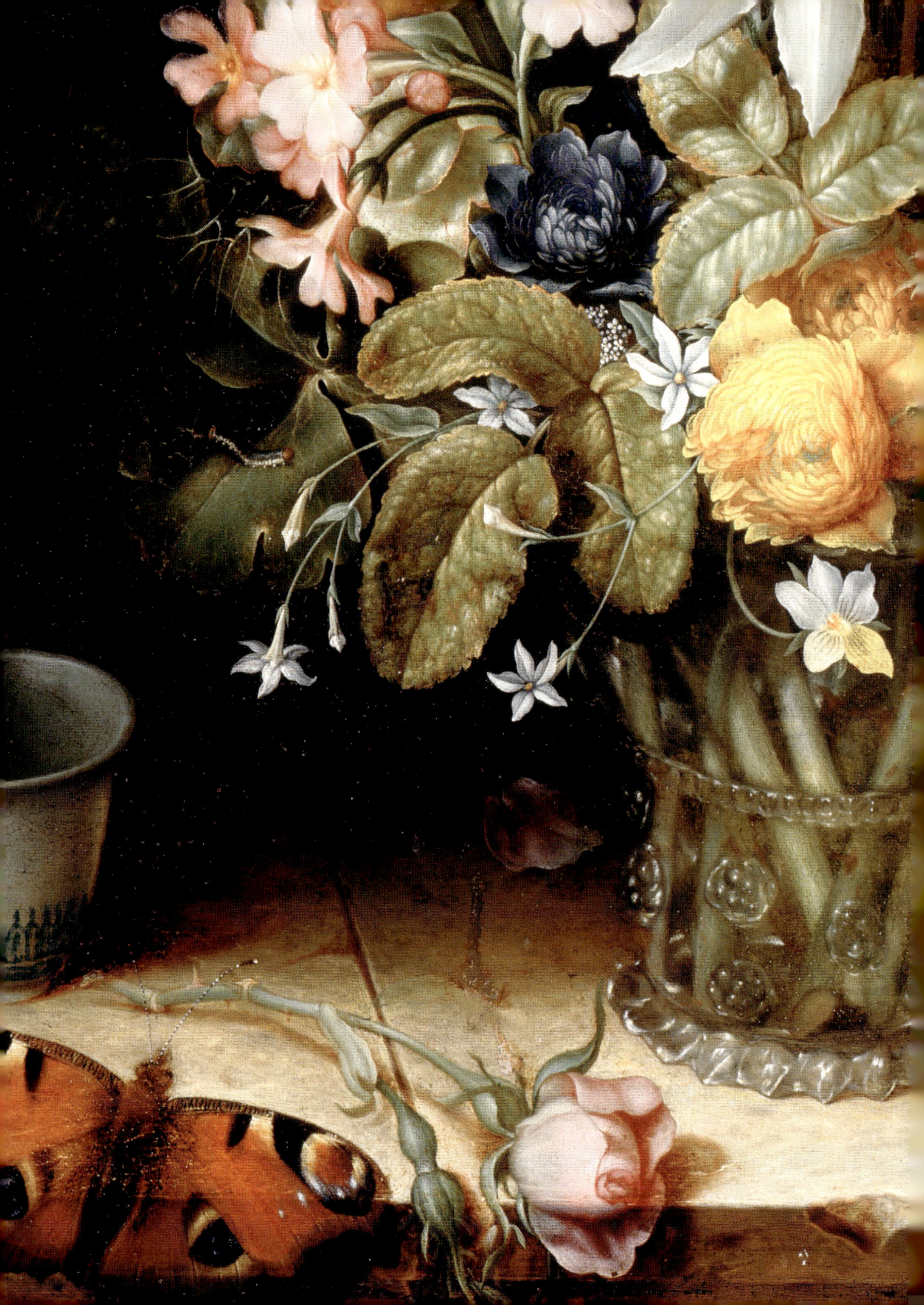

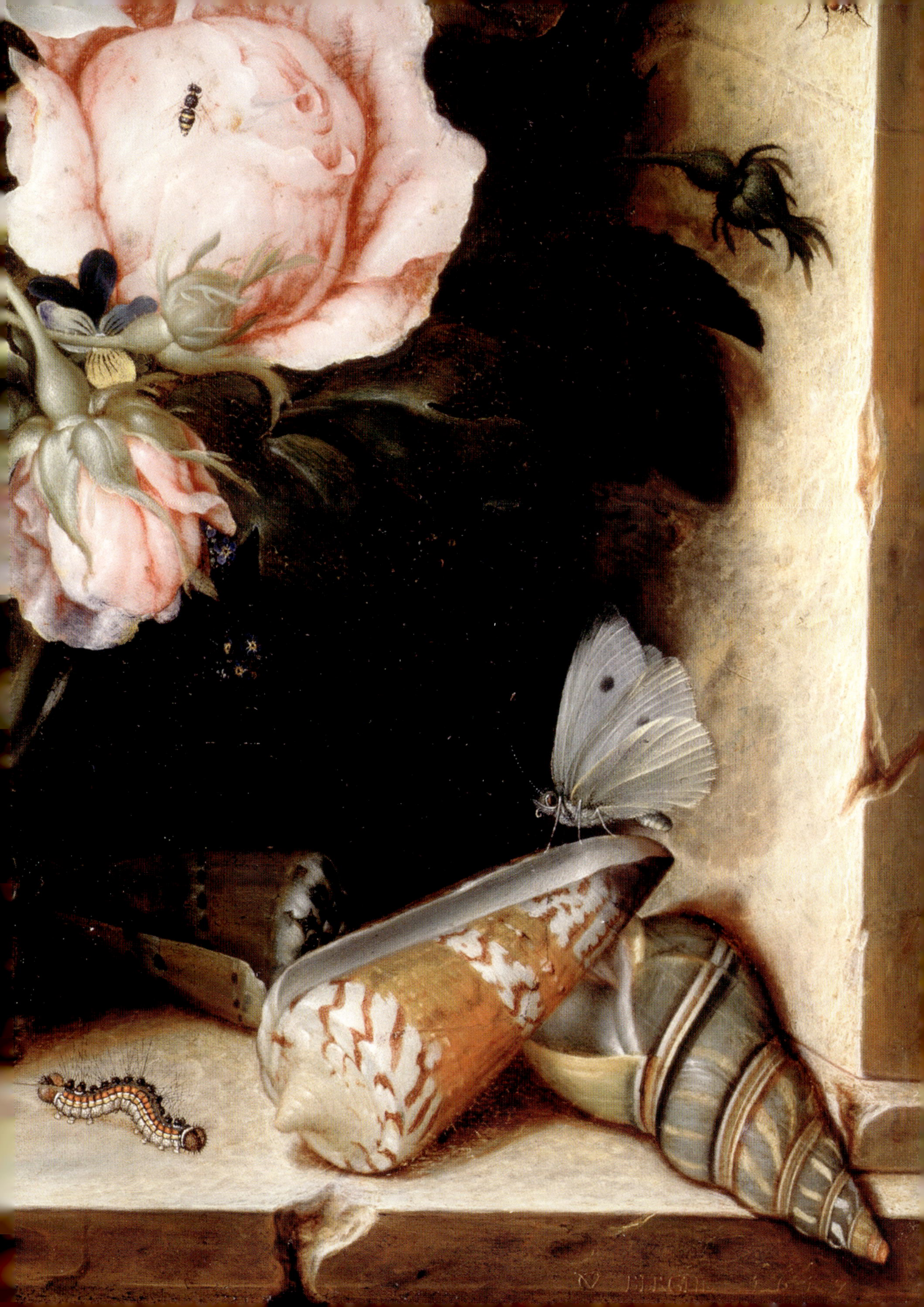

# CONTRIBUTORS

**Marisa Anne Bass** (PhD, Harvard University) is Professor of Northern European Art (1400–1700) in the History of Art Department at Yale University, and an Affiliate of Yale's Program in the History of Science and Medicine. Her scholarship focuses on intersections between visual and intellectual culture in the early modern period, addressing topics that range from natural history to antiquarianism to the politics of public art. Her publications include *Jan Gossart and the Invention of Netherlandish Antiquity* (Princeton University Press, 2016), *Insect Artifice: Nature and Art in the Dutch Revolt* (Princeton University Press, 2019), and the coauthored exhibition catalogue *Beyond Bosch: The Afterlife of a Renaissance Master in Print* (Saint Louis Art Museum, 2015).

**Stephanie Dickey** (PhD, Institute of Fine Arts, New York University) holds the Bader Chair in Northern Baroque Art at Queen's University in Kingston, Canada. Her research and publications explore the art of the early modern Netherlands, especially the work of Rembrandt van Rijn and his circle, and aspects of the history and collecting of prints, portraiture as a cultural practice, and representations of gender and emotion. Among her books and edited volumes are *Rembrandt: Portraits in Print* (John Benjamins, 2004), *The Passions in the Arts of the Early Modern Netherlands (Netherlands Yearbook for the History of Art*, 2010), *Rembrandt and His Circle: Insights and Discoveries* (Amsterdam University Press, 2017), and *Rembrandt in Amsterdam: Creativity and Competition* (Yale University Press, 2020).

**Anne Goldgar** (PhD, Harvard University) is the inaugural Garrett and Anne Van Hunnick Professor of European History at the University of Southern California. Her research focuses on the intersections of history, art history, economic history, and the history of science, particularly issues of collecting, science, and trade in the early modern period. Among her published works are *Impolite Learning: Conduct and Community in the Republic of Letters, 1680–1750* (Yale University Press, 1995) and *Tulipmania: Money, Honor, and Knowledge in the Dutch Golden Age* (University of Chicago Press, 2007). Her current project focuses on Dutch identity from 1598 to the present through the lens of representations of the Nova Zembla *overwintering* of 1596–97.

**Anna Grasskamp** (PhD, Leiden University) is Assistant Professor at the Academy of Visual Arts of Hong Kong Baptist University. Her articles on artistic and material exchange between Asia and Europe have appeared in the *Rijksmuseum Bulletin*, *World Art*, and *Res*. She coedited *EurAsian Matters: China, Europe, and the Transcultural Object, 1600–1800*

(Springer, 2018) and is the author of *Objects in Frames: Displaying Foreign Collectibles in Early Modern China and Europe* (Dietrich Riemer Verlag, 2019). Her current book project examines the relationships among artists, collectors, materiality, and thingness in the transculturally connected art worlds of early modern Eurasia, with a special focus on maritime material culture, craftsmanship, and the female body.

**Hanneke Grootenboer** (PhD, Visual and Cultural Studies, University of Rochester) is Professor of the History of Art at Radboud University Nijmegen. In her research she examines early modern visual culture through the lens of contemporary art and theory, focusing on topics that include intimacy, interiority, introspection, and miniaturization. Her publications include *The Rhetoric of Perspective: Realism and Illusionism in Seventeenth-Century Dutch Still Life Painting* (University of Chicago Press, 2005), *Treasuring the Gaze: Intimate Vision in Late Eighteenth-Century Eye Miniatures* (University of Chicago Press, 2012), and *The Pensive Image: Art as a Form of Thinking* (University of Chicago Press, 2021).

**Claudia Swan** (PhD, Columbia University) is the Mark S. Weil Professor of Art History & Archaeology in the Department of Art History and Archaeology at Washington University in St. Louis, and a scholar of early modern northern European visual culture with a concentration in the Dutch world. She has published numerous articles and book chapters on early modern art, science, and collecting, and on Dutch visual culture. Recent publications include a special issue of *NUNCIUS* on early modern geometries; a study of Dutch *liefhebberij* as a market sensibility, and *Rarities of These Lands: Art, Trade, and Diplomacy in the Dutch Republic* (Princeton University Press, 2021).

**Róisín Watson** (PhD, University of St Andrews) is Departmental Lecturer in Early Modern European History at the University of Oxford. Her research explores the dynamics of religious change in early modern Europe and the development of confessional identities. She focuses in particular on the relationship between religion and visual culture in Reformation Germany. Her forthcoming book *The Beauty of Belief: Decorating the Württemberg Church during the Reformation* (Brill, 2021) explores church interiors in southwest Germany. She has also published work on the connections between funerary monuments and the construction of confessional memory.

# INDEX

**Bold** page numbers indicate figures.


Aar, Jan Govertsz van der, 27, 31, 34, 123, 181n22; Goltzius portrait of, 12, 27, **30**, 31, 40, 42, 43, 47, 93, 94
*Achilleid*, 58
Achilles, 53, 57, 58
aesthetics, 35, 47, 129; and art and nature, 42; and Black bodies, 180n2; and Borsselen, 27; and cabinet arrangements, 10; and collectors, 8–9; European, 14; and Gersaint, 165, 166, 172; and God, 9; and marbled items, 7–8; and natural knowledge, 16; and nature, 16; and Porret collection, 40; and religion, 11; and Rembrandt, 155; in G. E. Rumphius, 25, 47; and self-representation, 10; and sensual appeal, 14; and slave trade, 7; and Valéry, 17. *See also* beauty
Africa, 6, 65, 70
African men, 68
agate, 7
Albrecht VII, 132
Alciato, Andrea, 152
Aldegrever, Heinrich, 161
Aldrovandi, Ulisse, 65; *De reliquis animalibus*, **62**, 62–63, 182n31, 183n43
Alkmaar, island of, 26
allegory, 21, 68, 113
amateurs, 2, 79, 161, 167, 174, 180n7. *See also* connoisseurs
Amboina wentletrap, 122. See also *Epitonium scalare* (precious wentletrap)
Ambon, 2, 7, 22, 26, 45, 163, 180n4, 185n13
Ambras Castle, 49
Amphitrite, 40, 53, 181n37
Amsterdam, 35, 111, 170; collections in, 5, 9, 34, 106, 107, 122, 141, 161; as market for cowrie shells, 40; market in, 161–62; Oortman's residence in, 104; sales in, 6; and trade, 163; and Valentyn, 2
Andromeda, 87, 89, 90, 128
Angola, 40
animals: and Aldrovandi, 62, 65; and Aristotle, 3, 65; connection of shells with, 15; figures of, 51, 63, 64, 68, 145; and Galatea, 70; made from shells, 145; and processing, 45, 169; and G. E. Rumphius, 7. *See also* birds
Anna Maria Louise de' Medici, 111
Anonymous: *Cameo medallion with portrait of Prince Maurits, Stadholder*, 35, **38**; *Ewer from Burghley House, Lincolnshire*, 63, **63**; illustration from Amerigo Vespucci, *Diß Büchlin saget*, **68**, 69
Anonymous (Dutch goldsmith), *Golden Cradle with Child*, 111, **111**
Anonymous (Flemish?), *Scene of Shells and Birds*, 99, **100–101**
Antilles, 5
antiquarians, 93, 94, 98, 99
Antwerp, 53, 59, 62, 93, 182n18
Aphrodite, 89. *See also* Venus
Aphrodite-Arsinoe, 98
apothecaries, 1, 9, 162
Appel, Jacob, *Painting of the Dollhouse of Petronella Oortman*, 113, **114–15**, 116
architecture, 15, 16; and Furttenbach, 138; garden, 132; and grottoes, 3, 128, 129, 132, 133, 138, 139, 140, 141, 142, 147, 148, 153; and Montaigne, 123; and nature, 99; of shells, 17, 75, 93, 95
*arcimboldesques*, 10
*Argonauta argo* (paper nautilus), 3
aristocracy, 1. *See also* elites; nobility
Aristotle, 3, 26, 42, 45, 65, 123, 181n39
Arnold, Joseph, *The Wholesaler and Ironmonger Dimpfel Family Curiosity Cabinet*, 59, **60–61**, 62
art: and Borsselen, 27; collection as, 9; commodification of, 58; early modern connoisseur's love of, 58; and Gersaint, 16; and Kaltemarckt, 49; knowledge about, 16; and nature, 4, 9–10, 15,

art (*cont.*)
26–27, 31, 42–43, 46, 104, 132, 155, 161, 164, 174, 175, 181n39; and Oortman, 119; price of, 159; reflections on, 14; and science, 10; and Valéry, 99
art cabinets. *See* cabinets, art; *Kunstkammern*
artifice, 112, 153; divine, 84, 150, 175; and divine nature, 87; and Furttenbach's grotto, 140; human, 79, 147, 161; human vs. natural, 123; and nature, 46, 79; of shells, 95, 99
*artificialia*, collection of, 5, 26, 104
artisans, 53; and grottoes, 128–29, 133, 138, 140, 141, 142, 147, 150; and Oortman's dollhouse cabinet, 107
artists, 12, 16, 159, 162
Arundel, Thomas Howard, Earl of, 172
Asia, 35, 62, 63, 70
Asian ceramics, 58–64
Asian sea snails, 53
Asian shells, 49, 53, 58, 59, 64, 70, 71
Ast, Balthasar van der, 84, 187n8; *Still Life with Fruit and Flowers*, 81–84, **82–83**, 90; *Study of Two Shells ("Herts Horen" and "Geel Belleken")*, 159, **159**, 169
Astley, Edward, 171
auctions, 7, 122, 162, 164, 165–68
automata, 127–28, 133, 150

baby shells, 104, 106, 107, 117, 120, 123
Bachelard, Gaston, 17, 87, 104, 106, 119
Baldung, Hans (Hans Baldung Grien), *The Witches*, 183n42
Banks, Joseph, 189n78
banquet scenes, 79
Barnard, John, 171
Bartsch, Adam, 171
Bassé, Jan, 161, 162
beauty, 1, 3; and Ast, 84; and Bachelard, 87; and Borsselen, 27, 84, 159; and Brinck, 34; and Camphuijs, 26; and Francken, 58; and Furttenbach, 141; and Gersaint, 167, 168, 169–70, 172; and Grotta Grande, 132; and Kornblum, 69, 70; and Linard, 98, 99; and marbled items, 7–8; and Palissy, 95; and Rembrandt, 175; and G. E. Rumphius, 9, 25, 43, 45; and Spinoza and Beudeker, 122. *See also* aesthetics; eroticism; sexuality
Beekman, E. M., 7, 181n43
Beham, Hans Sebald, *The Peasants' Feast or the Twelve Months*, 182n7
Bella, Stefano della, *View of a grotto in the grounds of the Villa Pratolini*, 138, **139**
Bellekin, Cornelis, *Nautilus Cup*, 182n7
Bellekin, Jan, *Carved Nautilus (The Sloane Nautilus)*, 35, 99, **101**
Bellekin family, 35, 99
Belon, Pierre, 7, 59, 94–95
Berghe, Christoffel van den, *Still Life with Flowers in a Vase*, 31, **32**, 34, 79
Beudeker, Christoffel, 122–23
Beyeren, Leendert van, 187n16
Bible, 149
biography: of objects, 2; of shells, 14. *See also* social life, of things
birds: carvings of, 63; images of, 49, 62, 63, 99; made from shells, 145; mechanical, 127. *See also* animals
black amber, 140
Black bodies, 7, 21, 22, 46–47, 180n2. *See also* body; people of color
Blanc, Charles, 162
Blijenburgh, Cornelis van, 8
Bloemaert, Abraham, *Tronie of a Black Man with a Conch*, **20**, 21–22, 181n44
Boboli gardens, Grotta Grande, 132, 149
Bocskay, Georg, and Joris Hoefnagel, *Mira calligraphiae monumenta (fol. 118r)*, **86**, 87
body: anatomical features of, 52; of beholder, 64; as bound to soul, 15, 120; of Christ, 59; enslaved, 14; foreign, 7, 64, 65, 70, 71; interaction with, 14; laboring, 14, 46; male, 53, 57, 58; and Oortman, 106; parts of, 52–58; and Pliny the Elder, 87; in Shakespeare, 89, 90; shells as substitutes for, 12. *See also* Black bodies; eroticism; nakedness/nudity; sexuality
body, female, 14, 57, 65, 71; and childbirth, 111, 116, 119, 120; of dragon, 150; in Erasmus, 119–20; genitals of, 53, 59; and Kornblum's cup, 69; and mind or soul, 119–20; and porcelain skin, 69, 70; and Rembrandt, 174; in Wtewael, 87, 89, 90. *See also* eroticism; sexuality; women
Bologna, Giovanni da, 141
Böner, Johann Alexander, 127; Interior of Magdalena Sibylla von Württemberg's grotto, **126**, 127, 128, **131**, 145, 149; Obelisks in Magdalena Sibylla von Württemberg's grotto, **130**
Bonnac, Marquis de, 164
Borsselen, Philibert van, *Strande (The Beach)*, 7–8, 9, 27, 35, 84, 123, 159
Bosch, Hieronymus, 76
Bosschaert, Ambrosius, *Bouquet of Flowers on a Ledge*, 79, **158**, 159, 169, 187n8
Boucher, François, 165, **165**
Brandt, Hendrina, 113, 116
Brandt, Johannes, 110, 111, 113, 117
Breda, Anthony van, 9
Breda, Johanna van, 9, 10, 107
Brinck, Ernst, 34, 35, 40
British Museum, 163, 164, 171
Brune, Johan de, 118
Brussels, 132, 133, 141
Buckingham, Duke of, 34
Buonanni, Filippo, 7, 166, 171; *Heads*, 107, **109**; *Ricreatione dell'occhio*, 174, 183n43
Buontalenti, Bernardo, 132
Burke, Jill, 71
Buton, 43
Buysen, Andries van, after Romeyn de Hooghe, *Visitors in the Natural History Cabinet of Levinus Vincent in Haarlem*, **8**, 9
by-works (*parerga*), 63

cabinets, 1, 188n64; arrangements in, 10; and Bachelard, 104; and Oortman's dollhouse, 106; preservation in, 170; for prints, 188n64; and Quentin de L'Orangère, 168. *See also* dollhouses
cabinets, art, 103, 111. See also *Kunstkammern*
cabinets, of curiosities, 16, 103, 116; and Beudeker, 122; and contemplation, 120; and *Conus marmoreus*, 159–61; and Courten, 164; in Gheyn, 94; and grottoes, 16; in Luyken, 117; natural objects vs. human-made artifacts in, 104; and Oortman, 103, 119; and rarities, 168; and Rembrandt, 155; of Vincent, 165. See also *Kunstkammern*
Caligula, 3
calligraphy, 87
Callimachus, 98
Callot, Jacques, 168
cameos, 93, 161

Camphuijs, Johannes, 26
Cannstadt, 141
Cape of Good Hope, 5
Caribbean, 2, 89, 159
Carruthers, Mary, 106
Cartesian thought, 120
carving/carvings, 3, 25, 27, 35, 63, 99
*Cassis tuberosa* (striped king helmet), 98
casts, 95
catalogues, 10, 171; and aesthetics, 8; annotated, 171; and Duchess of Portland, 164; and Gersaint, 165, 166, 167, 168, 170, 171, 172; and Johann Friedrich's grotto, 141; for Porret, 40; and Seba, 152; and Worm, 162
Cats, Jacob, 112; "Children's Play" ("Kinder spel"), 117–18; *Houwelick (Marriage)*, 117–19
Caus, Salomon de, 133, 140, 142, 148; *Hortus Palatinus*, 132, **134**; *Les Raisons des Forces Mouvantes*, 133, **138**
ceramics, 58, 59, 62, 64, 95. *See also* porcelain
*Charonia lampas* (trumpet shells), 76–78, 183n5
*Charonia tritonis* (Triton's trumpet), 31, 58, 183n5
*Chicoreus sauli* (spiny snail shell), 93
children, 116, 117–18, 120
Chinese emperors, 64
Chinese porcelain, 59, 62, 71
Christian spirituality, 120
Christopher, St., 75, 76, 77, 78
church treasuries, 51
Cicero, 3, 11; *The Orator*, 120
*Cittarium pica* (magpie shell), 93, 183n13
classification, rise of, 3
Clusius, Carolus, 27
*Cochlea lunaris major*, 43, **44**
*Cochlea lunaris minor*, **44**
coconuts, 42, 49
coins, 79, 81, 93, 94, 99, 159, 161
collection, 4, 58, 70, 94, 121; in Amsterdam, 161; of *artificialia*, 5, 26, 104; and curiosity, 159; democratization of, 159; of Duchess of Portland, 7; and Gersaint, 172; and God, 9; history of, 3; of *naturalia*, 5; in Netherlands, 7; and Rembrandt, 155, 161, 162; and G. E. Rumphius, 25–26, 180n8; of Seba, 10; and social relationships, 10–11; and Valentyn, 10–11; in J. Zucchi, 68
collector(s), 34, 93, 171; and aesthetics, 8–9; cabinets of, 94; character of, 12; and Cicero, 120; and colonial exploitation, 14; and condition of specimens, 169; and culture, 4–5; and Doerr, 21; in Dordrecht, 1–2, 11, 171; and East India ships, 6; and household effects, 103; and Linard, 95, 98, 99; of multiples, 171; Netherlandish, 51; Oortman as, 123; portraits of, 21; of prints, 159, 161, 162, 163, 164, 165, 167, 168, 169, 170, 171, 174; of Rembrandt's etchings, 171; and G. E. Rumphius, 22, 25, 46; and sensuality, 14; and thought/reflection, 120; and Valentyn, 16
colonialism, 3, 5, 14, 17, 46, 63, 64; and commerce, 6; exploitation in, 14; and sexuality, 12; shells as currency in, 14; and trade, 4. *See also* imperialism
commerce, 3, 6, 14, 16, 17, 25, 40, 162, 168, 169. *See also* trade
commodification, 58, 69
commodities, 12, 47, 79; and emblem books, 113; exotic shells as, 14, 34, 35; and Rembrandt, 155, 167
*Concha murrha* (porcelain shell), 27
*conchas legere*, 120. *See also* gathering, of shells
conchophilia, defined, 1
conch shells, 53, 63, 84, 161, 164
connoisseurs, 1, 16, 169, 171; and Dordrecht collectors, 11; of prints, 174; and prints by Rembrandt, 171; and term *liefhebber*, 2; trade between, 171. *See also* amateurs
*Conus aulicus* (princely cone), 98
*Conus cedonulli* (admiral cone), 170
*Conus generalis* shell (remains of *Witte Leeuw*), **33**, 34
*Conus litteratus* (lettered cone), 34, 98
*Conus magnificus* (magnificent cone), 84
*Conus marmoreus* (marbled cone, Rembrandt's cone), 107, 155, 159, 161, 167, 169, 171, 172
*Conus marmoreus Linneaus 1758*, **154**, 155
*Conus* species (cone shells), 31
Coorte, Adriaen, *Shells on a Stone Plinth*, **92**, 93
coral, 58, 59, 65, 68, 76, 95, 111, 127, 133, 138, 139, 140, 148
Cornelisz van Haarlem, Cornelis: *Allegory of the Arts in a Time of Peace*, 31, 181n22; *Neptune and Amphitrite*, 40, **41**, 42, 43, 45, 47, 181n37
Cosimo III de' Medici, 26, 122, 163
Cospi, Ferdinando, 26
Costa, Emanuel Mendes da, 164, 189n78
Costa, Francisco Gomes da, 4
Coudenberg Palace, 132
Courten, William, 164–65
courtly celebration, and grottoes. *See* grottoes
cowrie shells, 6, 35, **39**, 40, 59
Cracherode, Clayton Mordaunt, 164, 170–71
craftsmanship, 9–10, 62, 63, 87, 104, 106, 112, 172
Critz, Thomas de (attributed to), *John Tradescant the Younger with Roger Friend and a Collection of Exotic Shells*, **6**, 7
Crucifixion, sculpture of (*Kunstkammer* from 1668), 59
culture, 4–5, 46, 63, 104, 146
Cupid, 53, 181n37
curiosity, 1, 4, 47; and collecting, 159; and exoticism, 5; and grottoes, 128, 129, 146, 147, 148, 149, 153; and Rembrandt, 155; in G. E. Rumphius, 25
curiosity cabinets. *See* cabinets, of curiosities
currency, shells as, 3, 6, 14, 35, 40
Cybele, 46
*Cymbiola vespertilio* (bat volute shell), 31
*Cypraea arabica* (cowrie) shell (remains of *Witte Leeuw*), 35, **39**
*Cypraea tigris*, **38**

Daulby, Daniel, 171
death, 81, 149; contemplation of, 12, 16; and Erasmus, 119–20; and grottoes, 127, 128, 129, 149, 150; and liminality, 4, 12; and Oortman, 116–17, 120; in Shakespeare, 89, 90; and still life, 78; and Wtewael, 89. *See also* lifelessness/lifeless things; reanimation
Deidamia, 57, 58
Delft, 2, 59, 122
*denkbeelden* (thought-images), 112–13, 116
Descartes, René, 119
Dimpfel family, 59
Diomedes, 53
Doerr, Arthur, "The Shell Collector," 21
*The Dollhouse of Petronella Oortman*, 15, **102**, 103–23, 185n26; baby shells in, 104, 106, 117, 120, 123; birth depicted in, 116; *Christ Approached by the Children*, 117; cost of, 110; death depicted in, 116–17, 120; dolls in, 113, 116; *The East-Indian Art Cabinet with Miniature Shell Collection*, 103, **105**, 106, 118, 120, 121, **121**;

*The Dollhouse of Petronella Oortman* (cont.)
inventory of, 112; lying-in room of, 107, **110**, 110–12, **115**, 116, 117; and miniaturization, 15, 104, 106, 107, 110, 111, 112, 113, 118, 119, 123; outer case of, 110, 111; outer frame of, 111; rooms of, 110–11; tapestry room of, 103, **115**, 116, 117
dollhouses, 12, 15; and control of environment, 106; as *denkbeelden*, 112–13, 116; Dutch vs. German, 111; as ego-documents, 106, 184–85n6; lying-in rooms in, 107, 111; as mental retreats, 106; and miniaturization, 103; as play for adults, 117
dolphins, 63
*doodverf*, 43
Dordrecht, collectors of, 1–2, 11, 171
dragons, 63, 87, 89, 150
Dresden, 59, 133
Dresden *Kunstkammer*, 59, 62
Duflos, Claude, **165**
Dürer, Albrecht: *Adam and Eve*, 164; portrait of Frederick the Wise, 161; *Witch Riding Backwards on a Goat*, 183n42
Durlach, 140
Dutch East India Company (VOC), 5, 22, 26, 31, 35, 40, 77, 107, 171
Dutch West India Company (WIC), 5, 77

East Indian shells, 35, 46
East Indies, 2, 35, 45, 107
Edam, island of, 26
elites, 7, 51, 128, 133. *See also* aristocracy; nobility
emblem books, 5, 81, 112–13, 118, 120
emblems, 11, 34, 104, 127, 152
embroidery, 9, 10, 104, 111
emu eggs, 42
England, 159
English slave trade, 6
Enkhuizen, 34, 40
*Epitonium scalare* (precious wentletrap), 15, 107, 120–23, **121**, 162, 172
Erasmus, Desiderius, 3, 8, 120; *Colloquies*, 119–20; "The New Mother," 119–20
eroticism: and foreign oceans, 65; in Francken, 58; and Indian shells, 14; and *Kunstkammer*, 52, 64, 70–71; and shell cups, 52, 53, 64; and Weston, 70; in Wtewael, 89. *See also* beauty; body; body, female; sexuality
etching, on shells, 2, 10, 27
Europe, 4, 7, 10, 12, 14, 16, 63
Europeans, 21, 22, 25
*Euspira nitida*, 93
exotica/exoticism, 1, 31, 35, 78, 79; in Bloemaert, 21; and bodies, 12; and Cosimo III de' Medici, 26; and F. Costa, 4; and foreign trade, 5; and market, 169; and Oortman, 104; of origins, 14; in G. E. Rumphius, 25; and slave trade, 6, 7. *See also* rarity/rarities

faces, 10, 107, 139, 145, **145**, 147, 150, 152
Faille, Johan de la, 8, 122, 170
Farnese Palace, 146
Ferdinand II, Archduke, 49
Fibonacci sequence (golden section), 11–12
fish-tailed figures, 63, 64, 65
Flinck, Govert, 162
Florence, 59, 65, 132, 149
flowers, 2, 10, 150; and Aar, 27; in Ast, 81; and Berghe, 31; and Beudeker and Spinoza, 122; in Borsselen, 84; and *Conus marmoreus*, 159; images of, 62, 63; made from shells, 139, 142, **144**, 145, 146, 147; in Peeters, 79; and Rijswijck, 35; variety of, 147; and Venne, 84, 87
flower still-life paintings, 78, 79, 84, 159
foreignness, 14, 70, 71
foreign oceans, 65
foreign peoples, 12, 16, 69
Fortuna, 49, 65
France, 35, 140, 159, 165
Francesco I de' Medici, 65, 132
Francken, Frans, II, *A Curiosity Cabinet*, **160**, 161
Francken, Frans, II (workshop), *Ulysses Recognizes Achilles Dressed as a Woman among the Daughters of Lycomedes*, 53, **56**, **57**, 57–58, 62, 63, 182n15
François I, King of France, 94
Frederick Hendrick, 162
Frederick the Great, 133
Friedrich, Johann, 152
Friedrich V, 140
Fugger, Octavian, 182n5
Furttenbach, Joseph, the Elder, 128–29, 132, 133, 138–40, 141–42, 146–49, 150; *Architectura civilis (Civil Architecture)*, 140, 142, **144**, 148; *Architectura privata (Domestic Architecture)*, 140; grotto of in Leutkirch, 139, **139**, 140

Galatea, 69, 70, 71, 133
gardens, 84, 112, 141, 152; and ancient Romans, 3; design of, 15; and Furttenbach, 140; and grottoes, 128, 132, 133; Italian, 3; of northern Europe, 3; and Schickhardt, 138; and Stuttgart's ducal residence, 128; of Versailles, 147; of Zwinger, 133
gathering, of shells, 4, 12, 16–17, 21, 22, 25, 26, 27, 45. See also *conchas legere*; labor; preparation and processing, of shells
gems, 27, 90, 93, 94, 99
genitals, 53, 57, 58, 59, 65, 71. *See also* body; sexuality
Genoa, 139
German courts, 132–33
German print culture, 65
Gersaint, Edmé-François, 7, 16, 164–68, 169, 170, 171, 174, 185n44; *Catalogue raisonné de Coquilles, Insects, Plantes marines, et autres Curiosités naturelles*, **165**, 165–66, 167, 172, 189n78; *Catalogue raisonné de toutes les Pièces qui forment L'Oeuvre de Rembrandt*, **166**, 166–67
Gesner, Conrad, 7
Geyer, Elias, *Nautilus Cup supported by a Vintner*, 27, **50**, 51, 52
Gheyn, Jacques de, II: *Neptune, Amphitrite and Cupid with a Nautilus Shell*, 47, 53, **55**, 57, 70, 181n37; *Portrait of Abraham Gorlaeus at Age 52*, 35, 93, 94, **94**
Gheyn, Jacques de, III, 163
Glomy, Jean-Baptiste, 166, **166**, 167
Goa, 3
God, 12; and aesthetics, 9; and artifice, 84, 87, 150, 175; and Beudeker, 122–23; and Borsselen, 9, 123; and collection, 9; creation of, 1, 11; design of, 11, 12; and Dordrecht collectors, 11; and grottoes, 129, 149, 150, 152; and Hoefnagel, 87; ingenuity of, 155; and nature, 9; nature of, 11; relationship to, 16; and Rembrandt, 155, 175
Godewijck, Margaretha van, 111
Goldney, Thomas, II, 142
goldsmiths, 49, 52, 64, 70
Goltzius, Hendrick, *Portrait of the Merchant and Shell-collector Jan Govertsz van der Aar*, 12, 27, **30**, 31, 40, 42, 43, 47, 93, 94, 123
Gontard, Carl von, 133

Gorlaeus, Abraham, 35, 93, 94, **94**
Gossart, Jan, *Neptune and Amphitrite*, 183n5
Gothic cathedrals, 95, 99
Greeks, ancient, 94, 98
Green, Toby, *A Fistful of Shells*, 6
Grimaldo, Leodegar, *Drinking Automata with Neptune as Vintner*, 51, **51**, 52
Groningen, Jan Swart van, *Christ in a Landscape*, 183n8
grotesques, 10, 133, 139, 145, **145**
grottoes, 3, 7, 12, 15–16, 127–53; and ancient Romans, 3; in Bible, 149; and conversation, 90; and courtly celebration, 128, 129, 146, 148, 153; creation of, 10; and curiosity, 128, 129, 146, 147, 148, 149, 153; and death, 127, 128, 129, 149, 150; decoration imitating, 133; fashion for, 128; and heavenly realm, 127, 149; as *Kunstkammern*, 150, 152; light in, 148; mannerist, 133; of natural theology, 11; natural vs. artificial, 132; and nature, 129, 132, 146, 149, 150, 152, 153; plural responses to, 129, 132, 146; Renaissance courtly, 132; and sensuality, 12; and sexuality, 12; as shell in macrocosm, 16; subterraneous, 132; theatricality of, 146; and thought/reflection, 12, 16, 127, 128, 129, 132, 146, 148, 149, 152, 153; water features in, 127, 128
Grotto of the Nativity, 149

Haarlem, 2, 9, 165
Hades, descent into, 90
The Hague, 2
Hainhofer, Philip, 181n39
Halle, church treasury of, 182n5
Hameel, Alart du, *St. Christopher*, 76–77, **77**
*Harpa amouretta* (minor harp shell), 31
*Harpa doris*, 183n13
*Harpa harpa*, 183n13
Hay, Jonathan, 64
Heidegger, Martin, 106, 119
Heidelberg, 132, 133
Helle, P.C.A., 166, 167
Henriette Adelaide, Electress of Bavaria, 133
Henriëtte Catherine of Nassau, 133
Henry III, King of France, 146
hermit crabs, 75, 81, 90
*Hexaplex rosarium*, 183n13
Heyden, Jan van der, 185n17
Hinloopen, Michiel, 170
*Hippopus hippopus* (bear paw clam, horse's hoof clam, or strawberry clam shell), 181n22
Hirst, Damian, *Treasures from the Wreck of the Unbelievable*, 17
Hoefnagel, Joris, **86**, 87
Hoffmannswaldau, Christian Hoffmann von, "Description of Immaculate Beauty," 183n49
Hollar, Wenzel, 172, 189n83; *Conus imperialis*, 172, **173**, 174
holotypes, 171
Holy Roman Empire, 132
Hooch, Pieter de, 116
Hoogstraten, Samuel van, 162
Hooke, Robert, 93
Hornick, Erasmus, ewer, 53, **54**
horseshoe crabs, 161
Hortus Palatinus, Heidelberg, 132
Houbraken, Arnold, 169, 174
Houbraken, Jacob, 164, 169, 171
Hulsen, Esaias van, 141, 142, 145
humanism, 5, 11, 120
hydraulics, 133, 138

Idstein, grotto at, 142
imperialism, 3, 10. *See also* colonialism
India, 6
Indian Ocean, 25, 45, 107
Indians, 63
Indian women, 69
Indies, 10, 16, 71
indigenous peoples, 2, 3, 7, 183n25
Indonesia, 2, 5, 22
Indo-Pacific region, 31, 42, 84, 98
Indo-Pacific snail, 155
ingenuity, 76, 112, 127, 147, 150, 152, 155, 161
insects, 81, 84
interiority, 15, 104, 106, 113, 123, 149. *See also* thought/reflection
intimacy, 15, 31, 104, 106
Isabella Clara Eugenia, 132
Italian *studiolo* collections, 65, 68
Italy, 3, 128, 138, 140, 141

Jamaica, 163
Jamnitzer, Wenzel, 52–53
Jansz, Volckert, 172
Japanese lacquerware, 43, 45
jasper, 140
Jeannin, Pierre, 35
Jesus Christ, 127, 128, 149
Johannes, Count of Nassau, 142
Johann Georg II von Anhalt-Dessau, 133
Johann Wilhelm von der Pfaltz-Neuberg, 111
joke cups *(Scherzbecher)*, 52. *See also* vessels
Jong, Jan de, 107
Joyeuse, Duke de, 146

Kalf, Willem, *Pronk Still Life with Nautilus Cup*, **x**, 3, 79
Kaltemarckt, Gabriel, 49, 51
Kemp, Martin, "Wrought by No Artist's Hand," 46
Kepler, Johannes, 11
*konstkamers*, 51. See also *Kunstkammern*
Kornblum, Marx, *Cup*, 27, **48**, 49, 51, 52, 63, 64, 65, 69
kraakware, 58, 59
Kraus, Johann Ulrich, *Description de la Grotte de Versailles*, 145, **145**, 147
*Kunstkammern*, 104, 146; of Dimpfel Family, 59, 62; Dresden, 59, 62; as eroticized spaces, 52, 64, 70–71; and Furttenbach, 140; and Godewijck, 111; and grottoes, 16, 150, 152; as microcosm, 52; and Oortman's dollhouse, 103; and Ovid, 69; of Rembrandt, 162; vessels in, 49, 51; and J. Zucchi, 68. *See also* cabinets, of curiosities

labor, 6, 14, 16, 22, 25, 26, 27, 34, 46, 47. *See also* gathering, of shells; preparation and processing, of shells
lacquerwork, 4, 35, 43, 45, 121
*Lambis chiragra chiragra* (spider conch), 98
Later, J. de: after Jan Goeree, Frontispiece in Georg Eberhard Rumphius, *D'Amboinsche rariteitkamer*, 22, **23**, 46; after Paul Augustus Rumphius, *Portrait of Georg Eberhard Rumphius at the Age of 68*, 22, **22**

Lauffen am Neckar, 141
Leiden, 162
Leonardo da Vinci, 129, 147
Leonhard, Karin, 4
Lesser, Friedrich Christian, 141, 147, 149, 150
Leucopetra, 90
Leutkirch, 133
Leyden, Lucas van, 162; *Young Man with a Skull*, 161
*liefhebberij*, 25
*liefhebbers*, 2, 11, 22, 25, 34, 45, 117
lifelessness/lifeless things, 22, 90, 98, 104. *See also* death
Lightfoot, John, 164
*Liguus fasciatus*, 187n8
*limaçon de Pascal*, 122, 186n48
liminality, 4, 12, 149
Linard, Jacques, 94; *Still Life with Shells and Coral*, 95–99, **96–97**
Linnaean nomenclature, 181n44
Lister, Martin, 16, 164, 165, 170, 171, 172
Lister, Martin, and daughters, *Historiae conchyliorum*, 7, 164
*Livona pica* (West Indian top shell), 31
Lonitzer, Adam, 7
Louis XIII, King of France, 94
Ludwigsburg, 141
Lutherans, 149
Luyken, Jan: "Het Kabinet," **117**; *Instructive Household Effects (Het Leerzaam Huisraad)*, 113, 117, **117**, 120
Lycomedes, daughters of, 53, 58, 182n15
Lycomedes of Skyros, 53, 57, 58

Maar, Dora, *Hand-Shell*, 17
Maldives, 6
male body. *See* body
Man, Cornelis de: *The Curiosity Seller*, 12, **13**, 104; *Merchant in an Interior*, 185n17
Mander, Karel van, 99
marble, 7, 140
marbled items, 7–8
Marguerite de Vaudemont, 146
Marie de' Medici, Queen of France, 35
market, 35, 40, 79, 159, 161–62, 170; and Gersaint, 7, 16, 164, 165, 166, 167; for prints, 16, 159, 162, 164, 165, 166, 167, 168–69; and Rembrandt, 159, 168–69; and G. E. Rumphius, 7; and Visscher, 34
Martirano, Bernardino, 90, 184n39
masks, 93, **145**, 146. *See also* faces
Masques de coquillages et de rocailles, **145**
Masson, Bonnier de la, 185n44
Master of the Holy Figures, *Triptych of St. Christopher, St. Jerome, and St. Anthony*, **74–75**, 75–78, **76**
materiality/material culture, 2, 3, 4, 12, 113
mathematics, 11–12, 122, 133, 186n48
Maurice of Hesse-Kassel, 34
Maurits, Prince of Nassau, 35
medals, 93, 94, 159, 161
Medici, Duke of, 62
Medici collections, 62
Medici family, 59
Melanesian Muslims, 21
memento mori, 149. *See also* death; skulls
merchants, 34, 51, 162
*Mercure de France*, 165
Merian, Maria Sibylla, 180n4
Merian, Matthaus, *Grotto in the Hortus Palatinus*, 132, **134**
mermaids, 49, 65, 69–70, 71
mermen, 69, 71, 76
metaphors, 11, 12, 16, 64, 70, 93; as borrowed residence, 123; and dollhouse items, 106, 113, 118, 120, 121, 123
microcosm, 12, 52, 149
Midas, 128
Middelburg, 84
middle-class, 159, 162
Middle Stone Age, 3
Milton, John, 90
Ming dynasty artifacts, 64
*Mitra mitra* (episcopal miter), 31, 84
*Mitra papalis* (Papal miter), 42
*Mitra stictica* (Pontifical miter), 42
*moeite en verdriet*, 42–47
mollusks, 7, 16, 25, 53, 75, 123
Moluccas, 5, 22, 26
Montaigne, Michel de, 120, 123
Montconys, Balthasar de, 122
morality, 5, 81, 104, 152
mother-of-pearl: baby figure made of, 111, **111**; and black skin, 7; in Bloemaert, 21; in Borsselen, 35; in Brinck, 34; in Francken, 63; and Furttenbach, 142; in Goltzius, 27; in Cornelisz van Haarlem, 42; and Kornblum, 49; and Oortman, 107; and porcelain, 185n11; in G. E. Rumphius, 9, 43, 45; and Württemberg *Kunstkammer* shell cup, 150
moths, 35, 161
multiples, 11, 15, 16, 77, 132, 152, 159, 166, 167, 169, 171
Munich, 132, 147
murex, 3
*Murex radix* (murex), 84
mythology, 10, 14, 68, 71

nakedness/nudity, 22, 42, 49, 64, 65, 69, 70, 71, 128, 174. *See also* body
Namban lacquerwork, 35
Naples, 90, 141
natural history, 3–4, 9, 22, 25, 46, 93, 94, 174
natural history objects, 6, 9, 10
*naturalia*, 5, 26, 34, 94, 104, 161, 164, 167, 169, 172, 174
naturalists, 11, 21, 79, 84
natural philosophy, 7, 120
natural theology, 9, 11
*natura morta*, 78, 81
nature, 127, 161; and aesthetics, 16, 42; and amateurs, 79; and architecture, 99; and art, 4, 9–10, 15, 26–27, 31, 42–43, 46, 104, 155, 161, 164, 174, 175, 181n39; and artifice, 46, 79, 123; and Bellekin family, 35; and Borsselen, 9, 84; boundaries of, 15; bounty of, 147; fertility of, 159; generative powers of, 64; and Gersaint, 172; and God, 9; and grottoes, 129, 132, 146, 149, 150, 152, 153; improvement of, 10; and Linard, 99; and Mander, 99; as microcosm of heavenly realm, 149; nobles' authority over, 147; playfulness of, 11; real vs. represented, 51; and Rembrandt, 175; reproductive energies of, 64; in G. E. Rumphius, 47; and Spinoza, 122; wonders of, 14, 26
*Nature's Theatre of Wonder* (*Wondertoneel der Nature*), 107
nautilus cups, **x**, 14, 49, 51, 79. *See also* Bellekin, Cornelis, *Nautilus Cup*; Geyer, Elias, *Nautilus Cup supported by a Vintner*; Kalf, Willem, *Pronk Still Life with Nautilus Cup*

*Nautilus major sive crassus and its snail* (in Rumphius, *D'Amboinsche rariteitkamer*), **24**, 25
*Nautilus pompilius* (pearly nautilus), 25, 27, 35
nautilus shells, 3, 69, 70, 99; in Aldrovandi, 62–63, 65; in Arnold, 62; in Callimachus, 98; in de Gheyn, 53; in Dresden *Kunstkammer*, 59; as drinking vessels, **x**, 3, 4, 14, 49, 51, 79; and Fibonacci sequence, 11–12; in Francken, 58; and Oortman's dollhouse cabinet, 107; G. E. Rumphius on, 9; and Weston, 70. *See also* vessels, drinking
Nautilus shell with gilded silver mounts, 27, **28**
Neapolitan poets, 90
Neptune, 21, 40, 42, 43, 46, 52, 53, 128, 181n37
Nereids, 21, 40, 69
Netherlands, 4, 5, 159; auctions in, 165; collecting in, 7; and colonial trade, 5; and Gersaint, 165; print culture in, 65; and slave trade, 6, 40; and trade, 7
Neuburg Palace, Bavaria, 133
Neues Palais, 133
nobility, 34, 128, 146, 147. *See also* aristocracy; elites
Nuremberg, 52
nymphs, 21, 128, 133

Oortman, Petronella, 103, 110; and body, 106; and creation of shells, 123; and death, 116–17, 120; death of, 113; dollhouse of (see *The Dollhouse of Petronella Oortman*); and domestic sphere, 119; and exotica/exoticism, 104; monogram of, 111; and self, 119; and thought/reflection, 106, 112–13, 116–17, 119, 123; Warmoesstraat residence of, 104, 110, 121
Oranienbaum, 133
Orléans, Gaston, Duke of, 94
Orpheus, 128
ostrich eggs, 42, 49
Ovens, Jürgen, 122, 162
Ovid, 40, 132; *Metamorphoses*, 58, 69

Pacific Ocean, 1, 25, 107
Palissy, Bernard, 27, 95
Palissy, Bernard (attributed to), *Oval Basin with Life Casts of Reptiles and Sea Creatures*, 95
Paludanus, Bernardus, 34, 40
Paris, 7, 10, 164
parrots, 4, 68, 99
Pascal, Étienne, 122, 186n48
pearls, 42, 65, 68, 107, 161
Peeters, Clara, *Still Life with Flowers, Goblets, and Shells*, 79–81, **80**
Pegasus, 87
people of color, 65, 68. *See also* Black bodies
Perseus, 87
Petrarch, Francesco, *The Life of Solitude*, 120
Pfalz, Philipp Wilhelm von der, 133
Philippi, Gerhard, 140–41, 142, 148
philosophy, 10, 15, 16, 22, 106, 107, 112, 113, 119, 122, 123
photography, 17, 70, 174
*Pinnidae* (pen shell), 141
Pliny the Elder, 87, 147; *Natural History*, 3
polished, incised, and inked *Turbo marmoratus*, 27, **29**
polishing, of shells, 2, 4, 6, 9, 10, 16, 25, 34, 42, 43, 45, 62, 63, 142, 150, 161, 170. *See also* preparation and processing, of shells
politics, 12, 15, 35, 64, 147
Polo, Marco, *The Travels of Marco Polo*, 59
*Polymita picta*, 187n8
Pond, Arthur, 164, 171
Pöppelmann, Matthäus Daniel, **137**
porcelain, 14, 58–64; in Ast, 81; in Berghe, 31, 33; and Borsselen, 27; Chinese, 59, 62, 71; European hard-paste, 59; in Francken, 63; and multiples, 159; and Oortman's dollhouse, 104, 111, 185n11; in Peeters, 79; as term, 59; and women's skin, 69, 70. *See also* ceramics
*porcellane* (as term), 59
Porret, Christiaen, 40
Portland, Margaret Cavendish Bentinck, Duchess of, 7, 164, 168, 170–71
Portland Vase, 164
Portugal, 77
Portugal, King of, 170
Portuguese slave trade, 6
Pratolino, Villa di, 132, 138, 141, 146
preparation and processing, of shells, 2, 12, 42, 49, 161, 169–70; and Bloemaert, 21–22; and colonial exploitation, 14; and Furttenbach, 141–42; labor involved in, 6, 14, 16, 21–22, 27, 45; and Oortman's dollhouse, 107; in G. E. Rumphius, 43, 45–46, 47; by slaves, 6. *See also* gathering, of shells; polishing, of shells
prints: collectors of, 159, 161, 162, 163, 164, 165, 167, 168, 169, 170, 171, 174; condition of, 169–70; and Dordrecht collectors, 11; and Gersaint, 166, 172; market for, 16, 159, 162, 164, 165, 166, 167, 168–69; as multiples, 159, 166, 167, 169, 171; price of, 159; and Quentin de L'Orangère, 168; by Rembrandt, 155, 159, 164, 166, 168–69, 170, 171, 174; and Sloane, 163
Privy Gardens, Whitehall, 164
Prosperetti, Leopoldine van Hogendorp, 4
provenance, 49, 58, 62, 167, 170–71

Qing dynasty artifacts, 64
Quentin de L'Orangère, Louis, 167–68, 170

rarity/rarities, 31, 46, 167, 168–69; and aesthetics, 9; and collectors, 31; in Francken, 58; and Porret, 40; and prints, 162, 167, 168–69, 170, 171; and G. E. Rumphius, 22, 25–26, 45, 47; and sexuality, 12. *See also* exotica/exoticism
reanimation, 12, 15, 16
Recco, Giovanni Battista, *Still Life with Fish and Oysters*, 90, **91**, 93
Regensburg, 59
Regensburg *Kunstkammer*, 62
religion, 11, 15, 17, 78, 132
Rembrandt van Rijn, 155, 187n16; as collector, 155, 161, 162, 174; and Cosimo III, 163; etchings by, 155, 162, 164, 167, 168–69, 171, 172, 174–75; finances of, 162; and Gersaint, 164, 166–67, 174; and Hollar, 172, 189n83; *The Hundred Guilder Print*, 162; prints by, 159, 164, 166, 168–69, 170, 171, 174; *The Rat Catcher*, 164; *The Sampling Officials of the Amsterdam Drapers' Guild, known as 'The Syndics'*, 172; *Self-Portrait*, **166**; *The Shell (Conus marmoreus)*, 162, 164, 166–67, 170, 172, 174–75, 187n22, 188n62; *The Shell (Conus marmoreus), first state*, 155, **156**, 169; *The Shell (Conus marmoreus), second state*, 155, **157**, 172; *The Shell (Conus marmoreus), third state*, 155, **157**
Remy, Pierre, 164
*Reuse-ooren* (Giant's ears), 43
Rice wine cup, Jingdezhen, Wanli-period, porcelain (remains of *Witte Leeuw*), 31, **33**
Rijswijck, Dirck van, *Tableau with Flower Still Life*, 35, **37**

Rizio, Paolo, 139
Roberval, Gilles de, 186n48
Roeters, Ernst, 122
Romans, ancient, 3, 94
Rondelet, Guillaume, 7
Roos, Anna Marie, 4
Rothé, Sara, 118, 120
Rotterdam, 2
Röver, Valerius, 162
Royal Society of London, 163
Rudolf II, 34, 87
Rumphius, Georgius Eberhard, 9, 161, 163, 165, 171, 181n13, 181n43, 181n44, 185n13; *Het Amboinsche kruidboek* (The Ambonese herbal), 22, 181n13; and collection, 22, 25–26, 46, 180n8; *D'Amboinsche rariteitkamer (Ambonese Curiosity Cabinet)*, **22**, 22–26, **23**, **24**, 43, **44**, 45–47, 122, 170, 172, 180n4, 180n8; and natural philosophy, 7; as Pliny of the Indies, 22; preparation and processing in, 43, 45–46, 47; rarity in, 22, 25–26, 45, 47; slaves of, 6; and Valentyn, 2
Rumphius, Paulus Augustus, 181n13; *Portrait of Georg Eberhard Rumphius at the Age of 68*, **22**

Sandrart, Joachim von, 78
Sannazaro, Jacopo, *Piscatorial Eclogues*, 90
Saxony, court of, 51
Scaliger, Joseph, 34
Scaliger, Julius, 59
Schickhardt, Heinrich, 138
Schijnvoet, Simon, 122, 170, 172, 180n4
Schloß Neuburg an der Donau: *große Grottenhalle*, 133, **135**; *Grotte des Hirschen*, 133, **136**; *Raum der Tierkreiszeichen*, 133, **136**
Schmidt, Nikolaus, 52–53
science, 4, 10, 25, 26, 133, 141, 165, 174
scientific publications, 22
Scot, Reginald, *The discoverie of witchcraft*, 65
seahorses, 161
Seba, Albertus, 9, 10, 163; *Kunstkammer*, 152
Segwaart, Bartholomeus van, 1–2
sensuality, 7, 12, 14, 15, 16, 17, 22, 40, 42, 47, 58, 64
sexuality, 12, 17, 64; and alcohol, 52; and female body, 14; and foreign people, 69; in Gheyn, 53; in Cornelisz van Haarlem, 42; knowledge of, 52; and Kornblum, 69; and peasant feasts, 182n7; and seduction fantasies, 71; and snails, 65; and Weston, 70; and witches, 65. *See also* beauty; body; body, female; eroticism
Seychelles nuts, 42
Shakespeare, William, *King Richard III*, 89, 90
shell cups. *See* vessels
shell-flowers. *See* flowers
Shell goblet with dragon foot, 150, **151**
shell life, 15, 78–87
shell money/wampum. *See* currency
silversmiths, 42
*Sinnelickheden* (sensualities), 40
Six, Jan, 171
skin, 52, 58, 65–71. *See also* Black bodies; body
skulls, 89, 90, 93, 149
slavery, 6–7, 14, 26, 40
Sloane, Hans, 163–64
snails, 35, 43, 98, 122, 161; Asian sea, 53; gilded silver, 49; golden, 65, 159; and Grimaldo, 51; and grottoes, 141; and Heidegger, 119; Indo-Pacific, 155; and Kornblum, 49, 51, 65; and sexuality, 65; and soul, 120, 185n34; vineyard, 51
social life, of things, 2–3, 10. *See also* biography
social relationships, 6, 10–11, 14, 15–16, 17, 22, 170, 171
Solander, Daniel, 164
soul, 15, 113, 120, 123, 185n34
South America, 107
Spain, 35
Spanish Netherlands, 132
Spary, Emma, 4
spider conch, 150
spiders, 81
Spinoza, Baruch, 122
spiny murex, 161
spirals, 11–12
spiral staircase, 123
*Spondylus princeps* (vibrant spiny oyster), 98
Sprankhuysen, Dionysius, 119
Steen, Jan, 118
still life, 7, 15, 75, 78, 79, 81, 84, 99, 113, 116–17, 174
striated mollusk, 87
*Strombus* (screw shell), 141
*Strombus gigas*, 89
Stuttgart: court of, 148; ducal residence of, 128, 132, 133, 140, 141, 142; grottoes in, 145
Suetonius, 3
Sulawesi, 43
Sustris, Friedrich, 132

Tansillo, Luigi, 184n39
*Terebra subulata* (auger shell), 31, 93
terrapin shells, 51
Thetis, 132
thought/reflection, 1, 2, 14, 75, 78, 87, 99, 104; aesthetic, 161; and Cicero, 3; and collectors, 120; as craftsmanship, 106; and decoration, 15; and dollhouses, 15; and grief, 15, 16, 149; and grottoes, 12, 16, 127, 128, 129, 132, 146, 148, 149, 152, 153; on identity, 14; of life and death, 12, 16; and liminality, 4; and Magdalena Sibylla, 129; on mathematics, 11–12; and Oortman, 106, 112–13, 116–17, 119, 123; and shell life, 15; and social relationships, 11; as theme in art, 17; with things, 106. *See also* interiority
time, 98, 99, 150, 152
Tobin, Beth Fowkes, 4
trade, 3, 5–6, 7, 10, 14, 43, 77, 99, 159; and Aar, 31; and Amsterdam, 163; and colonialism, 4; and cowries, 40; Dutch colonial, 5; in East and West Indies, 35; and exoticism, 5; and Gersaint, 170; and imperialism, 64; in Linard, 98; and market, 169; and rulers, 147. *See also* commerce; slavery
tridents, 52
Trobriand Islands, 3
Trojan War, 53
trumpet shells, 76–78, 90, 183n5
tulips, 5, 7, 169
turban shells, 11, 31, 40, 53, 59, 62, 63, 150
*Turbo chrysostomus* (turban shell), 81, 84
*Turbo marmoratus* (marbled turban shell), 21, 25, 27, **29**, 31, 40, 42, 43, 45, 98
*tyd en moeite*, 25, 26, 45

Uffenbach, Zacharias von, 107, 113, 116, 185n16, 185n26
Ulm, 128, 138, 140
Ulysses, 53, 58
*Umbilicus Veneris* (Venus's navel), 43
uncanny effect, 90
uncanny things, 15, 75, 78
United Provinces, 35
Utrecht, 142

Valentyn, François, 1, 2, 8, 10–11, 16, 185n13; *Treatise on Seashells and Sea Creatures*, 2
Valéry, Paul, 17, 99, 104, 119
vanitas still life, 113, 116–17
Venice, 140, 141
Venne, Pieter van de, *Still Life with Shells and Insects*, 84, **85**, 87
Venus, 49, 52, 65, 181n37. *See also* Aphrodite
Venus-Fortuna, 65, 69, 70
Versailles, palace of, 145, 147
Vertue, George, 172
Vespucci, Amerigo, 69
vessels, 182n5; and body, 53; human and natural design in, 46; for meaning, 16, 17; and Palissy, 95; porcelain, 63–64; in G. E. Rumphius, 47; shells and skulls as, 149
vessels, drinking, 42, 52, 104, 150; and art and nature, 42–43; in banquet scenes, 79; and Borsselen, 27; and elites, 7; *Kunstkammern* display of, 49, 51; and myth, 14; nautilus shells as, x, 3, 4, 14, 49, 51, 79; in Peeters, 79
Vincent, Levinus, 9, 10, 107, 163, 165, 170; *Shell Cabinet (Tab. II)*, 107, **108**
vintner figures, 51
Visscher, Roemer, 34, 112; *Sinnepoppen*, 5, **5**, 34, 81
Volkertsz, Jan, 162, 172
Vondel, Joost van den, 3

Walker, Kara, *Fons Americanus*, 17
Walther, Johann, *Interior of Count Johannes of Nassau's grotto at Idstein*, 142, **143**
Wan-li vase, 81
Watteau, Jean-Antoine, *L'Enseigne de Gersaint*, 164
Weismann, Ehrenreich, 127
Wely, Jan van, 34
West Africa, 6, 77, 107
West Indies, 1, 35
Weston, Edward: *Shells*, 70, **71**; *Shells* series, 17, 70
Wilhelm V, Duke of Bavaria, 132, 141, 147
witches, 65, 69
*De Witte Leeuw*, 31, **33**, 34, 35, **39**, 171
Wittelsbach residence, Munich, 132, 147
women, 1, 65, 128; and Cats, 117–18, 119; as collectors, 103; and desire, 58; and dollhouses, 103; Indian, 69; and miniaturization, 103, 107; non-European, 69, 71; and ownership over artifacts, 107; porcelain-skinned, 69, 70; and transgressive behavior, 65; wealthy, 103. *See also* body, female
women of color, 65. *See also* Black bodies
Worm, Ole, *Musei Wormiani Historia*, 162, **163**
Wtenbogaert, Jan, 162–63
Wtewael, Joachim, *Perseus and Andromeda*, 87–89, **88**, 93
Württemberg, Duchy of, 128, 138
Württemberg, dukes of, 141, 150
Württemberg, grottoes of, 129, 140
Württemberg, Johann Friedrich von, 128, 129, 132, 133, 140, 141, 142, 145, 146, 147, 148, 149–50
Württemberg, Magdalena Sibylla von, 127–28, 129, 132, 133, 145, 147, 148, 149–50, 152
Württemberg, Wilhelm Ludwig von, 127
Württemberg *Kunstkammer*, 150

Zeeland, 2, 84
Zucchi, Jacopo, *The Treasures of the Sea (Coral Fishers)*, 65, **67**, 68
Zucchi, Lorenzo, after Matthäus Daniel Pöppelmann, *Grote Qui Est, dans un Sallon a L'Orangerie Royale*, 133, **137**
Zwinger, Dresden, 133
Zwolle, 2

# ILLUSTRATION CREDITS

Amsterdam Museum (fig. 11)
ARTIS—Collections of the University of Amsterdam (fig. 8, fig. 9, fig. 20)
Artwork in the public domain (fig. 51, fig. 52)
© Ashmolean Museum, University of Oxford (fig. 3)
© Bayerische Schlösserverwaltung, Maria Custodis, München (fig. 65)
© Bayerische Schlösserverwaltung, Ulrich Pfeuffer, München (fig. 64)
© Bayerische Staatsbibliothek, Munich (fig. 69, fig. 73)
© Bibliothèque Nationale de France, Paris (fig. 70)
© bpk Bildagentur / Staatliche Kunsthalle, Karlsruhe / Photo: Wolfgang Pankoke / Art Resource, NY (fig. 37)
© bpk / Staatliche Kunstsammlungen Dresden (fig. 22)
© Center for Creative Photography, Arizona Board of Regents / Artists Rights Society (ARS), New York (fig. 33)
© Detroit Institute of Arts (fig. 39)
Digital image courtesy of the Getty's Open Content Program / J. Paul Getty Museum, Los Angeles (fig. 40, fig. 45)
© Duke of Northumberland (fig. 81)
© Fondation Custodia, Frits Lugt Collection (fig. 80)
© Gallerie degli Uffizi, Florence (fig. 54)
Getty Research Institute (fig. 84)
© Herzog Anton Ulrich-Museum Braunschweig, Kunstmuseum des Landes Niedersachsen (fig. 6)
© KIK-IRPA, Brussel (fig. 34, fig. 35)
© Kunsthistorisches Museum, Vienna (fig. 21, fig. 47)
© Landesmuseum Württemberg, Stuttgart / H. Zwietasch (fig. 74)
© Los Angeles County Museum of Art (fig. 79)
Metropolitan Museum of Art (fig. 10, fig. 16, fig. 30, fig. 63, fig. 67)
© The Montreal Museum of Fine Arts / Photo: MMFA, Christine Guest (fig. 46)
© Museum Boijmans van Beuningen, and Studio Tromp (fig. 12)
© Museum für Kunst und Gewerbe Hamburg (fig. 23)
© Museum Ulm, Photo: Museum Ulm (fig. 28)
© Nationalmuseum, Stockholm / Photo: Erik Cornelius (fig. 42)
Philadelphia Museum of Art (fig. 13)
© Photo Scala, Florence (fig. 31)
Rijksmuseum, Amsterdam (fig. 2, fig. 4, fig. 7, fig. 14, fig. 15, fig. 17, fig. 18, fig. 38, fig. 43, fig. 44, fig. 49, fig. 50, fig. 53, fig. 55, fig. 56. fig. 57, fig. 58, fig. 59, fig. 76, fig. 77, fig. 78, fig. 85)
© RMN-Grand Palais / Art Resource, NY (fig. 41)
© RMN-Grand Palais (Musée du Louvre) / Hervé Lewandowski (fig. 26, fig. 27)
© Robin Halwas, London (fig. 83)
© Sächsische Landesbibliothek—Staats- und Universitätsbibliothek, Dresden (fig. 60, fig. 61, fig. 62)
© SLUB Dresden / Deutsche Fotothek (fig. 66)
© Smithsonian Libraries (fig. 82)
© Sotheby's 2020 (fig. 5)
© Staatliche Museen zu Berlin, Kunstbibliothek. Photo: Dietmar Katz (fig. 24)
© Staats- und Universitätsbibliothek Göttingen (fig. 29)
© Statens Museum for Kunst (fig. 1)
Stichting P. en N. de Boer, Amsterdam (fig. 19)
© The Trustees of the British Museum (fig. 36)
© The Trustees of the Natural History Museum, London (fig. 48, fig. 75)
© Universitätsbibliothek, Heidelberg (fig. 71, fig. 72)
© Universitätsbibliothek Freiburg i.Br. (fig. 32)
© Victoria & Albert Museum, London (fig. 68)
© Wallraf-Richartz-Museum & Fondation Corboud / Photo: Rheinisches Bildarchiv Cologne, Meier, Wolfgang F. (fig. 25)